Intelligent and Sustainable Cement Production

Intelligent and Sustainable Cement Production

Transforming to Industry 4.0 Standards

Edited by

Anjan Kumar Chatterjee

CRC Press
Taylor & Francis Group
Boca Raton London New York

CRC Press is an imprint of the
Taylor & Francis Group, an **informa** business

First edition published 2022
by CRC Press
6000 Broken Sound Parkway NW, Suite 300, Boca Raton, FL 33487-2742

and by CRC Press
2 Park Square, Milton Park, Abingdon, Oxon, OX14 4RN

CRC Press is an imprint of Taylor & Francis Group, LLC

ISBN: 978-0-367-61840-7 (hbk)
ISBN: 978-0-367-61843-8 (pbk)
ISBN: 978-1-003-10679-1 (ebk)

DOI: 10.1201/9781003106791

Typeset in Times New Roman
by SPi Technologies India Pvt Ltd (Straive)

Contents

Preface vii
Editor xiii
Contributors xv
Notation xvii

Chapter 1 Contemporary Cement Plants: Scale, Complexity, and
Operational Variables 1

Anjan Kumar Chatterjee

Chapter 2 Transforming Cement Manufacturing Through
Application of AI Techniques: An Overview 39

Xavier Cieren

Chapter 3 Process Automation to Autonomous Process in Cement
Manufacturing: Basics of Transformational Approach 79

Anjan Kumar Chatterjee

Chapter 4 Electrical Systems for Sustainable Production in
Cement Plants: A Perspective View 99

Peddanna Shirumalla and Anjan Kumar Chatterjee

Chapter 5 Data-Driven Thermal Energy Management
Including Alternative Fuels and Raw Materials
Use for Sustainable Cement Manufacturing 141

Ashok Dembla and Matthias Mersmann

Chapter 6 Control of Cement Composition and Quality:
Potential Application of AI Techniques 199

*Mohsen Ben Haha, Maciej Zajac, Markus Arndt, and
Jan Skocek*

Chapter 7 Asset Performance Monitoring and Maintenance
Management in Cement Manufacturing 225

*P.V. Kiran Ananth, K. Muralikrishnan, and
Anjan Kumar Chatterjee*

Chapter 8   Digital Twin and Its Variants for Advancing
Digitalization in Cement Manufacturing 263

Anjan Kumar Chatterjee

Chapter 9   Developments in Application of Sensors to
Sustainable Manufacturing of Cement 293

Kamal Kumar, Anupam, and Anjan Kumar Chatterjee

Chapter 10 Integrated Enterprise Resource Planning in
Sustainable Cement Production. 323

Rameshwar Dubey

Chapter 11 Implementation of Digital Solutions in Cement
Process and Plants 347

*Sriram Seshadri, Jeyamurugan Kandasamy, and
Manikandan Rajendran*

Chapter 12 Technological Forecasting for Commercializing
Novel Low-Carbon Cement and Concrete Formulations 405

Sadananda Sahu

Epilogue 455

Index 461

Preface

In the title of the book three attributes have been used to describe the emerging shape of cement manufacturing – 'intelligent', 'sustainable', and 'Industry 4.0', and it is highly pertinent to understand what such a combination leads up to. The intelligent manufacturing systems are already familiar to industry at large. Such systems were introduced with the goal to integrate the abilities of humans, machines, and processes to achieve the best possible manufacturing outcome. It is designed to cover the entire process from raw materials to finished products and seeks to achieve optimal utilization of manufacturing resources with minimal process waste and to add value to the entire business. While the traditional manufacturing works on the existing knowledge and experience of operators, the intelligent manufacturing systems must involve learning from the past production data, understand all complexities, and find better alternatives. Such systems as designed and applied, therefore, have already been a combination of sorts between the operation technology (OT) and information technology (IT). Depending on the level of IT, a system may have only involved digitization with the help of computers, or may have been more extensive with computerization and internet communication, or in some cases may have further advanced into basic applications of artificial intelligence (AI). In broad terms, the systems have been essentially inward-looking and focused on efficient production and maintaining consistent quality in products.

Sustainability in essence has a different dimension. Decades back in the Brundtland report of the World Commission on Environment and Development, sustainability was defined as '....meeting the needs of the present generations without compromising the ability of future generations to meet their own needs'. Later, there was an attempt to define more specifically what the 'environmental sustainability' would mean and in the EU MDG7 report it was defined as '.....meeting current human needs without undermining the capacity of the environment to provide for those needs in the long term'. The industry's progress towards adhering to the principles of general and environmental sustainability continues to be a baffling issue. Even in 2017, the report of Circle Economy, a not-for-profit organization, showed that the world was using up more than 100 billion ton of natural resources per year. It is predicted that by the middle of the 21st century the consumption level may rise up to 170–184 billion ton. At the same time the resource circularity, which was 9.1% in 2015, started showing a further downward trend. In addition to the resource burden, the environmental sustainability is essential from another perspective of greenhouse gas emissions, which were globally about 50 billion ton in 2016 with energy-related and direct process-related emissions being more than 75%.

In keeping with the demands of resource and energy conservation and also of climate protection, the manufacturing industries are in a race to reach a new technological height designated as Industry 4.0. It may be recalled that the industrial culture dawned in the human society in the late 18th century with the first commercialization of water- and steam-powered machines and devices, which is now marked as Industry 1.0. In other words, the introduction of mechanical production facilities had ushered in the industry culture. Subsequently, the beginning of the 20th century marked the start of the second industrial age or Industry 2.0 as we recognize it today, when the electrical energy became the primary source of power. Mass production of goods using assembly lines became the standard practice. The next industrial age, designated as Industry 3.0, is associated with the invention of PLC (programmable logic controllers) in 1960s and the application of interconnected computer networks in 1980s. In reality, the advances in electronics and invention of transistor technology brought in the third industrial revolution. Although it took about 100 years to move from Industry 1.0 to Industry 2.0, the Industry 3.0 tag was reached in about 50 years. The progress to Industry 4.0 revolution, which is happening now, has occurred even faster with the advent of internet and telecommunication facilities in 1990s. Although the transition from Industry 3.0 to Industry 4.0 was gradual and imperceptible in the last decade of the 20th century, there is a sudden spurt in the related technological developments in the last two decades, which appear disruptive in character. The core concepts are extensive digitalization and the merger of the boundaries of the physical and virtual world, giving shape to cyber-physical systems (CPS). Apart from internet-of-things, the transforming technologies include cloud computing and storage, big data analytics, machine learning (ML), and other technologies of AI. These developments have influenced the manufacturing industries to such an extent that they are now able to think about achieving new heights that were unthinkable earlier.

Where does the cement industry fit into this broad frame? It is a question that is being seriously and extensively pondered over by the cement community. In fact, it is a valid concern as the Portland cement manufacture is considered a mature industry. It was patented almost two centuries back in 1824 in the United Kingdom, and it has shown an unprecedented growth all over the world in this span of about 200 years, and more specifically, in the last two decades. The production of Portland cement crossed 2.0 billion tons per year in 2003, 3.0 billion tons in 2009, and 4.0 billion tons in 2013. Even a conservative prediction of the global annual demand of cement by the International Energy Agency indicates that it will be more than 4.68 billion tons in 2050. Up until the 1980s, growth was driven by scale and cost of production. After the 1980s, energy conservation and pollution control became the prime drivers. From the beginning of this century it was realized that sustainability would have to be of paramount importance to increase further capacity of cement production. Another unavoidable barrier of increasing the cement capacity is the compulsion to move to a less-polluting regime in terms of CO_2 and other gaseous emissions. With all the above challenges, the cement industry is in a constant endeavor to reshape itself with energy efficiency, resource conservation, and sustainability. The same trends of industrial restructuring are more prominent now, but the solace is that the barriers may be overcome and the goals achieved, if the industry earnestly takes to the path of digital maturity.

The technological history of the cement industry shows that during 1970s and 1980s automation was used as a means to maintain the plant operations in safe and continuous mode. During 1990s, the process control and information system of the then vintage were integrated with the plant operation, leading to higher productivity, better efficiency, lower manufacturing cost, and improved product quality. During the above period, emphasis was on adopting computer-integrated manufacturing systems with different levels of computing functionality. The endeavor was to elevate from equipment control systems to supervisory personnel interaction systems, then to plant-wide information management systems, aiming further toward divisional integration, followed by corporate integration systems.

In the late 1990s and in early years of the 21st century, the cement industry became more and more market-driven from its earlier orientation of being production-driven, which prompted a change in the operation of plants and management of enterprises. This was done by integrating the enterprise resource planning (ERP), the manufacturing execution system (MES), and the plant control system (PCS). While the ERP system included the whole range of management such as the financials, materials management, procurement, marketing, sales and customer services, the MES covered production scheduling, process management, material tracking, and quality management. This integration was possible due to progressive advances in the plant process control technologies. The capability of widely installed DCS (distributed control systems) with PLC at sub-systems levels, further strengthened with SCADA (supervisory control and data acquisition) systems, was enhanced with 'Expert Systems' first, and then with 'Production Optimizers'. There was a perceptible shift toward making the processes more and more data-driven by adopting MPC (model predictive control) approaches, based on neural networks and computer vision techniques for production optimization.

With adoption of all the above technological advances in an integrated manner, the contemporary cement plants and enterprises have so far been able to be competitive, cost-effective, market-responsive, and relatively sustainable. But the technology effects seem to be flattening out, while the cement demands are galloping, barring some cyclic slow-downs. The expert systems and production optimizers are not able to survive a large number of disruptive variables and disturbances in the cement process. They also need substantial maintenance to adapt to the frequently changing operating conditions including natural raw materials. Furthermore, the industry has only limited success in moving toward circular economy and in reducing greenhouse gas emissions as necessary for climate protection. Perhaps, the solution lies in making the cement process and plant operations predictive and autonomous. A paradigm shift is foreseen in the cement industry to make it intelligent and sustainable, which can only happen through digitalization toward Industry 4.0.

The cardinal requirement for causing this paradigm shift is to strengthen the IT base which must be capable of undertaking data analytics. It is estimated that monitoring only one cement production line involves approximately 700 data samples collected at intervals of every 30–60 seconds, which may mean a billion data samples per year. The process data is collected from different sensors located throughout the production line. A beginning has also been made to develop and apply 'soft sensors', because many parameters in cement manufacturing are not directly measurable by instruments. The vast majority of contemporary cement plants operate simultaneously multiple production lines. In addition, the cement enterprises consist of multiple locations in different parts of the world. Further, for each production line at each location, a large volume of historical data is stored. Thus, the data sources are different, the data quality is varied, and the datasets may require specialized storage and processing tools. The data needs to be time-stamped and be on a time-series. Moreover, such databases require special data compression algorithms to ensure storage without loss and to facilitate fast retrieval. The larger the network for gathering information, the more valuable it can become by combining the experience and expertise of different but relevant segments. These networks are effective, only if the real-time and historical information is available in easy and flexible format everywhere across the network with reliability and accuracy.

It is relevant to mention here that cloud computing, i.e., the on-demand availability of computer system resources, especially data storage and computing power, is a very promising technological development that is expected to push the cement industry to level 4.0. Since cloud computing relies on sharing of resources to achieve coherence and economies of scale and since it helps to avoid or minimize up-front IT infrastructure cost, it will certainly facilitate the cement enterprises to get their applications up and running faster with improved manageability and less maintenance. Competence in cloud engineering encompassing systems, software and web on one side, and performance, security, and risk on the other is now more widely available. Minimization of the risk of internal outages is no more a difficult proposition. Thus, the application of cloud computing is seen as a booster for data management in the cement manufacturing facilities at large.

Data analytics and cloud computing will pave the way toward the application of ML, which is a form of AI. It enables a system to learn from data by using a variety of algorithms that iteratively learn from data to improve, describe data, and predict outcomes. As the algorithms ingest training data, it becomes possible to produce more precise models based on the data. ML requires the right kind of data that can be applied to the learning process. It is important to note that an organization, to start with, does not have to have 'big data' in order to use ML techniques, though big data always helps to improve the accuracy of ML models. In addition to ML, the database can be used for developing 'digital twins' to mirror the entire production process through a digital model or a 'twin' of the physical assets, processes, and systems of the plant spanning over its entire life. The digital twin technology has immense potential and will benefit the cement industry in many ways, though its implementation is a complex process requiring multiple technologies and tools.

It is pertinent to note that an integrated cement manufacturing plant of today is a fitting example of an industrial IoT (industrial-of-things). It is a system of systems with networked equipment and devices with sensors. The overall infrastructure is basically amenable to digitalization, which is also in a happening phase, but the destination is still some distance away. The cement operations will have to adopt AI involving a set of ML and deep learning techniques. This strategy will be essential to predict operational and correctional failures to optimize production processes, to practice predictive maintenance, to undertake remote operation in digital twin mode, to get into new product design, to manage product quality in high-performance mode, and to institutionalize smart supply chain. In parallel, the cement plants will have to strongly focus on sustainability, more than ever before, because of its energy- and resource-intensity, and large environmental impacts.

To the best of my understanding, the progress in digital transformation that has happened in the cement enterprises and that are in offing has not been chronicled systematically in a single publication so far. The published information on the subject is scattered in different types of promotional and trade literature, apart from the scientific publications in the field of IT, which are not specific to cement. A few of the multinational cement companies, which are apparently pursuing the course of digital transformation, have been releasing short communications on the progress of their efforts from time to time, but the publications are essentially promotional in nature. The proposed book on the subject is perhaps a prime effort to consolidate the past, present, and future of digital transformations in the cement industry.

The proposed book is designed to be a repository of the present state of knowledge regarding the practices in vogue and new possibilities in development to make cement plants data-driven, predictive, and autonomous. It will consist of 12 chapters as given in the table of contents. Chapter 1 is intended to recapitulate the unit operations of the cement plants, process variables and operational interdependence and complexity from raw materials exploitation to packing and dispatch of finished products. Chapter 2 will provide an overview of the present state of the digital transformation happening in the cement industry with the help of ML. Chapter 3 will deal with the transformational journey of automatic processes to being autonomous in operation. Chapter 4 will focus on the sustainable use of electrical energy with the help of IT. Chapter 5 will deal with the advances in thermal energy management systems and software including the adoption of more effective means of using alternative fuels and raw materials. Since cement is a composite product with multiple variables, the present practices of quality control and assurance often fall short of the needs and targets. Hence, Chapter 6 is intended to delve into predictive quality analytics. In cement plants it is crucial to look for patterns in asset data, how the asset is being used, and the environment in which it is operating. There is a specific need to focus on identifying risks and optimizing the asset reliability, which requires the adoption of the strategy of data-driven predictive maintenance. This topic will form the core of Chapter 7. As mentioned earlier, the digital twin technology is being seen as an important tool for process optimization and training. Chapter 8 is intended to deal with these aspects. It is understood that no AI techniques can be introduced without proper data collection, data processing, and data storage. These operations will depend on the sensor technology and its progress toward soft sensors for measuring the process parameters. These aspects will be covered in Chapter 9. As the operations in cement enterprises extend much beyond the process of manufacture, the ERP has become an integral part of the business. This topic will be dealt with in Chapter 10. A sense of extensive implementation of digital technologies in cement plants can be obtained from Chapter 11. Our effort will remain incomplete if we do not cover the commercial prospects of novel low-carbon cement and concrete formulations. This topic will be dealt with in Chapter 12. Finally, an attempt will be made to summarize the scope and trend of digital transformation in cement production as the form of an Epilogue.

It is obvious from the above description of chapters that topics are highly specialized and cannot be handled by a single author with authority and justice. Hence, the proposal is to build up the book with chapters contributed by selected authors. Since I have to my credit a book entitled *Cement Production Technology: Principles and Practice*, published by CRC Press, Taylor & Francis Group, in 2018, I found the application of digital technologies and advances in IT to cement manufacture as a theme in sequence to what I had covered in the previous book. I have a very long and intimate association with the cement sector spanning over almost five decades and the previous book was borne out of this association. I am fortunate to continue my association with the cement and concrete world and I find newer directions of advancements in the industry. The pace of introduction of such novel techniques and technologies needs acceleration. The proposed book owes its origin to such a thought process. I believe that the proposed book under my editorship with chapters contributed by me and a few other erudite experts will be a very practical and useful handbook on digital transformation for the cement fraternity.

Anjan Kumar Chatterjee

Editor

Anjan Kumar Chatterjee is a Materials Scientist and has spent almost five decades in the cement and concrete industry as a research scientist and a corporate executive. He holds a doctorate degree from the Moscow State University, Russia, based on his research studies on electro-re-melting slags at the Baikov Institute of Metallurgy in Moscow. He also carried out research work at the Building Research Establishment in the United Kingdom. His professional activities encompassed the cement manufacturing process, blended and special-purpose cements, refractory products, ready-mixed concrete production, cement and concrete materials science, and chemistry. He is an alumnus of Presidency College (now a university) in Kolkata.

Dr. Chatterjee started his career as a faculty member of the Materials Science Center of Indian Institute of Technology, Kharagpur, India, and later moved to Cement Research Institute of India (now National Council of Cement and Building Materials), New Delhi, and then to Associated Cement Companies Limited (now ACC Limited), Mumbai, from where he retired as its Whole-time Executive Director. During his tenure in ACC, he was responsible for the company's R&D, new cement projects, and all diversified business activities. In the post-retirement period, he has been engaged in providing technical support services to the cement, concrete, and mineral industries in India and abroad in various capacities. In addition, Dr. Chatterjee has been on various international assignments with UNIDO and other organizations. He continues to be the Chairman of Conmat Technologies Private Limited, a service-oriented outfit, in Kolkata.

Dr. Chatterjee is a fellow of the Indian National Academy of Engineering, Indian Concrete Institute, and Indian Ceramic Institute. He is on the editorial boards of international journals such as 'Cement and Concrete Research' and 'Cement Wapno Beton'. He has continued to be a member of the steering committee of International Congress on the Chemistry of Cement for the last four decades. He has been conferred lifetime achievement awards by the Indian Concrete Institute, Association of Consulting Civil Engineers, Confederation of Indian Industries, and Cement Manufacturers Association of India. Among his numerous publications, the most note-worthy is the book on 'Cement Production Technology: Principles and Practice', published by the CRC Press of Taylor & Francis Group. He has many awards to his credit.

Contributors

Markus Arndt
HeidelbergCement AG
Leiman, Germany

Anjan Kumar Chatterjee
Conmat Technologies Private
 Limited
Kolkata, India

Anupam
Process Optimization and
 Productivity
National Council of Cement and
 Building Materials
Ballabgarh, India

Xavier Cieren
Research Scientist
LafargeHolcim Innovation Center
 (LHIC)
St. Quentin Fallavier, France

Ashok Dembla
Humboldt Wedag India Private
 Limited
New Delhi, India

Rameshwar Dubey
Liverpool Business School
Liverpool John Moores University
Liverpool, UK

Mohsen Ben Haha
HeidelbergCement AG
Leiman, Germany

Jeyamurugan Kandasamy
Group Digital
F.L. Smidth India
Chennai, India

P.V. Kiran Ananth
Confederation of Indian
 Industries
CII – Sohrabji Godrej Green
 Business Centre
Hyderabad, India

Kamal Kumar
Process Department
Holtec Consulting Private
 Limited
Gurgaon, India

Matthias Mersmann
KHD Humboldt Wedag
 International AG
Cologne, Germany

K. Muralikrishnan
Confederation of Indian
 Industries
CII-Sohrabji Godrej Green Business
 Centre
Hyderabad, India

Manikandan Rajendran
Assets Insights
Group Digital
F.L. Smidth India
Chennai, India

Sadananda Sahu
Solidia Technologies
Piscataway, NJ, USA

Sriram Seshadri
Smart Products
Group Digital
F. L. Smidth & Co.
Copenhagen, Denmark

Peddanna Shirumalla
ERCOM Engineers Private Limited
New Delhi, India

Maciej Zajac
HeidelbergCement AG
Leiman, Germany

Jan Skocek
HeidelbergCement AG
Leiman, Germany

Notation

- In a few chapters standard cement chemistry notations have been used such as $A=Al_2O_3$, $C=CaO$, $F=Fe_2O_3$, $H=H_2O$, $M=MgO$, $K=K_2O$, $N=Na_2O$, $S=SiO_2$, $\bar{S}=SO_3$.
- C-S-H denotes a variable composition of the hydrate phase.
- Calcium monosulfate-type structure, also known as AFm, and ettringite-type structure, also known as Aft, denote a solid solution range.
- 'Ton' has been used for capacity/product mass without arithmetic correction of 'tonne'.

1 Contemporary Cement Plants

Scale, Complexity, and Operational Variables

Anjan Kumar Chatterjee

Conmat Technologies Private Limited, Kolkata, India

CONTENTS

1.1   Introduction 1
1.2   Scale and Scatter of Production 3
1.3   Complementary Role of Material Chemistry and
       Process Engineering... 5
       1.3.1   Key Features of Raw Materials Influencing the Process 8
       1.3.2   Criticality of the Clinker-Making Stage................................. 10
       1.3.3   Material Chemistry in Clinker Grinding................................ 12
1.4   Resource Efficiency and Material Flows in the Production
       Process... 13
1.5   Thermal Energy Performance of the Kiln Systems 16
1.6   Cement Kilns for External Waste Management............................... 18
1.7   Electrical Energy Performance 19
1.8   Pollutions and Emissions in Cement Manufacturing....................... 20
1.9   Characteristic Features of Portland Cements 24
1.10 Modelling, Simulation, and Advances in Process and
       Quality Control Systems.. 25
1.11 Integration of Business, Management, and Production 30
1.12 Integrated Features of Contemporary Cement Plants 32
1.13 Concluding Observations 33
References... 36

1.1   INTRODUCTION

For all practical purposes, the production of Portland cements has evolved into a mature industry, although the rudimentary process of manufacture for the basic product was patented only about two centuries back in 1824 in the United Kingdom. Since then, the process has undergone sustained development, the demand for the diversified group of products has continuously risen, and the industry has flourished. More specifically, the growth of the industry has been unprecedented in the last two decades all over the world. The production of Portland cements crossed 2.0 billion tons per year in 2003, 3.0 billion tons in 2009, and 4.0 billion tons in 2013. With high growth rate in the previous decade, the industry was expected to cross the mark of 5.0 billion tons of production per year in 2019, but it did not happen due to the impacts of a global economic recession. According to some forecasts, the industry is back on its growth path and will reach 5.8 billion tons by the year 2027 [1].

Fundamentally, the production of Portland cements is dependent on natural raw materials and involves several unit operations of varying efficacy and environmental impacts. These unit operations cause both physical and chemical transformation of the in-process materials and are interfaced with each other in an integrated plant. The quality parameters of the finished products are maintained within a narrow range of composition and microstructural features for effective applications. Notwithstanding these rigours, the industry has grown phenomenally as mentioned above. Up until the 1980s, the growth was driven by scale and cost of production. After the 1980s, energy conservation and pollution control became the prime drivers. From the beginning of this century, it was realized that sustainability would have to be of paramount importance in increasing further capacity of cement production. Another unavoidable environmental requirement that has emerged in further expanding the cement capacity is the compelling move to a less-polluting regime in terms of CO_2 and other gaseous emissions. Because of the above change drivers, the cement industry has always attempted to reshape itself with energy efficiency, resource conservation, and sustainability. The same trend of industrial restructuring is more prominent now, further enhanced by the adoption of various digital solutions.

Tracing back the technological history of the cement industry, one may observe that during 1970s and 1980s automation was used as a means to maintain the plant operations in safe and continuous mode. During 1990s, the process control and information system of the then vintage were integrated with the plant operation, leading to higher productivity, better efficiency, lower manufacturing cost, and improved product quality. During the above period, emphasis was on the computer integrated manufacturing system (CIMS) with different levels of computing functionality. In late 1990s and in early years of the 21st century, the cement industry became more and more market-driven from its earlier orientation of being production-driven, which prompted a change in the CIMS model. The execution of manufacture was no more limited to only the plant control system but necessitated the integration of the enterprise resource planning (ERP). Such integration was possible because the plant process control could be brought to a more precise state due to various developments including the use of more sensitive and accurate sensors.

The extent of process evolution that has thus taken place in the plants producing various types of Portland cements is significantly large and consistent with the scale of production, which has surpassed that of other commercially produced construction materials such as steel, aluminium, or plastic. This chapter is aimed at reviewing the technological status of contemporary cement plants with a view to understanding the transformational progress it has made towards achieving the industry 4.0 standards.

1.2 SCALE AND SCATTER OF PRODUCTION

The recent trends of demand and per capita consumption of Portland cement have been compared on a global basis with those of steel, aluminium, and plastic in Table 1.1 [2–5].

From the above table it is evident that the present dimension of the cement industry is much larger than that of other competitive manufactured products. At the same time, it should be borne in mind that the Portland cement is seldom used in the form it is produced. It acts as an essential ingredient in a range of value-added products such as mortars, plasters, renderings, repair materials, and concrete, the global volume of which is many folds higher than that of cement itself. Since the application of cements is significantly diverse with substantial expansion of volumes, the production facilities are generally equipped with machinery and expertise to support the post-production services.

The Portland cement is produced in a large number of countries in various regions of the world. It is reported that the global installed capacity of cement production reached 6.28 billion ton per year in 2016, although the total capacity of the companies listed in the Global Cement Directory in 2018 totalled 4.47 billion ton [6]. Irrespective of the capacity data, the consumption of cement in the world increased fourfold from about 1.0 billion ton in 1990 to about 4.0 billion ton per year in 2017. There has been some contraction in

TABLE 1.1
Recent Demand Trends of the Major Manufactured Construction Materials

	Global Production (Million Metric Ton)			
Year	Portland Cement	Crude Steel	Primary Aluminium	Plastic
2019	4200	870	63.7	—
2018	4100	1808	64.4	359
2017	4050	1732	63.4	348
	Per capita consumption (kg)			
2018	521	225	11	28

TABLE 1.2

Major Cement-Producing Countries of the World and their Recent Production Trends

Serial No.	Country (Number of Facilities in 2017)	Average Capacity per Facility in 2017 (t/d)	Production of Cement (Mn t)		
			2017	2018	2019
1	China (861)	7000	2220	2200	2200
2	India (264)	3000	290	380	320
3	Vietnam (89)	2400	79	90	95
4	USA (105)	2200	86	87	89
5	Egypt (25)	5800	53	81	76
6	Indonesia (42)	4200	65	75	74
7	Iran (86)	1700	54	58	60
8	Russia (63)	2400	55	54	57
9	Brazil (98)	1500	53	53	55
10	S. Korea (24)	6400	56	57	55
11	Japan (31)	4900	55	55	54
12	Turkey (78)	2800	81	72	51

cement production between 2015 and 2018 due to the global economic downturns, but the market forecast presents a positive outlook [1]. The major cement-producing countries of the world are listed in Table 1.2 [7]. The growth pattern indicates that three Asian countries top the list with the volume of production in China accounting for almost 56% of the global production of cement. Another development trend is important to note. The installed capacity of only ten cement companies accounted for over 40% of the total capacity of the listed entities as indicated in Table 1.3.

The data presented in Tables 1.2 and 1.3 help making the following observations on the recent trend of cement production:

- The capacity of individual production facilities in the major cement-producing countries varies approximately from 1500 t/d to 7000 t/d.
- The major cement-producing companies have integrated facilities having annual capacity approximately ranging from about 1.0 to 3.0 million ton of cement, although there are single kilns having capacities of even 5.0 million ton per year.
- The average annual capacity of the individual stand-alone grinding plants ranges from about 0.6 million ton to over 1.0 million ton.
- The production facilities even under one entity have multiples locations to run the business.

The above trends are only indicative for appreciating the technology dimensions of the cement industry. The statistics would vary, depending on the regions and plants considered, as the listed operating facilities in the world are in excess of 2500 [7].

TABLE 1.3

Production Capacities of the Top Ten Multinational Cement Enterprises in 2018

Serial No.	Country of Primary Operation (Name of the Enterprise)	Total Annual Capacity (Mn.t)	Share in the Total Listed Capacity (%)	Approximate Average Annual Capacity per Integrated Plant Facility (Mn.t)	Approximate Average Annual Capacity per Stand-alone Grinding Facility (Mn. t)
1	China (CNBM & Sinoma)	521.0	11.6	NA	NA
2	Switzerland (LafargeHolcim)	356.0	8.0	1.92	0.81
3	China (Anhui Conch)	335.0	7.5	NA	NA
4	Germany (Heidelberg)	187.8	4.2	1.56	0.67
5	Mexico (CEMEX)	95.6	2.1	1.63	0.73
6	India (Ultratech)	93.5	2.1	2.90	1.03
7	China (China Resources)	83.3	1.8	NA	NA
8	Brazil (Votorantim)	70.9	1.6	1.40	0.66
9	Taiwan (Taiwan Cement)	69.0	1.5	NA	NA
10	Ireland (CRH)	63.3	1.4	1.07	0.57
Total		1875.4	41.2	–	–

1.3 COMPLEMENTARY ROLE OF MATERIAL CHEMISTRY AND PROCESS ENGINEERING

'Material chemistry' and 'process engineering' together constitute the foundation of cement manufacturing. Over the decades, however, the engineering advances have been phenomenal to convert the chemical concepts into very large, fast reacting, energy-efficient, and pollution-controlled plant systems. In the course of over 190 years of existence of the generic Portland cement the fundamentals of manufacturing chemistry have not undergone any disruptive change, while considerable engineering advances have been made in the hardware and software of cement manufacture in order to achieve optimum cost and quality. These advances reflect in the following:

- The capacity of a single kiln system has reached 14,500 tons per day or almost 5.0 million ton per year [8].
- With automation, instrumentation, and computer-aided controls, the man-hours per ton of cement produced has come down to one or even less, thereby reducing the application of human discretion and increasing the dependence on electronic gadgets.

- The variations in the hardware design cause differences in the chemical performance, thus necessitating precision in the hardware design.
- The raw material characteristics and their impurities strongly influence the system performance, thus making it imperative for the hardware to be compatible with raw materials characteristics.

Before one goes into the specifics of the above developmental features, it might be relevant to recapitulate the broad frameworks of interaction between the multifarious materials and processing steps, they are subjected to, at different stages of cement production from cradle to gate. The frameworks of the individual stages are presented schematically in Figures 1.1–1.4 and the typical process layout of an integrated manufacturing facility is shown in Figure 1.5 [9].

All the four stages are sensitive to a large number of physical properties and chemical composition of the materials including the minerals present in the raw state or compounds formed in them due to subsequent reactions. Since the stages are well integrated, the output of the previous stage serves as the input for the next processing step, which makes the total production facility highly complicated. Further, the principal pieces of equipment for mining, crushing, grinding of raw materials, blending, high-temperature processing, clinker grinding, and finally storing and packing of cement essentially depend on a very large number of auxiliaries, which include electrical motors, pumps,

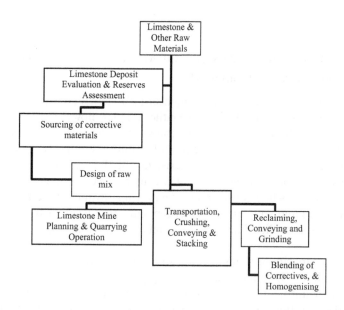

FIGURE 1.1 Materials and process steps from raw materials exploration to kiln feed preparation.

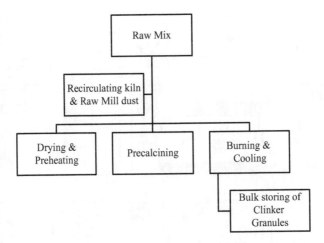

FIGURE 1.2   Processing steps for transforming the raw mix to clinker.

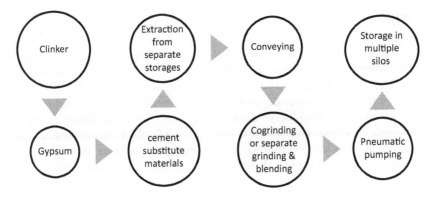

FIGURE 1.3   Processing of clinker into various types of Portland cements.

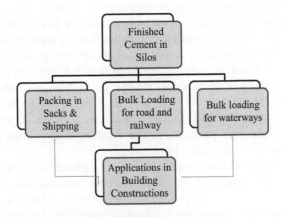

FIGURE 1.4   Steps in cement packing and shipping.

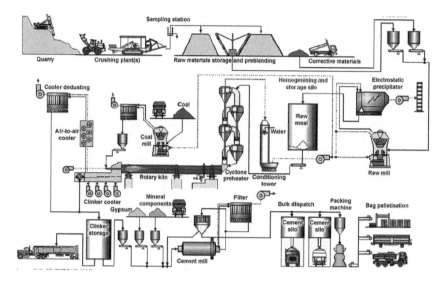

FIGURE 1.5   A typical process layout of a modern cement plant [9].

compressors, transformers, fans, blowers, conveyors, cooling towers, bag filters, electrostatic precipitators (ESP), lighting systems, transporting vehicles, etc. The achievement of steady-state operation and uninterrupted campaign of the production facility depends on the electrical and mechanical reliability of the auxiliary machinery. Coupled with the above requirements are the demands for energy conservation, resource sustainability, environmental protection, and low-carbon footprint. As a result, the technical management of cement production facilities has turned out to be a challenging task.

1.3.1   KEY FEATURES OF RAW MATERIALS INFLUENCING THE PROCESS

The process and the stoichiometric requirements indicate that a cement-grade limestone should have at least about 44.0% calcium oxide with magnesia not exceeding 3.5% and certain other impurities such as manganese oxide, alkali oxides, sulphur determined as sulfate, and phosphorus pentoxide individually below 0.5%. A limestone containing more than 0.15% Cl⁻ is also unsuitable for cement making under normal circumstances. Another important consideration is to ensure that the free silica or quartz content in the limestone should not exceed 8%. [10]. Moisture and clay contents in limestone often pose feeding and flow problems at the crushing stage, which operationally often turn out to be of crucial importance.

Limestone duly proportioned with other supplementary raw materials are subjected to milling in three alternative systems: ball mills, vertical roller mills (VRM), and hydraulic roll presses (HRP) often in combination with ball mills. A variant of roller mill in horizontal disposition, horizontal roller mills (HRM), has recently come into commercial application. New grinding installations are primarily VRMs. Roll presses are used in upgrading existing ball mills for increased production or decreased specific power consumption. All raw grinding systems are close-circuited with separators (also known as classifiers) for efficient grinding. It should be borne in mind that the productivity of the raw milling systems depends on feed size to mill, grinding behaviour of the feed material, drying capacity of the system, product fineness required, and the grinding power available. The above milling systems are also used for clinker grinding but with noticeable difference in performance, which has been discussed later.

While in general the raw milling is controlled for the stoichiometric requirements by the three important oxide ratios, i.e., Lime Saturation Factor (LSF), Silica Modulus (SM), and Alumina Modulus (AM), the homogenization of raw meal (which in effect is the ground raw mix) prior to pyro processing has always been a very important step in clinker manufacture. The most common homogenization system is the pneumatic one, which is based on air fluidization method. This system can either be discontinuous (or batch type) or continuous. The batch mode is adopted only in special cases where very high order of homogenization is required. It may have a blending factor of 20 or more. The most common practice is to have pre-blending of crushed limestone and continuous homogenization of ground raw mix. In plant practice, the homogeneity is determined on the basis of n numbers of hourly spot samples (normally 24 samples) and the targets are set as follows:

$$S_{LSF} \leq 1\%$$

$$S_{CaCO_3} \leq 0.2\%$$

$$S_{SM} \leq 0.1 \text{ unit}$$

where S is the standard deviation.

In some plants, the homogeneity is measured with four or eight hourly samples. These samples are analysed for major oxides and these data are converted to potential C_3S or LSF. A kiln feed should typically have an estimated standard deviation of less than 3% C_3S or 1.2% LSF. Although the homogeneity of kiln feed is expressed in terms of standard deviations of certain parameters, it should be borne in mind that it has a limitation in the sense that the standard deviations of any parameter does not distinguish between a steady trend and constant fluctuation.

There are no rigid standards for raw meal fineness. It is determined empirically and should be as coarse as a given kiln system can tolerate. Typically, a raw meal is ground to about 15% residue on 88-μm sieve and correspondingly to 1.5%–2.5% residue on 212-μm. With improved burning systems and techniques, the residues can be as high as 30% on 88-μm and 6% on 212-μm sieves, as maintained in some plants with large preheater-precalciner kilns. It should be borne in mind that an optimum but narrow particle size distribution is required, as fines tend to increase the dust losses by entrainment in exhaust gases, while the coarse particles are harder to react in the kiln, resulting in high free lime or high fuel consumption.

From the considerations of burning behaviour, it is advantageous to limit the top sizes of the following mineral phases in the raw meal as follows:

- Silica minerals: 44 μm (e.g., quartz, chert, acid insoluble residue, etc.)
- Shale particles: 50 μm
- Silicate minerals (e.g., feldspar): 63 μm
- Carbonate minerals (e.g., calcite, dolomite): 125 μm

How to achieve the target parameters in the raw material preparation section of the cement manufacturing facility is a complex issue because of close interactions of material chemistry, comminution process, and size reduction equipment. In the modern plants, the practice of automatic sampling and online analysis has made the quality control more effective, but it still has operational limitations. Certainly, there are unexplored opportunities for application of artificial intelligence (AI) techniques in this segment of the production process.

1.3.2 CRITICALITY OF THE CLINKER-MAKING STAGE

The success of cement production, however, hinges more on the raw materials behaviour and productivity at the clinker-making stage. The entire plant is designed and built, keeping the clinker-making stage at the centre. Reactions that occur in the course of clinker making are shown in a simplified form in Figure 1.6 [10].

The reactions are strongly dependent on the intrinsic reactivity of the materials and the empirical burnability parameter of the kiln feed prepared from the raw materials. The formation of clinker, irrespective of the system in use, involves the sequential occurrence of multiple chemical reactions at short intervals during the passage of the kiln feed through the kiln system. For illustrating the trend of formation of the major clinker phases along the length of the kiln, a model-derived diagram is presented in Figure 1.7 [11].

The important chemical transformations that occur in the modern kiln systems consisting of multistage cyclone preheaters, precalciners, rotary kilns, and grate coolers for the chemical reactions to take place can be summarized as follows:

- Calcining or decarbonation of limestone to the extent of 85% or more in the preheater-precalciner tower outside the kiln
- Solid state formation of compounds by the oxides released from both the carbonate and other components of the kiln feed upstream in the kiln
- Melt formation and liquid-phase sintering in the downstream burning zone inside the kiln

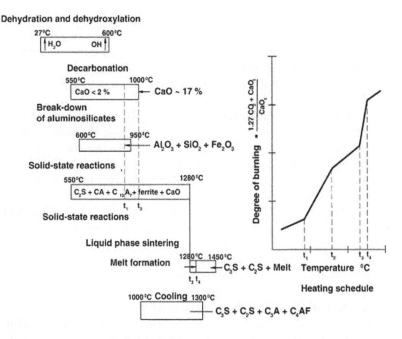

FIGURE 1.6 Approximate reaction sequence in clinker formation at almost constant rate of heating [10].

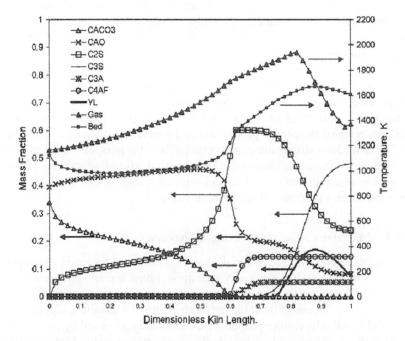

FIGURE 1.7 Formation of the major clinker phases along the kiln length [11].

- Cyclic occurrence of volatilization and condensation of alkalis, sulfates and chlorides between the preheater and the rotary kiln
- Stabilization of the clinker phases and microstructure in the cooler
- Combustion of fuels in the kiln and the precalciner
- Emissions of exhaust gases

In addition to the reaction kinetics, the decarbonation or calcination of raw meal depends on several systemic parameters such as the precalciner temperature, residence time of solids inside the precalciner, efficient separation of gas and solids, effect of dust circulation, etc. In fact, the entire kiln system has to deal with 10%–20% dust cycles at various sections, viz., between the cooler and the kiln, between the burning zone and the back end of the kiln, between the kiln and the preheater, between the preheater cyclones, and losses from the system. The liquid-phase sintering of the kiln feed at the burning zone is extremely critical for clinker formation. Obtaining an optimum quantity of the melt phase at the burning zone, which ranges from about 24% to 32%, and characterizing the melt properties such as viscosity and surface tension are crucial for the formation of clinker phases and in achieving the required granularity of the bulk clinker. The process of clinker cooling is not only to make the clinker physically amenable for transportation but also for phase stabilization, freezing of clinker microstructure, and preheating the combustion air. It is essential to understand that all the critical physico-chemical transformations of the in-process materials in the kiln system occur in quick sequence. The material residence time in the preheater tower is less than 30 seconds, in the rotary kiln section about 20–30 minutes, depending on the L/D ratio of the kiln, and finally in the clinker cooler the material passage time is about 30 minutes. Targeting the set points for most of the above reactions, monitoring the trends in operation, and course correction by process intervention in real time, with a view to maximizing the kiln productivity and minimizing the energy consumption, are the utmost complexities in clinker production. Further, it is important to understand that the correct preparation of raw meal upstream of the clinker production and converting clinker into the specified cements downstream are equally crucial from the perspectives of materials chemistry. Since, as already mentioned, a production line consists of four stages, one cascading into another, and since each stage has several units, the real-time control becomes all the more intricate.

1.3.3 MATERIAL CHEMISTRY IN CLINKER GRINDING

Clinker, being a multiphase material, shows significant variation in its properties due to changes in its chemical composition, phase assemblage, and microstructural characteristics. The phase composition variations are caused by the presence of foreign elements in the lattice structure of C_3S (alite), C_2S (belite), C_3A, and C_4AF. The clinker properties are ultimately governed by the relative proportions of the above four major phases, the polymorphic forms of alite

TABLE 1.4

Relative Performance of the Different Clinker Grinding Systems

Serial No.	Parameters	Ball Mill	Vertical Roller Mill	High Pressure Rolls	Horizontal Roller Mill
1	Achievable specific surface area (Blaines, cm²/g)	>6000	4500	4000	4000
2	Range of particle size distribution in terms of n value in RR plot	0.85–1.10	0.65–1.10	1.0–1.10	1.05–1.10
3	Typical specific energy consumption with respect to Bond's work index, kWh/t	33.5 (BWI:14.4)	27.5 (BWI: 19.6)	20.6 (BWI: 14.7)	17.5 (BWI: 14.4)

and belite, alkali-modified crystal symmetry of C_3A, and the solid solution state of the C_4AF phase [10].

The microstructural feature defined by the size and shape of alite and belite grains, characteristics of C_3A and C_4AF as the interstitial phases, clustering tendency of the silicate phases, mode of occurrence of free lime and free magnesia (periclase), clinker nodule porosity, clinker granulometry, etc. govern the grinding characteristics of the clinker. It has been observed that clinkers with fine granulometry, ill-formed microstructure, low porosity, high belite content, large crystal size, and high sulfate content show harder grinding properties [10].

Clinker is ground into cement by employing different milling systems, such as, ball mills, VRM, high-pressure grinding rolls (also known as HRP), and HRM. The relative performance of these mill systems is shown in Table 1.4 [10, 12]. The ball mills are the oldest and most prevalent ones, although they are highly energy inefficient. The energy considerations and the large throughput capacities required for large plants have driven the development of VRM and roller presses. These new systems are progressively surpassing the popularity of ball mills. HRM, or horomills as they are called, are the latest development in the field of grinding. It seems to combine the advantages of both the ball mills and VRM. The present limitation of horomills is the relatively low capacity for a single unit (180–425 t/h, depending on the feed material). Selection, system design, operation, and energy conservation of the grinding process are interlinked with each other and are influenced by the characteristics of materials to be ground.

1.4 RESOURCE EFFICIENCY AND MATERIAL FLOWS IN THE PRODUCTION PROCESS

It needs no reiteration that the production of cement is material-intensive. While, of course, the specific consumption of materials varies widely, depending on the scale, process, equipment, and regulatory measures in practice, it is

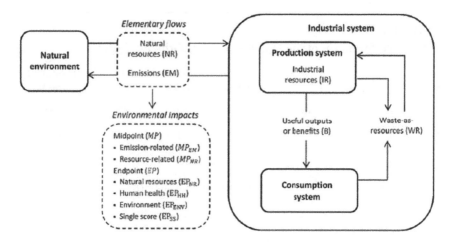

FIGURE 1.8   Schematic diagram of resource efficiency in a production system [13].

important to be equipped with the data on material flows and resource efficiency for a given facility. The fundamental concepts of using the natural resources and the consequent environmental impacts in a production system are shown schematically in Figure 1.8 [13]. While the material balance is computed from the input of natural resources to a production system and the corresponding emissions from it, the natural resources are converted into industrial resources in the production system with a material efficiency that is computed from the ratio of useful outputs to waste-as-resources returned to the system in a cyclic manner. The elementary flows of resources and emissions result in various environmental impacts, which can appear at mid-points or end-points of the flows and can be assessed with the help of specific models. The eco-efficiency of the resources is determined from the ratio of intended benefits to the assessed impacts. This approach for determining the resource efficiency and eco-efficiency of resources can be applied at different scales of production, micro to mega. When the concept is applied to a constituent unit of a production facility, the material efficiency of this process segment can be evaluated.

The material efficiency of the cement plant machinery and the constituent processes has not been studied as much as their energy efficiency and environmental impacts. Considering the production scales at various levels of the cement industry, this parameter deserves to be investigated more extensively for evolving proper resource policies. A specific investigation, reported for a plant in China having a five-stage preheater-precalciner kiln of average 2500 t/d clinker production, illustrates the specifics of such a study. The material flow profile of this plant, based on actual measurements, is given in Figure 1.9 [14]. The figure shows that 1.21 t of limestone, 5.6 t of air and 0.40 t of other materials including coal are required to produce a ton of cement of a given specification. Waste gases amounting to 4.79 t are discharged into the atmosphere. High- and low-temperature gases, 1.5 t in volume, are recycled to

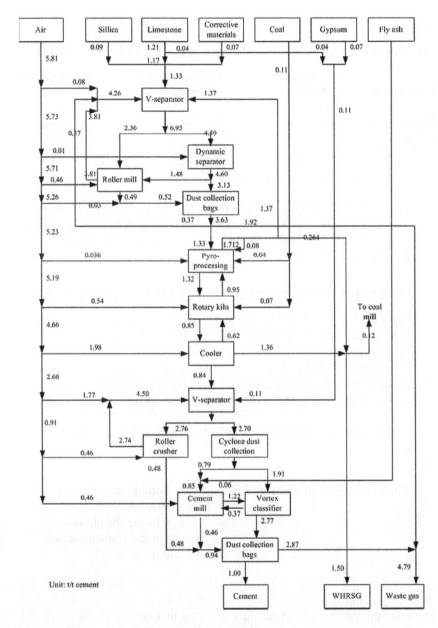

FIGURE 1.9 Flow of materials as measured in a cement plant in China [14].

generate electricity, and 0.12 t of hot air is reused for drying coal. The material efficiency parameters for the constituent units of the production facility are given in Table 1.5. The material efficiency values range from about 26% for the cement grinding to about 40% for the rotary kiln, the efficiency of other units remaining within this range. It is evident, therefore, that the

TABLE 1.5

Material Efficiency Parameters as Determined in the Chinese Plant [14]

Unit Processes	Material Efficiency (%)	Waste-as-Resource Generation (%)	Recycle Ratio of Waste (%)
Raw mill	36.69	63.31	16.33
Preheater tower	34.24	65.76	81.98
Rotary kiln	39.24	60.76	100.00
Clinker cooler	29.76	70.24	99.53
Cement mill	25.88	74.12	0.00

application of data analytics to understanding the perspectives of material efficiency in the cement production process of individual facilities as well as for groups of facilities is important for resource planning and also for formulating the resource policies.

1.5 THERMAL ENERGY PERFORMANCE OF THE KILN SYSTEMS

Broadly speaking, the thermal performance of a kiln system is assessed by its productivity and specific heat consumption. Although these parameters are generally estimated from the quantity of fuel consumed, its calorific value, and the amount of clinker produced during a given campaign, a more meaningful and comprehensive approach is to monitor the 'heat and mass balance' of the kiln system, based on the fundamental principles:

$$\sum \text{Mass of all input streams} = \sum \text{Mass of all output streams}$$

and

$$\sum \text{Total Energy Input} + \text{Energy Released*} = \sum \text{Total Energy Output}$$
$$+\text{Energy Absorbed}$$
$$* [*\text{During the physico}$$
$$-\text{chemical transformations}$$
$$\text{in the process}]$$

For monitoring the 'input' and 'output' streams, it is essential to recognize the 'boundary limits' of a system. The boundary limits, for example, for a typical preheater-precalciner kiln system are depicted in Figure 1.10, in which the inputs and outputs of energy are shown in the top and the inputs and outputs of materials are presented at the bottom.

A number of design improvements have already been introduced in the kiln systems for reducing their specific heat consumption. The improved inlet geometry of cyclones, configuration of immersion tubes for low gas velocities, spiral entry of gases into the cyclones, and increased number of cyclones in the preheater facilitated effective heat exchange resulting in low exit gas

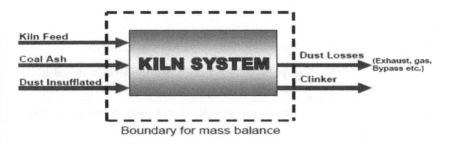

FIGURE 1.10 Schematic diagram of the kiln system for heat and mass balance.

temperature of 260°C–270°C and pressure drop not exceeding 600 mm of water gauge even with six stages. The modern precalciners are capable of accepting low-grade fuels, waste-derived fuels, and lumpy materials including shredded tyres with reduced NOx emissions. In addition, the design of burners has attained significant improvement with efficient air stream distribution and mixing of combustion air and swirl air. The air pressures in the burners can be manipulated for various contradictory demands such as high flame temperature, low NOx emission, stable kiln operation, and low power consumption. In the contemporary cement plants, multichannel burners with flame momentum of 2000% m/s and above are in use for hard-to-burn low-volatile fuels such as the petroleum coke.

For efficient clinker cooling, use of air beams, static grate with moving cross bars for clinker transport, and walking floors are some examples of design improvements in the new-generation grate coolers. These coolers have made a remarkable breakthrough in improving the heat recuperation efficiency. The cooling air input is reduced to a level of 1.7 Nm³/kg of clinker with minimum volume of air vented through the stack. These features have reduced the heat loss by about 30 kCal/kg of clinker. High-efficiency coolers operate on the principle of controlled supply of cooling air to the individual plates on the grate resulting in high thermal efficiency (>75%). The specially designed grate plates also prevent fall-through of clinker dust and in many cases are equipped with air-flow regulators to match the flow of air with the clinker bed height on the plate. Installation of energy-efficient fans and motors has further improved the energy performance of the modern plants.

Another effective energy conservation measure in the cement plants is the waste heat recovery from the exhaust gases for power generation. While the most common water–steam cycles operate at heat source temperatures as low as 300°C, for heat recovery from still lower temperatures, the systems based on Organic Rankin Cycle, utilizing organic compounds as process flows or the Kalina Cycle, using a water–ammonia solution, are now available for implementation in cement plants.

For the purposes of illustration, the thermal efficiency and productivity parameters as measured for two groups of preheater-precalciner kiln systems, one with five-stage cyclones and another with six stages, are given in Table 1.6 [15].

TABLE 1.6

Thermal Performance of Some Contemporary Kiln Systems [15]

Parameter	Five-Stage Preheater-Precalciner Kilns		Six-Stage Preheater-Precalciner Kilns	
	Best Achieved	Average of 10 Plants	Best Achieved	Average of 7 Plants
Kiln volume load t/d/m^3	7.0	5.7	7.1	6.0
Kiln thermal load mkCal/h.m^2	5.8	4.0	4.33	4.0
Specific fuel consumption, kCal/kg clinker	707	730	686	702
Preheater exit gas volume, NM^3/kg clinker	1.39	1.53	1.45	1.50
Preheater exit gas temperature, °C	260	311	245	272
Preheater exit gas draught, mmWG	380	542	450	576
False air across the preheater,%	2.0	7.0	2.0	7.6
Cooler cooling air Nm^3/kg clinker	1.70	1.92	1.60	1.80
Clinker discharge temp. °C	120	145	120	145
Heat loss through the preheater exit gas kCal/kg clinker	126	165	127	141
Heat loss: cooler vent + water spray, kCal/kg clinker	85	104	96	106

1.6   CEMENT KILNS FOR EXTERNAL WASTE MANAGEMENT

The cement kilns are well proven to thermally treat the combustible waste materials, in many cases better than the conventional incinerators. This is because of the fact that the cement kilns are suitable for safe destruction of hazardous and toxic organic substance for providing an alkaline and oxidizing environment, for trapping heavy metals in clinker, and certain other operational merits. These advantages of the cement kiln systems opened up an avenue to solve the problems of massive quantity of waste generated outside the boundary of the industry from manufacturing and material processing plants, agriculture and municipalities. A severe problem is the increasing generation of urban solid waste, which is more than 2 billion ton in 2016 and is projected to touch 3.4 billion ton in 2050. The unrecyclable combustible part of this waste can be treated in cement kilns after necessary pre-processing

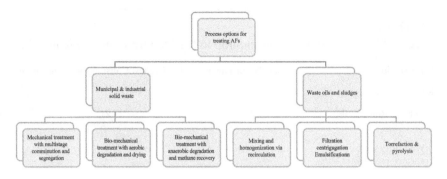

FIGURE 1.11   Pre-combustion treatment options for wastes [16].

(Figure 1.11) [16]. The technological and engineering developments in this respect have been significant.

The purpose of all such pre-treatment processes is to remove undesirable components present in the alternative fuels, increase their calorific value, homogenize the fuels, and convert them into a form that can be fed into the pyroprocessing systems more easily. It is pertinent to note that an effective and efficient pre-treatment facility for solid waste should have primary and secondary crushers/shredders, ferrous and non-ferrous metal separators, air sifters, as well as the essential feeding systems, conveyors, and discharge belts. The facilities are sometimes equipped to mix combustible liquid or sludge with segregated solid fractions in the production of refuse-derived fuels (RDF). The pre-treatment facilities are increasingly becoming parts of the total logistics in cement production. The contemporary plants, therefore, are now designed and operated with adjoining pre-treatment facilities or firm and regular supply links from distantly located units.

1.7   ELECTRICAL ENERGY PERFORMANCE

All the major machineries in the cement production facility are continuous in operation with a large number of moving parts. Further, their functions depend on a series of conveyors and transportation systems. Hence, the energy efficiency of the major equipment and, more particularly, of the grinding installations as well as the electrical energy used by a plethora of motors, drives, fans, blowers, compressors, etc. are crucial for sustainable operation of the cement production facility. The salient features of energy conservation in the modern plants include:

- Replacing ball mills by vertical or HRM and high-pressure roll presses
- Use of variable frequency drives for all process fans, high-capacity bag filter fans, compressors, etc.
- Mechanical conveying in place of pneumatic facilities

TABLE 1.7

Specific Electrical Energy Consumption in the Modern Cement Plants [15]

Plant Section	Best Performance	The Highest Consumption Monitored
Raw material grinding with		
• Vertical Roller mills	13.3	21.2
• Ball Mills	16.5	26.3
(kWh/t of raw meal) Preheater-precalciner kiln system with		
• Five stages of cyclones	16.3	32.0
• Six stages of cyclones	23.7	27.5
(kWh/t of clinker) Cement grinding system with		
• Ball Mills		
• Ball mills with pregrinding	27.2	45.2
VRM	23.8	30
• Vertical Roller Mills (stand-alone)	21.0	42
(kWh/t of cement) Total plant (kWh/t of cement)	67.2	88.0

- Low pressure drop in preheater and other systems
- Efficient power supply and distribution systems

The status of the specific electrical energy consumption in the modern plants as measured for a cluster of 17 units is illustrated in Table 1.7 [15].

1.8   POLLUTIONS AND EMISSIONS IN CEMENT MANUFACTURING

The cement industry is responsible for discharge of a large quantum of pollutants into the atmosphere. The contribution is not limited to the real-time industrial operation of a cement plant, but it also encompasses a whole range of activities covering the backward and forward linkages. The upstream stage is essentially the stage of obtaining raw materials and energy inputs, which gradually leads to depletion of natural resources – ecological disturbances like soil erosion and deforestation, noise, and emissions of gases like carbon dioxide. At the process stage, the problems are due to emissions of particulate matters and gases affecting the ambient air quality including the greenhouse gases. The downstream stage essentially relates to the process-derived wastes, the disposal of which, in the absence of a viable waste management system, would result in the loss of valuable land on one hand and increasing accumulation of wastes and pollution of environment on the other. The concept of sustainable production involves elimination or minimization of all the above problems through prevention, containment, and waste utilization and benign disposal.

The most important polluting substances within the meaning of the directive of most of the countries in the world are as follows:

- Process-generated and fugitive dust
- sulphur dioxide and other sulphur compounds
- Oxides of nitrogen and other nitrogen compounds
- Organic compounds, in particular hydrocarbons (except methane)
- Heavy metals and their compounds
- Chlorine and its compounds
- Fluorine and its compounds

The abatement techniques for dust emission primarily include the use of fabric or bag filters, ESP, and gravel bed filters, often supported by cyclones as pre-collection systems. The primary techniques for reducing the generation of nitrogen oxides are based on the use of low-NO_x burners or the use of staged combustion in the modern precalciner kilns. The secondary methods suitable for the cement production process are the selective 'non-catalytic' and 'catalytic' reactions (SNCR and SCR). The emission of sulphur dioxide is not a major issue in the cement plants. In special situations, where the sulphur dioxide emissions are unusually high, the technology of online injection of calcium hydroxide or end-of-the-pipe scrubbing is adopted. The cement plants need to be provided with monitoring systems for both dust and gaseous pollutants. For dust emissions the corona power based system for monitoring the performance of ESPs is installed. Sometimes the opacity meters are used for continuous monitoring of stacks. Although the continuous monitoring systems are preferred, they have their limitations and problems. Sometimes manual methods of gas sampling and measurement of pollutants are taken resource to. The abatement techniques and the pollution-monitoring methods have been described in detail by the author in [10].

In cement plants, there is always a problem of noise pollution. It occurs at all stages of the process and originates from the machinery such as crushers, grinding mills, fans, blowers, compressors, conveyors, etc. Generally, the noise level in cement plants varies from 80 to120 dBA (decibels) [10]. There are national standards in different countries for maintaining the ambient noise below the specified limits during the day and night times, and for time-dependent noise exposure limits for the plant operating personnel. The reduction measures for noise pollution involves three elements, viz., the source, the transmission path and the receiver, i.e., the working population. The noise control at source is tackled generally by using vibration damping pads under the machine base or isolating the vibrating components from the main body of the machine. The choice of the measures is based on the frequency spectrum of noise. The noise control in the transmission path is accomplished by erecting barriers or enclosures in between the source and the workers. In the construction of such structures, sound-absorbing materials are used, the selection of which is site-specific and depends on the acoustic properties. Still the noise prevention measure at the receiver end involves the use of personal gears of ear protection.

As far as the greenhouse emissions are concerned, the global cement industry is responsible for about 8% of the total anthropogenic CO_2 emissions. Hence, the carbon dioxide emission continues to remain a major concern for the Portland cement manufacturing process. The concern emanates from the release of about 535 kg CO_2 per metric ton of clinker from the limestone calcination and about 330 kg CO_2 per metric ton of clinker from the fuel combustion, resulting in direct emission of 835 kg CO_2 per metric ton of clinker. The corresponding figure for cement would vary, depending on the quantity of clinker used in making a tonne of cement and the grinding technology adopted. The industry at large has adopted several measures to reduce the CO_2 emission level at the clinker-making stage [10]. The effective ones include the following:

- Use of alternative non-carbonate calcium-rich raw materials
- Enhancing use of alternative fuels in place of conventional coal
- Making unit operations more energy efficient
- Generating electricity with waste heat
- Increase in the use of renewable energy

However, it is doubtful if the above measures will be enough to reach the targeted 50% reduction in CO_2 emission per metric ton of cement produced by 2050 with respect to the emission level of 1990. The broad emission scenario is depicted in Figure 1.12 [17], which traces the overall 450-ppm CO_2 mitigation path with a sharp fall in CO_2 emissions from 2025 to 2050. On the contrary, the global cement production in the business-as-usual mode is predicted to show a 260% increase in specific CO_2 emissions, instead of 50% reduction. It would perhaps be imperative to adopt the additional measure of carbon capture and storage (CCS) in addition to the measures indicated above. At the same time, it is understood that the financial implications of adopting CCS are so adverse for the cement industry that it may not turn out to be viable on its own. The economic feasibility of CCS will depend not only on the viable trading of CO_2 but also on the sale of value-added products that could be developed with sequestered CO_2. This strategy, therefore, brings in a focus on carbon capture and use (CCU) in addition to CCS itself.

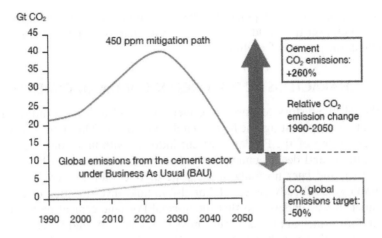

FIGURE 1.12   Trends of CO_2 emissions under different conditions.

In general, the cement industry has so far attempted to make use of the best available technologies to prevent pollution and emissions from the process but more technological advances will be necessary to meet the future requirements of sustainability. The present status of the environmental sustainability efforts can be summarized as follows:

- Quarry reclamation through conservation, stockpiling and use of top soil and overburden, re-contouring of slopes to minimize erosion and run-off, and planting of native vegetation
- Stack emissions for dust brought down to 30 mg/Nm^3
- Stack emissions of SO_2 contained to less than 100 mg/Nm^3 (unless special dispensations are called for due to pyrites content in raw materials exceeding 0.25%)
- Stack emissions of NO_x contained to less than 1000 mg/Nm^3 in preheater-precalciner kiln systems
- HCl emission not to exceed 10 mg and HF emission to 1 mg/Nm^3
- Total Organic Carbons (TOC) emission below 10 mg/Nm^3
- Emissions of Hg, Cd, and Tl and their compounds not to exceed 0.05 mg/Nm^3
- Heavy metals (Sb + As + Pb + Co + Cr + Cu + Mn + Ni + V and their compounds) to be limited to 0.5 mg/Nm^3
- Dioxins and Furans emission not to exceed 0.1 ng/Nm^3 at 10% O_2
- Maintaining the ambient air quality in terms of yearly mean for dust and gaseous emissions below 100 μg/m^3
- Control of noise pollution below safe limits
- High level of wastes recycling
- Recycling of cooling water
- Reducing the CO_2 emission

Thus, sustainable development has become an integral part of cement production all over the world. However, the road map is complex, regional, and is still in the process of evolution.

1.9 CHARACTERISTIC FEATURES OF PORTLAND CEMENTS

Although the history of Portland cement is spread over two centuries from the date of its patenting, the first seven decades of the 20th century saw massive expansion of the Portland cement industry with numerous engineering innovations and deeper understanding of cement chemistry. Changes were seen then and later not only in the scale of operation and energy-efficient technology, as already discussed, but also in huge diversification of the basic product. It was understood that the normal Portland cement performs well when four principal phases, for example, C_3S, C_2S, C_3A, and C_4AF appear in the composition in a given proportion. It was also understood that the family of Portland cement can be expanded by adopting several process steps at the time of manufacture as shown in Figure 1.13.

By virtue of these manufacturing strategies, the normal Portland cement has added to its family rapid hardening cement, sulfate resisting cement, low heat cement, and white cement, in which only the relative proportions of the

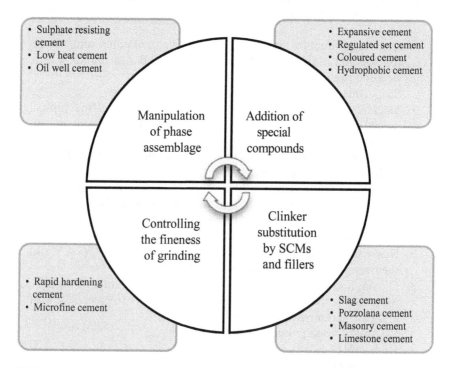

FIGURE 1.13   Process steps for producing different varieties of cements.

phases changed with or without the adjustment of fineness. Another class of products such as expansive cement, regulated set cement, and coloured cement are produced by adding special compounds to the basic cement. The third group of cements, called blended cements, are manufactured by substituting clinker partly with granulated blast furnace slag, fly ash, natural pozzolans, limestone powder, etc. All these products are manufactured in accordance with standard specifications adopted in almost all the major cement-producing countries. The production of cement, therefore, cannot be undertaken without an elaborate laboratory infrastructure for characterization and testing of properties. The important properties such as consistency, setting time, compressive strength, soundness, and their relationship with chemical and phase composition of cements are still empirical. The 28-day compressive strength of cement mortar is the most important quality parameter, the compliance of which in a production facility entails prolonged storage of cements before shipping. Numerous attempts have been made to predict the 28-day strength from the early age properties by statistical analysis and mathematical modelling and the validity is mostly local [18]. Further, the application of the products and their field performance depend on hydraulic reactivity, the hydrate phases formed and the microstructure that develops in the hydrated cement paste. Since the cement paste forms the core of concrete and is primarily responsible for its performance, the cement performance is closely interlinked with durability of concrete and with its environmental impacts during and after use in construction. The life cycle assessment of cement is crucial in this context. These concepts have grown widely in the industry but their objective implementation is still arbitrary in practice.

1.10 MODELLING, SIMULATION, AND ADVANCES IN PROCESS AND QUALITY CONTROL SYSTEMS

The cement production is a multi-input and multi-output process with strong couplings and inherent non-linear behaviour. Hence, its automation is a challenging proposition and in most plants, the unit processes are designed with conventional control loops augmented by various expert knowledge systems with empirical correction factors, which have been derived from extensive modelling studies. The process optimization studies in the initial stages used to rely extensively on physical models based on air–water or acid–alkali mixing techniques. Subsequently, simplified mathematical models of precalciner kilns were developed based on empirical relations to describe the physical and chemical processes. Such models were and even now are useful in determining the overall heat and mass balance, sulphur and alkali cycles, raw mix proportioning and homogenization, etc. The models make use of plant-derived algorithms and are validated with more operational data from the stably running plants. Even then, the generalization of the findings is limited in application as the process models are essentially for steady-state conditions. Attempts were made to refine the models, for example, by numerical integration of ordinary non-linear differential equations [19], combining 1-D longitudinal and 2-D cross-sectional models to analyse spatial temperature distribution [20], simulating mass fractions along the kiln axis [21], and so on. A detailed modelling study based on energy and material balance for a kiln system was reported in [22]. The model was based on energy and material balances for the kiln section that had been divided into about 1000 balance locations. The most important chemical reactions calculated in these locations were the calcination of the limestone, the clinker phase formation, and the combustion of fuel. Most of the process engineering principles concerned the circulating dust systems, which included the repeated mixing and separation of gas and dust in the cyclone stages and the entrainment of dust in the kiln and the cooler. The particle size distribution of the clinker in the cooler, which can have a significant influence on energy consumption, was also taken into account. The differences between the calculated and measured parameters were significant in many locations. As an illustration, the differences between the measured and calculated values of the volatile constituents of the feed samples passing through the preheater section and the clinker from the cooler discharge are shown in Table 1.8.

Apart from such observed differences between the simulated and actual values of different parameters, the cement plant operations have become more complex over time with several time-dependent and interrelated processes, more particularly with additional fuel streams and sometimes, with unconventional raw materials. Some present approaches include computational fluid dynamics (CFD) and artificial neural network (ANN). The CFD models in particular enable the identification of fuel and air mixing regions, dust recirculation in gas streams, internal build-ups, development of reduction conditions, and formation of pollutants in the event of use of alternative fuels. Based on such studies, the present status of the plant control systems in the modern plants has evolved as outlined below.

Distributed digital control system (DCS) comprising separate large memory capacity and Programmable Controllers with high-speed computation are installed in the cement plants for each process section, replacing the conventional motor control and PID control systems. The control systems in the modern plants consist of human–machine interfaces, control software, and programmable logic controllers. They include data packages that can bring out trends of control parameters, alarm provisions, and even log details of shift operators. These packages have large flexibilities to change the graphics and control logic and the unit processes are controlled from a central control room. The process instrumentation has expanded considerably and computer models are used to operate complex processes. Fuzzy-type or rules-based logic gained wide popularity in the 1990s, and its use is continuing more extensively. The processes for kiln optimization and mill control are predominantly based on rules-based fuzzy logic. However, after being on the fringe for many years, the latest versions of neural net technology and model-based

TABLE 1.8
Calculated and Measured Volatile Constituents in the Feed, Cyclone Samples and Clinker

Constituents %	Preheater Inlet		Precalciner Inlet		Kiln Inlet		Cooler Discharge	
	Calculated	Measured	Calculated	Measured	Calculated	Measured	Calculated	Measured
K_2O	0.602	0.55	9.942	0.55	1.307	1.09	0.758	0.70
Na_2O	0.998	0.06	1.269	0.44	0.205	0.05	0.138	0.20
SO_2	0.262	0.32	8.688	3.65	1.799	0.60	0.758	0.62
S^{2+}	0.157	0.21	0.000	0.130	0.000	0.08	0.000	0.00
Cl^-	0.044	0.011	5.080	7.63	0.393	1.13	0.005	0.00

Source: Data extracted from [19].

predictive techniques are coming to the fore as competitive options. Expert packages with logical dynamic modelling tools and capable of integrating camera signals and soft sensors are some of the advanced systems in the market. The ramp-up in the market for expert systems in future would depend more and more on integration with high-quality soft sensors of in-process materials, camera signals, online particle-size analysers, etc. Further, many supervisors and laboratory managers have started making use of remote access software to communicate and to provide assistance to the plant. The next phase of control strategies seems to be heading towards intelligent field devices that use self-diagnostics and can electronically communicate specific instructions to the maintenance set-up of the plant. There is no doubt that technologically the plant control systems are progressing quite rapidly towards autonomous operation.

In the area of quality control, online Prompt Gamma Neutron Activation Analysers (PGNAA), robots coupled with X-ray fluorescence spectrometers, online free lime analysers, and particle size analysers have ensured stable uninterrupted operation of the kiln system to a large extent. Recent developments in the use of X-ray diffraction are changing the traditional methods of quality and process control, as they have the ability to measure mineral phases or compounds formed directly in real time. Cement and clinker production involves chemical reactions to produce precisely controlled blends of phases with specific properties. So far there has been overwhelming dependence on either offline or online oxide or elemental analysis of raw or in-process materials for QC. Methods and equipment are now available for continuous quantitative on-stream analysis of the mineral or phase composition of cement and clinker. The instrument is a stand-alone piece of equipment, which is installed at the sampling point. A sample for analysis is extracted from the process stream and after due preparation online the sample passes through the X-ray beam. The diffracted X-rays are collected over 0 to 120° by a detector. The Rietveld structural refinement technique is applied to analyse the resulting diffraction pattern. The analysis of the moving stream is done in close frequency of, say, once every minute. All analysis results are communicated directly to the plant PLC system. The real-time measurement of the mineral composition of cement and clinker for process control is a paradigm shift for the cement industry. The discernible benefits of using on-stream X-ray diffraction are the following:

- Control of kiln burner based on free lime, clinker reactivity, alkali and sulphur contents
- Control of cement mill separators and feed rates and proportions to achieve consistent cement strength at minimum power consumption
- Control of gypsum dehydration through cement mill temperature to give consistent setting times
- Control of mill weigh-feeders for different feed materials

The net advantages of implementation of such online QC systems are the optimum performance and cost, reduced risk of product failure, and consequent marketing benefits.

Another development in the on-stream analysis, apart from the widely used bulk analyser based on γ-radiation, is the application of infrared spectra that are provided by the stabilized white light source. The light illuminates the target bulk material to be analysed as it passes the unit on an existing conveyor belt. The infrared radiation excites vibrational oscillations of the molecular bonds in the material under test, which results in reflection and absorption spectra that are characteristic of minerals being analysed. The Near Infrared (NIR) ranges are applied for analysing limestone materials. It is claimed that the IR-based online bulk analyser shows better performance for the cement raw material constituents than the traditional γ-ray equipment. One additional advantage in this new development is the avoidance of potentially hazardous excitation sources.

Broadly speaking, the present state of development in the process and quality control is essentially for real-time control. The objective of autonomous operation is still some distance away. Developments pertaining to dynamic non-linear model predictive control (NMPC) are still to catch up. One such investigation for a 2400 t/d kiln system, based on a non-linear autoregression moving-average model with exogenous inputs (NARMAX), has been reported in [23]. The predictive control system is designed to achieve at the same time multiple objectives, such as minimizing the product quality deviations with respect to a reference, stable run of the process with predicted change of the disturbance mean value, meeting the production volume target and satisfying the input constraints. Consequently, the controller design turned out to be a complex optimization problem. Further, the minimization of the production volume error was augmented with a second output and an integrator. Since there were significant deviations in the production volume goals, an adaptive weighting of the cost of the function was introduced. The functionality of the designed NMPC was demonstrated with the help of a closed-loop simulation and three control approaches are compared in Figure 1.14, in which one may observe comparable fast transient response of

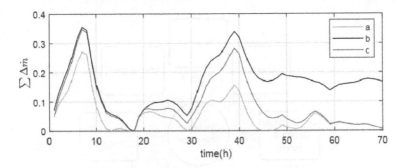

FIGURE 1.14 Cumulative production goal deviation: (a) mass flow reference tracking; (b) integration of production volume error; (c) adaptive weighting [23].

the controller with adaptive weighting (curve c) to that of mass flow reference tracking (curve a), while meeting the production goal over a finite time span.

It is important to note that the intrinsic nature of the cement production process demands non-linear dynamic model predictive controls in place of the widely prevalent steady-state MPCs. This advancement strategy obviously calls for more investigations involving data-driven modelling.

1.11 INTEGRATION OF BUSINESS, MANAGEMENT, AND PRODUCTION

In late 1990s and in early years of the 21st century, the cement industry became more and more market-driven from its earlier orientation of being production-driven, which unfolded a different paradigm for the CIMSs in practice in the modern cement enterprises. It became imperative to integrate all process information islands, such as maintenance, procurement, logistics, human resources, and commercials on one hand, and unit processes such as mining, comminution, pyroprocessing, and packing on the other, as this kind of integration only would support the decision making at operational, management, and execution stages. In multilocational units, the goals were to extend the integration of process control and automation systems with asset management and manufacturing operations at all levels and locations. In practice, the integration has taken place between ERP and manufacturing execution system (MES). While the ERP system included the whole range of management such as the financials, materials management, procurement, marketing, sales, and customer services, the MES covered production scheduling, process management, material tracking, and quality management. This integration has been possible due to the evolution of the plant process control to a more precise state and introduction of systems standardization in certain critical functional areas (Figure 1.15).

The functional areas indicated in Figure 1.15 are undoubtedly critical to operating a cement manufacturing facility and ensuring its economic viability. It was realised in 1980s that in bulk manufacture of chemical products like

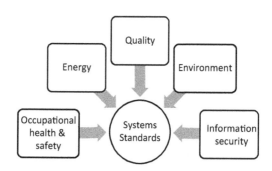

FIGURE 1.15 Critical functional areas for systems standardization.

cement it is practically impossible to physically test every gram produced and, therefore, it would be more effective to standardize the production process, in addition to testing periodic samples tested for compliance with the prescribed standard specifications. This consideration led to the practice of quality management systems, which culminated in the adoption of ISO 9000 family of standards in 1987. Within the next two decades, along with the periodic revisions of the standards, over one million organizations worldwide adopted ISO 9000 series in their operations. In keeping with the global trend, a large number of cement production plants in the major cement-producing countries adopted ISO 9001:2015 and they continue to practice seven quality management principles as shown in Figure 1.16 [24].

It is important to note that the QMS standard laid the path for the systems standardization in other critical functional areas shown in Figure 1.15. The ISO 14000 series for the environmental management came into practice in 1996 and the ISO 14001:2015 is widely adopted in the cement industry to minimize the adverse effects of operations on air, water, or land, to ensure compliance with applicable rules and regulations, and to improve continually the environmental performance of the plants. Following the benefits of QMS and EMS standards, the cement production facilities have progressively

FIGURE 1.16 Quality management principles in ISO 9001 standard.

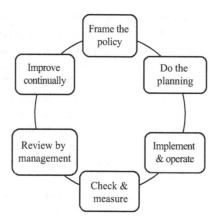

FIGURE 1.17 Expanded PDCA cycle.

integrated the energy management system standard (ISO 50001:2018) and occupational health and safety systems standard (ISO 45001:2018). With the vastly expanding information and database, the industry is preparing itself to embrace ISO 27000: 2018 series for information security management systems, which include financial information, employee data, intellectual properties, third-party information, etc.

Fundamentally, all the above standards rely on the same model. They are voluntary internal management tools with large documentation. The tools work on the principle of plan-do-check-act (PDCA) cycle (Figure 1.17). The standards become more effective to implement with authentic third-party certification. The system standards are essentially non-prescriptive with multiple interfaces. Hence, the operation of these management tools in the digital environment of the future cement manufacturing facilities would necessitate serious review.

1.12 INTEGRATED FEATURES OF CONTEMPORARY CEMENT PLANTS

Based on the foregoing discourses, a broad outline of the design and performance of the contemporary and near-future cement plants can be drawn up as follows, bearing in mind that the first and foremost criteria are the resource conservation, sustainability, and energy efficiency. It seems practical to forecast that the clinkering plants would continue to be installed with five–six stages of cyclones in preheater, low NO_x precalciners, ultra-short rotary kilns, multichannel burners, and advanced heat recuperation technology, which will offer specific heat consumption below 3 MJ/kg clinker. Apart from the high-performance grate coolers and tertiary air utilization, which are currently in practice, further opportunities that still exist for waste heat utilization will be harnessed to produce electrical power, industrial steam or hot water for local heating purposes. The concept of waste burning is now an accepted practice. Although the cement industry by itself is an insignificant generator of solid and liquid wastes, the cement kilns are likely to take up the increasing role of environment cleaning by utilizing alternative fuels and raw materials from other sectors without creating additional emission problems.

Upstream and downstream of the kiln systems, further advances will be seen in the substitution of ball mills by roller mills and HRP both in raw milling and finish grinding. The advances will be reflected in the process layout with more efficient separators, introduction of high-capacity HRM, improved design of process fans, shift from pneumatic to mechanical transport of materials, and use of high-efficiency electric motors and drives. The application of ultrasound techniques to separation and milling technology seems to be nearer to commercial adoption. Noise pollution reduction will be emphasized more in coming years. The future plants with their associated mining and crushing plants will undergo stricter regulatory controls, resulting in much quieter and cleaner manufacturing.

The process and quality controls in cement works of the future are likely to become more data-driven and more precise in real time. Since the end-properties of cement are dependent on chemistry and mineralogy, the online analysers for monitoring raw materials and ground raw mixes, the homogenizing process, the rotary kiln burning, and the finish cement grinding will not be limited to only the elemental or oxide composition and will extend to determining the minerals or compounds present. With such real-time data, the process parameters will be programmed to optimize the entire plant performance. Further, the operator response to process fluctuations, elimination of partial or no-load running of equipment, and higher levels of process automation will enhance the performance of plants. In addition, certain key elements of production such as energy, quality, environment, occupational health and safety, and information security will call for special attention including systemic standardization to the extent implementable.

In addition to the operational aspects, the visual impact of cement plants will turn out to be an important consideration. In most cases, the architecture of cement plants is monotonous and it creates discordance with the natural surroundings. Harmonization of the plant layout and architecture with the surroundings is a social requirement, which is hardly recognized. Some plants have adopted colour screening in order to camouflage the starkness of the kiln line against background of the surrounding countryside. Many others are trying to create visually appealing cement plant landscapes. Future plant designs should unquestionably develop such ideas further.

1.13   CONCLUDING OBSERVATIONS

The most extensive and diverse applications of Portland cements from tiny repairs to construction of massive structures in all climatic conditions have been possible due to the low unit value, domestic or regional production with local raw materials, and time-tested durability of the products. In keeping with the growth of the construction sector, the worldwide demand for cement has continued to increase; it crossed 2.0 billion ton per year in 2003, 3.0 billion ton in 2009, and 4.0 billion ton in 2013. Although the demand was forecast to exceed 5.0 billion ton in 2019, it did not happen due to the global economic recession. However, the demand is on a growth path now from a hovering level of 4.2 billion ton per year in 2019. The average per capita consumption of cement in 2018 was 525 kg, which is more than double that of steel and manifold higher than other primary construction materials. It is obvious, therefore, that the cement production is the largest factory-made materials industry in the world. Apart from the gigantic global production volume of cement, it is important to note that the average scale of production in the major cement-producing countries and in the top ten cement companies is very large, approximately in the range of 1.0–3.0 million ton per year. The largest operating kiln in the world today produces 14,500 tons per day or about 5.0 million ton per year of clinker. It is, however, important to note that despite the scale advantage, such large capacity kilns may not be the most frequently installed systems all over the world because of potential difficulties of logistics and cement evacuation. In all probability, the preferred capacity range will be 4000–8000 tons/day. It is important to consider that the product management capability of existing and new installations will be engineered in such a manner that multiple blended products manufactured from the same clinker may be stored, packed and dispatched simultaneously. Apart from the compulsion of economy of scale for the bulk production of clinker and market demand for different types of blended cement, the need for separate manufacturing facilities for niche products with specialized processing equipment as satellite plants may emerge. In this context, it is relevant to mention that the fluidized-bed clinkering plant of relatively small output is already in commercial operation.

The present practice of assessing the performance of cement plants is to use indicators such as the specific energy consumptions (thermal and electrical), emissions (dust, NO_x, and SO_2), and plant availability. It appears that the key performance indicators in the future plants will have to include additionally the following parameters:

- CO_2 emission (kg CO_2/t of cement)
- Resource efficiency
- Materials efficiency in the unit processes
- Data analytics performance

The worldwide focus and debate on CO_2 emissions will further intensify over coming years. In addition to the ongoing practices of clinker substitution, use of alternative fuels and raw materials, cogenerating electricity and energy conservation, the capture and recycling of CO_2 will move towards being viable through significant research programmes being carried out all over the world. Transformation of CO_2 into value-added binders or fuel or both are the directions of research and development, which the industry worldwide will continue to track and support.

Resource efficiency, signifying the quantity of inputs needed to produce a unit of output, will have to be central to the design of the production system. The parameter will depend on the ratio of wastes (including wastewater, air pollution, and greenhouse gases) to the inputs (including raw materials, energy, and water) per unit of output in the production process. In addition, the environment cleaning efficacy of the process in terms of utilization of industrial wastes and municipal refuse will be assessed. A corollary of this strategy will be the quantification of the extent of adoption of circular economy and partnership with other industrial units.

Material efficiency of the major cement machinery and the constituent processes, as defined earlier, is computed from the ratio of useful output to wastes-as-resources returned to the system in a cyclic manner. Monitoring and improving this parameter for each of the unit processes will help in achieving high efficiency of the integrated system.

The information management system (IMS) is at the core of transforming the cement manufacturing into Industry 4.0 standard. The present level of IMS is organized primarily for online process and quality monitoring in real time, while the cement business structure has evolved much beyond the traditional configuration of centralized groups to need-based networks. The larger the network, the more valuable it can become by combining the experience and expertise of different but relevant segments. These networks are effective, only if the real-time and historical information is available in easy and flexible format everywhere across the network with reliability and accuracy. A strong and effective IMS is more pertinent for the cement business due to its globalization as the information technology facilitates the optimization of resources across wider geographical and functional areas.

In summary, one has to reckon that the cement industry is characterized by high material footprint, large consumption of energy, and high CO_2 emissions, all of which have serious environmental implications. In 2016, the global cement industry, having an annual production volume of over 4.0 billion ton, consumed about 11 EJ of energy and emitted about 2.2 Gt of CO_2. Although there are no definite statistics, one may tentatively estimate that the industry would have consumed raw materials and cement blending materials in total quantity close to its production volume. These negative features of the industry are predicted to increase many folds by 2050, adding to the problems of climate change and unsustainable development. Hence, a technology roadmap for the cement industry to move towards a low-carbon production process has become indispensable. Such a roadmap has to include the path to Industry 4.0 standard having elements of big data analytics, machine learning algorithms, cloud computing, internet of things, virtual twins, and augmented reality.

REFERENCES

1. https://www.reportlinker.com/p05817680/Global-Cement-Industry.html?utm_source.GNW
2. https://www.worldsteel.org/media-centre/press-release/
3. www.world-aluminium.org/statistics
4. www.brinknews.com (>quicktake>plastic-production-on-the-rise)
5. https://www.statista.com/statistics/global-cement-production-volume
6. Armstrong T. (2018), *A review of the global cement industry trends*, International Conference 'Concreatech', Cement Manufacturers Association, New Delhi, India.
7. https://www.cemnet.com/global-cement-reports/country/the-next-20-years
8. https://www.cemnet.com/news/story/168425/
9. European IPCC Bureau BREF CLM 2013, Seville, Spain. https://eippch.jrc.ec.europa.eu
10. Chatterjee A.K., (2018), *Cement production technology: principles and practice*, CRC Press, Florida, USA.
11. Mujumdar K.S. and Ranade V.S., (2006), Simulation of rotary cement kilns using a one-dimensional model, *Chemical Engineering Research and Design*, 84, 3, 165–177.
12. Aydogan N.A. and Benzer H. (2011), Comparison of the overall circuit performance in the cement industry: high compression milling vs ball milling technology, *Mineral Engineering*, 24, 3–4, 211–215.
13. Huysman S., Sala S., Marcini L., Ardente F., Alvarenga R.A.F., Meester S de, Mathieux F., Dewulf Jo., (2015), Toward a systematized framework for resource efficiency indicators, *Resources, Conservation and Recycling*, 95, 68–76.
14. Gao T., Shen L., Shen M., Liu L., Chen F., (2016), Analysis of material flow and consumption in cement production process, *Journal of Cleaner Production*, 112, 553–565.
15. Confederation of Indian Industries and Cement Manufacturers Association, (2015), *Report on energy benchmarking for cement industry – version 2.0*, CII Sohrabji Godrej Green Business Centre, Hyderabad, India.
16. Chatterjee A.K. and Sui T., (2019), *Alternative fuels – effects on clinker process and properties*, Cement and Concrete Research: Special Issue: Keynote papers of International Congress on Cement Chemistry, Prague, Czech Republic, p. 105777.
17. www.feu.awsassets.panda.org/downloads/cement_blueprint_climate_fullenglrep_ir.pdf
18. Tsamatsoulis D. (2012), Prediction of cement strength: analysis and implementation in process quality control, *Journal of the Mechanical Behavior of Materials*, 21(3–4). 81–93.
19. Chen Y. Y. and Lee D. J., (1994), A steady-state model of a rotary kiln incinerator, *Hazardous Waste and Hazardous Materials*, 11(4), 541–559.
20. Boeteng A. A. and Barr P. V. (1996), A thermal model for a rotary kiln including heat transfer within the bed, *International Journal of Heat and Mass Transfer*, 39(10), 2131–2147.
21. Csenyei C. and Straatman A.G. (2016), Numerical modelling of a rotary cement kiln with improvements to shell cooling, *International Journal of Heat and Mass Transfer*, 102, 610–621.
22. VDZ Research Institute for the Cement Industry (2009), Modelling the circulating sulphur, chlorine, alkali systems in the clinkering burning process,

Part 1: comparison of measurement and calculation, Part 2: theory and discussion, *Cement International*, 7 (3 & 4), 75–87, 65–75.

23. Wurzinger A., Leibinger H., Jakubek S., Kozek M. (2019), Data-driven modelling and nonlinear model predictive control design for a rotary cement kiln, *IFAC PapersOnLine*, 52–16, 759–764.

24. https://www.iso.org/standard/62085.html

2 Transforming Cement Manufacturing through Application of AI Techniques

An Overview

Xavier Cieren
LafargeHolcim Innovation Center (LHIC),
St. Quentin Fallavier, France

CONTENTS

2.1 Preamble 40
2.2 Part 1: AI and Machine Learning Tools 40
    2.2.1 Preliminaries 40
    2.2.2 AI, Machine Learning, and Deep Learning:
    How Do They Differ?.. 42
        2.2.2.1 Machine learning... 43
        2.2.2.2 Deep Learning... 43
        2.2.2.3 How Does ML Work?... 44
        2.2.2.4 What Is a (Good) Algorithm? 45
    2.2.3 The Machine Learning Implementation Process at
    the Developmental Level... 46
        2.2.3.1 Categorize the Problem... 46
        2.2.3.2 Understand and Clean the Data 47
        2.2.3.3 Select the Best Algorithms and Optimize Them........ 48
    2.2.4 The Machine Learning Implementation Process at
    Production Level 48
    2.2.5 Is AI (or ML) a Future for Cement Manufacturing?.............. 50
        2.2.5.1 What Is Then the Potential of ML for the
        Cement Manufacturing Process?........................... 50
2.3 Part 2: "AI" Inside the Cement Production Process 52
    2.3.1 Main Components of a Cement Factory and
    Relevance of AI Applications... 52
        2.3.1.1 ML and Limestone Mining Operation...................... 53

DOI: 10.1201/9781003106791-2

2.3.1.2 ML and Raw Mix Design 54
2.3.1.3 ML in Clinker Production 56
2.3.1.4 ML to Build a Free-Lime-in-Clinker
Prediction Tool 58
2.3.2 Cement Grinding, Property Evaluation, and Hydration 62
2.3.2.1 Developing an ML Tool for Cement Compressive
Strength and Setting Time Prediction 62
2.3.2.2 AI in the Laboratory: Semi-automatic
Classification of Cementitious Materials
Using Scanning Electron Microscope Images 67
2.3.3 ML in Field: An Example of Sound Analysis 71
2.4 Concluding Observations 74
2.5 Perspectives 76
Acknowledgments 76
References 76

2.1 PREAMBLE

This chapter is divided into two main parts. Prior to the overview of some "Artificial Intelligence" (AI) techniques applied in the cement manufacturing process, the first part begins with some definitions and, challenges the concept of "artificial intelligence". It then focuses on the machine learning (ML) tools with a view to explaining what a good ML algorithm is, how it works, and how it is developed and implemented at industrial level.

The second part is built like a journey all along the industrial process, from the quarry to the cement, showing some potential benefits of "AI" tools with the help of various examples. Since an exhaustive list of all the possible techniques for the whole process cannot be presented here, a subjective selection has been made to fulfill the objectives of the chapter.

2.2 PART 1: AI AND MACHINE LEARNING TOOLS

2.2.1 PRELIMINARIES

AI could be defined in many different ways and various descriptions are found in books, papers, and on the internet. The basics are summarized below.

Digital computers appeared in the 1940s and were then able to process numerous mathematical operations and even crack some codes. The "Electronic Numerical Integrator and Computer" (ENIAC) was one of the first computers designed for the US army to calculate some ballistic properties in 1945 [1]. With time as the computation power increased essentially due to the speed of electronics and memory capacity, very complex and amazing tasks have been performed. For example, the Apollo Guidance Computer, weighing only 32 kilogram, provided computation that guided a spacecraft to travel to and from the moon [2]. Supercomputers are now used for

FIGURE 2.1   Advances in computing devices [Left: ENIAC computer [1], Middle: supercomputer FUGAKU [3], Right: a contemporary smartphone].

intensive calculation to model the world weather, simulate nuclear reactions or read some genetic code. The fastest supercomputer FUGAKU has appeared in June 2020 [3]. A modern smartphone possesses more memory and faster processing capability than the Apollo Guidance Computer is capable of recognizing human voice and face and can interact with objects. The landmark developments in computing devices are illustrated in Figure 2.1. It must, however, be understood that not all of these devices can be spotted as "AI inside".

In an "AI inside" device, the AI could be defined as "the ability of a computer or a robot (controlled by a computer) to perform tasks commonly associated with intelligent beings" [4]. This definition may further be qualified with "according to the human perception of intelligence". So, predicting the weather for the next week(s), driving and parking a car, sending a rocket into space (and getting it back…), playing the game of Go (and being its world master), diagnosing some human or earth disease(s), or even exhibiting human behavior or showing results at the level of a human expert by a computer is amazing but is definitely not a proof of intelligence. Today, despite extensive research work in the fields of hardware and programming, as well as the application of more surprisingly cognitive and social sciences, no programs/devices/machine can compete with human flexibility in even a limited domain space. So, the question arises: *are these systems "intelligent"*?

In his book entitled *"The artificial intelligence doesn't exist"* [5], Julia, the SIRI co-creator, questions, undoubtedly with a sense of provocation, the expression "Artificial Intelligence" with some known examples. The computer program AlphaGo succeeded in 2016 to beat the world Go master [6]. This was a challenge as the Go game has much more possible configurations and much more degree of freedom than chess. Intelligent? Julia explained that the program succeeded, but with the help of high energy consumption, hours of coding, costly equipment, and will fail to do anything else. Image recognition is another amazing achievement (think about real-time face recognition on video that your smartphone can do today!).

Classification of pictures of cats and dogs is a basic operation today, but it required a huge amount of various pictures (several thousand?) to reach an acceptable level of success. Intelligent? Well, a kid can do the job with a few explanations, in addition to a million other tasks. A vocal "assistant", powered by AI, is able to recognize your voice, to translate it into any foreign language, to indicate you your next appointment or the way to go home. Intelligent? Well, no way to ask about your feelings or to debate about, say, cement manufacturing.

A self-driven car is useful to transport you from A to B and eventually park your car, but not more. Some well-designed "AI" applications may predict an equipment failure, the characteristics of a manufactured product or the trends of the cement market but will fail because of a process or modifications in the world of rules!

Julia's point of view makes sense, and we will show later some examples of "AI-inside" tools that are doing the job but that are not "intelligent" at all. They rely on the ability of a standard computer to calculate highly complex operations to put out valuable predictions or provide sets of recommendations with a touch of human expertise. Many of these tools are in fact ML tools.

2.2.2 AI, MACHINE LEARNING, AND DEEP LEARNING: HOW DO THEY DIFFER?

To compete with human intelligence, AI would need to be a complex sum of various competencies as explained below and as displayed in Figure 2.2:

- Natural Language and Vision Processing and Understanding (accept sound and picture input and translate them into a computer format)
- Problem-solving (take decisions even with some incomplete or some false information)
- Knowledge representation (translate and store information in a form that speeds its access)

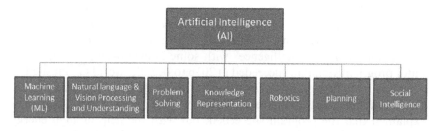

FIGURE 2.2 Diagram showing the components of the AI area.

- Robotics (respond to some request as a physical form)
- Planning (store information to draw conclusions close to real time)
- Social Intelligence (take into account the environment and its actions)
- ML (improve through experience)

Plus:

- Data for sure! then, the question of data quality is raised
- Hardware (sensors, processors, …)
- Coding with adequate algorithm choice strategy

In fact, the ML component is the most important as it is involved in almost all the others.

2.2.2.1 Machine Learning

One can see that ML is only a part of a system that is required to be qualified as an "AI-inside" system. ML is a set of statistical tools and algorithms allowing the building of a "prediction function" based on known data (historical data and then the "live" data). As ML is involved in almost all parts of the AI block and as it is probably the main component that can be used by itself in many applications, and more particularly in the cement industry, this will become the core of this chapter.

2.2.2.2 Deep Learning

Deep learning is a subset of ML that uses specific algorithms (Artificial Neural Networks, also named ANN or NN) to carry out the process of ML. A Neural Network (NN) is made of a single unit that mimics the neuron behavior. This unit gets some signals as input, sums them by applying specific weights, and passes this sum to an activation function that outputs a signal that can become itself an input for another unit. A single neuron, when trained, can compute simple functions such as "and" and "or" but also solve standard linear equations. A network of few neurons can model more complex mathematical functions, and, by enlarging the network to hundred neurons connected through several layers, one can solve very complex problems (think about playing chess or describe a picture!). The first and the last layer are called the input and the output layer, respectively. The layers in between, if any, are called the "hidden" layers. Networks with hidden layers are identified as "deep", giving its name to "deep learning". The input and output data can be multidimensional. For instance, a ML application may require a thermal picture of cement kiln as input and return a simple "false or true" data as output (that is translated into: "everything is ok" or "we are facing an issue").

The use of the term "artificial neuron" as the main component of deep learning algorithms and the fact that an artificial neural network is described as a human brain is for sure the source of the confusion in between the expressions "artificial Intelligence" and "machine/deep learning".

2.2.2.3 How Does ML Work?

The task of a ML algorithm is very simple: learn from experience. How? By building the best possible "prediction function" according to the data input and output quality, the selected algorithm and the training strategy applied by the developer. An algorithm uses computational methods to extract information from data without, in most cases, predetermined equations as a model. Depending on the data size, the algorithm selection, and the relationships between data, the model could be very complex, but essentially made of several matrices of coefficients and activation functions that calculate the output.

These activation functions are the core of ML models. For a single neuron, this is the function that let pass (or not) the signal to the next unit. They are generally nonlinear functions that render the mimic of very complex processes possible by their combination allowed by the network building.

This complexity and the fact that a ML model can solve almost every problem push some people to describe them as "black boxes". This is definitely not true. Whatever the algorithm type is, the structure of the model is perfectly known as well as the activation functions. The training is a question of successive partial derivations and is stopped, when all the necessary parameters to calculate the output are available, though sometimes very complex to apply. Actually, the "black box feeling" comes from the lack of "explainability", which may disturb the common sense of the scientists or experts. The principle of a ML model is described in a very synthetic way in Figure 2.3, which shows dataset splitting, training, and testing before operating.

As mentioned earlier, a model is built without predetermined equations (however, the algorithm type and the model architecture have to be selected in advance and data may need to be reshaped as well). Part of the data (the

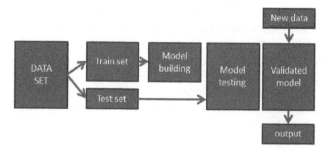

FIGURE 2.3 Principle of a machine learning application building.

"train" or the "training set" part) is firstly processed by the algorithm that will refine its parameters to produce the "best" prediction function, according to a chosen metric that estimates the difference between calculated and known results. This step is the "learning" step. Then the model is tested (not trained!) with the rest of the data, the "test set", to check if it is able to "generalize" the findings. If not, the model is probably facing an over- or under-fitting issue that should be solved by optimizing the dataset, modifying the model architecture, or changing the training parameters. This step is the testing step followed by the validation step. When the model is validated, it is implemented into the system and ready to produce its prediction.

The beauty of the ML models is that they normally improve their efficiency by learning, done by enriching the initial dataset by some validated output and retraining the model.

2.2.2.4   What Is a (Good) Algorithm?

As mentioned just above, an algorithm is a model able to output a value from an input. This model is not "created" by the computer from scratch (remember, no "intelligent" computer exists!) but is either selected from more or less complex functions or developed specifically. Two examples:

The **linear regression** is a ML algorithm. The predictive function is perfectly known:

$$output = b \times input + b,$$

and the learning process will refine a and b values in order to produce the best predictions. This very basic algorithm could be the best choice for some cases, but it shows some limitations as soon as some interactions exist in between several variables. For instance, building a linear regression model to predict the compressive strength of standard mortars as a function of the cement fineness would work, but only at very specific conditions and giving acceptable predicted values in a narrow range. A robust model would be obtained by using more complex algorithm (an example is described in the second part of this chapter).

The **Convolutional Neural Network (CNN)** is also a ML algorithm. CNN algorithms are fully adapted to image and sound classification or recognition, as these inputs are usually made of large size vectors or tensors. Unlike the linear regression algorithm, this prediction function is very complex. It consists of a series of convolutional layers (to filter the data), pooling layers (to reduce the dimensionality), and a fully connected neural network to resume the output. It is a very powerful algorithm that has to be used in specific cases: it could be the best choice to develop a model that predicts the amount of impurities in limestone samples from pictures, but definitely not a good one to estimate the free-lime content as a function of the kiln process.

Developing a ML application to predict an equipment failure from the analysis of vibration sensors is a totally different problem than to trigger an alarm when detecting on a video an unauthorized person in a restricted area. It is possible to get really similar results using different algorithms, and, on the contrary, get different result qualities using the same algorithm but with different parameters. Finally, the good algorithm is the one that gives the results at the expected level at the less possible cost, development time, and computing time. Details about the selection of ML algorithms and implementation of a complete solution are discussed in Section 2.2.3.

2.2.3 THE MACHINE LEARNING IMPLEMENTATION PROCESS AT THE DEVELOPMENTAL LEVEL

The first step is devoted to algorithm selection. This is done by looking at the data (type, size, available features) and the expected output. For instance, if the application purpose is to "recognize" and classify a cement material type by its chemistry or if the purpose is the prediction of the cement fineness as a function of the grinding process parameters, the requested amount and data type as well as the algorithm family will not be similar.

The usual route to develop ML can be described as follows:

a. Categorize the problem by looking at the available input and requested output
b. Look at and understand the data
c. Find the most suitable algorithm(s)
d. Implement the ML algorithms
e. Optimize hyperparameters

2.2.3.1 Categorize the Problem

If the input is labeled data, the problem is called "**supervised learning**". Labeled data means that data are tagged by categories such as "Clinker// Cement//Cementitious", "Safe//Unsafe", "60+ MPa // 45–60 MPa // <30 MPa"), or are numerical and continuous values such as "41.3 kWh", "56.3 t", "1425 K", "56.4 \$", "15.5 %").

Supervised learning is probably the most common form of ML. It presupposes a set of examples characterized by predictive variables for which the values of the target variable are known (or "labeled"). The learning process will refine the prediction function so that it generalizes the association between the explanatory variables and the target variable (assuming that the labeled training data is correct, without errors or bias). Algorithms such as "linear regression" or "support vector machine (SVM)" are two examples of supervised learning tools.

On the other hand, if data is "unlabeled", this is an **"unsupervised learning"** problem with a purpose to identify some structure or pattern into the data. This is the case for anomaly detection or clustering purposes for which the goal is to detect "different" signals or values compared to the others. For the cement manufacturing area, you may think about solving problems about predictive maintenance or alternative fuels chemistry analysis. "K-means" algorithm types are good for these purposes.

A third case is the **"reinforcement learning"** for which the goal is to learn from experience what were the best actions taken to reach an optimal state. The "Monte-Carlo" algorithm is a reinforcement algorithm.

If the output of a model is a number or some numbers, then the problem is one of **regression.** If the output is a class ("dog", "good", "yes"...), then the problem is one of **classification**. If the output is an ensemble of input groups, without labels, it's a **clustering** problem. Linear regression, logistic regression, and K-means are three adapted algorithms for these three problems, respectively. It is important to note that classification and clustering problems are supervised and unsupervised problems, respectively.

For selecting the most suitable algorithms, the ML developers are driven by several other considerations. For example, should the prediction function be linear or not? Is the prediction function parametric? Are the data available online? Is the data distribution known? Although these aspects are not further detailed here, it is evident that, as with all kinds of projects, it is essential to look at the starting materials (data in this case) and the objectives.

2.2.3.2   Understand and Clean the Data

Understanding the data plays a key role in the process of choosing the most adequate algorithm. The data size and type (categorical, numerical, multidimensional, etc.) are of great relevance, as some algorithms can work with smaller sample sets while others require huge sizes, some need numerical data, and some are devoted to the categorical type. It is also important to check the dataset "quality":

- Data distribution: if known, it helps to detect outliers (that have to be removed)
- Data size distribution for all the features (should be similar)
- Look at redundant or correlated features
- Apply the best strategy to replace the missing values

Data cleansing is an essential step (and the most time-consuming one) but should be done without excess in both directions. Identified errors should be removed with no exceptions, outliers (data that differs significantly from other) should be "marked", removed later if they clearly affect the quality of the model, and kept in order to increase the prediction values quality at extremes.

To sum up, the dataset quality should reflect the level that exists at the industrial level with its intrinsic uncertainty for building a ML tool based on the extraction of the top quality measurements of a dataset. On the contrary, inclusion of all the messy measurements will not deliver acceptable results with future input. One has to keep in mind the GIGO precept: "Garbage In, Garbage Out!"

2.2.3.3 Select the Best Algorithms and Optimize Them

The quality of a model can be figured out by the "distance" between the calculated values and the expected values, making it possible to compare different models or different strategies of the same model. This metric value is usually represented by the Mean Absolute Error (MAE) or the Root Mean Squared Error (RMSE), but some other parameters may be more suitable adapted depending on the data and objective.

After its selection, the algorithm has to be optimized. Assuming that the dataset is correct, the available levers are the so-called "hyperparameters". They are the core of the ML tool. For an Artificial Neural Network algorithm, some of the hyperparameters are the number of layers or the number of neurons per layer, the activation function type, the neuron weights initialization or the learning rate.

When optimized, the ML developer may focus on the output visualization optimization (the way to render the results).

Some other elements affecting the choice of a model are as follows:

- The accuracy of the model (set by the model, but to be challenged by the end user)
- The interpretability of the model (optional, it may be totally useless for the end user)
- The complexity of the model (same as the latter: if the complexity has no impact on calculation time or cost, complexity may have no consequence for the end user)

The steps d) and e) are part of the implementation at "production level"

2.2.4 THE MACHINE LEARNING IMPLEMENTATION PROCESS AT PRODUCTION LEVEL

Unless it is only meant to produce an impressive demo, a ML model implemented at production level is meant to show some benefits. There are different rules to follow to insure the success of a ML tool.

1. A clear and **well-defined use-case.** For instance, requesting an ML tool to predict the cement tonnage production or the electrical consumption

of a mill are not well-defined use-cases: what will you do with these numbers? The benefits should be estimated, for example, in terms of cost, time savings, waste reduction, and so on.

2. Check the **availability (and quality) of data**. Data is the core of the system. Having a huge amount of data to train a model is good, but being unable to feed the model "online" or identify data errors when implemented will lead to failure. In addition to adequate data availability (that supposes a proper data governance), data should be constant in its feature distribution (or at least the model should be able to alert if not).

3. Identify the **data value** location. Ensure a constant exchange between experts (those who are building the ML model and the "field" experts) to validate the various assumptions that are to be set, such as feature importance, outliers/errors detection, output class identification or identification of links in between historical data reading and real events. Expertise is also required if some feature engineering is required.

4. Model **building and validation**: Like for the previous step, a mutual exchange should occur in between model experts and the end user (assuming that the end user is expert as well). The end user validation is required to challenge (in both directions) the metrics type and value that will trigger the validation of implementation (at that stage, the model is just entering its automation and production rollout phase).

5. **Automation and production rollout:** As soon as the validated model is implemented and starts to deliver its output, one should find time to:
   • Check that the process is running without bugs in "real life" (data feeding, output visualization, etc.)
   • Check that the output delivered by the model is consistent with the expert expectations.

6. **Monitor and update the model**: Once installed, the model (in fact its output) should be continuously monitored and approved. The quality of the results provided by a ML tool is expected to increase with time by updating it with the new data. On the other hand, as the model was trained on historical data that were produced from a specific process, an important variation of the latter may affect the new data. This is the reason why the model should be continuously monitored and validated, so that a drift or an unexpected output can be immediately attributed to a model failure or seen as a real effect.

The algorithm selection, as stated above, is driven by several considerations. Ted Dunning, a renowned expert in ML and other big data solutions, has listed five of them (and with no specific order) [7]:

• Deployability: upscaling ability is essential, especially in the case of industrial implementation

- Robustness (real-life data may be incomplete, corrupted or exotic. The model has to "stay quiet"!)
- Transparency: the quality of the prediction should increase with time. If not, the system should be able to set an alert
- Skillset: the implementation and the use of the model should not require a very high level of expertise
- Proportionate: invested time or money should be proportionate to the expected gains

2.2.5 Is AI (or ML) a Future for Cement Manufacturing?

Let us first see what ML can offer today. Now when people think about AI, more realistically about ML, the following applications come to mind:

- Image analysis: dogs and cats or handwritten digit recognition are no more issues. Today, ML applications are able to do face recognition in real time on video
- Text and language processing: getting a translation of a word or a text is more than easy. Today, ML tools are able to classify texts according to their topics or their styles, write summaries or concatenate various sources to build a text. Real-time language translation is no more a dream.
- Combination of both: ML can describe a picture with the appropriate legend, or on the opposite, retrieve a picture from an oral request.
- Business: customer preference predictions, trends predictions, logistic chain optimization
- Health and safety: Medical prediction and diagnoses, identification of high-risk patients, recommendation of best possible medicine

This is a non-ending list.

2.2.5.1 What Is Then the Potential of ML for the Cement Manufacturing Process?

The cement manufacturing process is a very wide and complex sum of processes. "AI", through the ML tool abilities shown before, has a full potential to be implemented at almost every area of the cement manufacturing ecosystem: from the quarry to the final material dispatch, from the suppliers to the customers, from human resources to health and safety management. Some examples are described in Part 2 of this chapter.

The power of ML is its ability to combine and process features of various kinds. For instance, to predict the 28 days compressive strength of standard mortar (at the time of cement manufacturing, therefore 28 days ahead in time). For this purpose, at the cement dispatch output, the ML would require at least the chemical properties of the clinker and the added materials (chemical element concentrations and crystal phase ratios) as well as some physical properties such as the granulometry and the loss of ignition.

Another example is to switch to a continuous prescription tool that provides a parameter set to drive the cement mill in order to get the requested cement quality, with a goal of energy consumption reduction and raw material cost reduction; new data from process and logistic are required and would take part in the training process of the tool.

Although a very experienced engineer could integrate all of this information, doing so intensively, continuously, and in real time is impossible with traditional tools, making the use of the ML tool more than relevant.

In all the cases, the essential ingredients are the following:

- Data that is relevant and accessible at all time (including historical data)
- The possibility to use data from all the processes, including outsourced data
- The share of the expertise of the ML developers and the local people

It is sometimes said that the cement industry is "old-fashioned", but, in fact, cement manufacturers were early adopters of automation and control systems and they have been using them for many years. A visual comparison of a typical cement factory control room in the 1980s and in 2020 is presented in Figure 2.4. The figure is also a witness to the fact that the cement manufacturing process has been monitored for years.

Nowadays, the number of sensors in a plant can be over 10,000, delivering data every second for some of them. This huge amount of data should be used at a higher level than just making work easier for operators and process and quality engineers by offering them real-time data visualization or historical data graphs.

The LH plants (and in almost all cement works) are equipped with HLC ((High-Level Control) systems which support operators to control different processes via decision-making dashboards and/or automatic control, usually

FIGURE 2.4 A typical cement factory control room in the 80s (left; Photograph from Longmont Museum, [8]) and in 2020 (right, LH © LafargeHolcim 2020.

based on a simple linear model (and sometimes empirical model) using the data provided by the sensors. Despite the power of these tools, a holistic view is lacking. This is hard to achieve due to the complexity and numerous interactions all along the industrial process, from raw mix computation to cement quality validation.ML offers an opportunity to improve these control systems rather than to replace them.

The second part of this chapter is a journey along the cement manufacturing process.

Some basics of all the processes will be reminded together with the potential locations of some ML tools. Some are only mentioned, others are more detailed, the objective here being only to provide an overview both of different ML techniques and of the possible contributions to the cement process.

2.3 PART 2: "AI" INSIDE THE CEMENT PRODUCTION PROCESS

2.3.1 MAIN COMPONENTS OF A CEMENT FACTORY AND RELEVANCE OF AI APPLICATIONS

There are about 3000 cement plants in the world, dispatched in 160 countries and territories, run by more than 650 companies [9], with a global cement production of more than 4 billion ton a year [10]. All are different in size and process type but a typical full integrated cement plant process could be described as follows (Figure 2.5):

- Mining, extracting, and crushing raw materials that include limestone and clay, eventually correcting the chemical composition by adding materials such as bauxite (aluminum source) or iron ore. The output of this first step is the "raw meal"

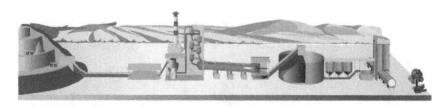

FIGURE 2.5 Diagram of a typical integrated cement plant. From the left to the right: the quarry ending with the raw material mill, then the chimney, the preheater and precalciner tower, the rotary kiln, the cooler and the clinker silo, then the cement mill with its added materials silos, and, at the end, the cement storage silos and the dispatch. (Picture from LafargeHolcim Maroc 2020).

- Recombine the chemical elements of the raw meal by heating it in a rotary kiln at high temperature (as high as 1450°C). Clinker is the output of this step, called pyroprocessing
- Mix and mill clinker with gypsum and eventually other material such as fly ash, slag, limestone to produce the final product: the cement
- Dispatch the cement

Let us enter into details of these different steps that are themselves built of numerous sub-processes. The objective is not to detail the cement production process neither to establish an exhaustive list of possible ML tools. Therefore, the following overview will highlight the huge potential of "AI" for cement manufacturing. A large part of the details about the industrial process are taken from [11]

Cement manufacture begins in the quarry as it is the location where most of the raw materials needed are extracted. Among them, the limestone is the primary raw material for cement production and is to be sourced far in advance of the cement plant conception.

Some ML tools exist to help geologists to identify and classify areas of potential interest but are not covered here, as they are outside the scope of "cement manufacturing". Ironically, these tools that were developed for studying very large scale areas (satellite pictures or maps with mile or km as order of magnitude) are very similar to the tools used at a very low scale to identify and classify crystal types and clusters of micrometer size from microscopy images. A detailed example is given in this section.

2.3.1.1 ML and Limestone Mining Operation

Limestone is predominantly mined from a quarry that is an open pit exposed to the surface. The limestone quality and location vary within the quarry. It is therefore necessary to blend rocks from different benches and faces to optimize material extraction. 3D modeling quarry management software has been around for years. However, the trend is to implement ML algorithms that are the best candidates for the image analysis, 3D modeling, rock volume estimation, front analysis (color and roughness are parameters revealing the nature and composition of the rocks), blasting video analysis, rock drilling sound analysis, etc.

The images are eventually captured by drones (themselves piloted with the help of ML tools). The Figure 2.6 is showing two examples of drone images processed by ML to estimate volume of the front or of some aggregate piles.

In contrast to drones, a different illustration is furnished in Figure 2.7 from an iron ore quarry site. The mining company Rio Tinto operates 73 of these titans hauling iron ore 24 hours a day at four mines in Australia's Mars-red northwest corner.

It is important to note the importance of research in the area of application of AI to mining. Below are listed the titles of some examples of scientific

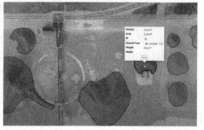

FIGURE 2.6 Drone pictures of a quarry front (left) and aggregate piles (right) that are processed as augmented reality images giving various valuable information (courtesy of DELAIR [12]).

publications found with the two keywords "machine learning" and "quarry" [14–17]:

- "Prediction of blast-induced flyrock in Indian limestone mines using neural networks"
- "Blast Vibration Analysis by Different Predictor Approaches- A Comparison"
- "Support vector machines approach to mean particle size of rock fragmentation due to bench blasting prediction"
- "Minimising rock break at the Dewan cement limestone quarry using an artificial neural network"

2.3.1.2 ML and Raw Mix Design

Calcium carbonate (limestone) and aluminum silicate (shale, clays, etc.) are basically required to produce the cement clinker. The typical Portland clinker composition, when expressed as oxides, is 62%–65% of CaO, 19%–21% of SiO_2, 4%–6% of Al_2O_3, and 3%–5% of Fe_2O_3 on a loss-free basis. This optimal composition may not be achieved by using only the two main raw materials. Hence, some corrective materials such as bauxite, laterite, iron ore, or sand may be added to the mix to achieve the adequate stoichiometric proportion, not forgetting some chemicals that are added to facilitate the grinding process (grinding aids) and the clinkering process (mineralizers).

At that stage, optimal raw mix composition is calculated using some oxide ratio and the empirical Bogue's equations that estimate the four major clinker phases and melt phase contents, but other features (that impact the clinker quality) have to be taken into account:

- Raw material impurity contents (clays, quartz)
- Chemistry of minor constituents

FIGURE 2.7 Movement of gigantic dump trucks, facilitated by MI tools, at West Angelas, the iron ore mining property of Rio Tinto in Australia [13].

- Burning behavior
- Fuel nature

Such a variety of parameters makes the raw mix design a complex operation, which makes it a good candidate for ML algorithms. As already said in the first part, most ML tools can deal with data from all types, therefore, one can imagine adding several other parameters to make them powerful predictive and prescriptive tools:

- The real raw meal composition to feedback the model
- Effective clinker quality (the one measured at the kiln output), to feedback the model
- Some kiln parameters (including fuel nature)
- Quarry operation cost
- Raw material and fuel availability and cost
- Crushing, milling, and homogenization process costs

The following titles correspond to some examples of publications about ML and raw mix design [18–21]:

- "Mathematical modeling of a cement raw-material blending process using a neural network"
- "Modelling and simulation of raw material blending process in cement raw mix milling installations"
- "Intelligent Control System for Cement Raw Mill Quality Based on Online Analysis"
- "Application of Monte Carlo Simulation for Cement Raw Material Blending Optimization"

2.3.1.3   ML in Clinker Production

The clinker production (pyroprocessing) is the central process of the cement plant. Its input is the kiln feed (the raw meal that was designed from raw materials and corrective materials). Its output is the clinker that is itself the essential component of cement.

This process starts at the kiln feed extraction from its storage location and concludes at the point where the cooled clinker is sent to its silo, covering a huge amount of sub-processes involving various equipment used in various environments.

Nowadays, this part of the cement manufacturing process is probably the one that is the most monitored and automated: it is a very complex process, designed to run non-stop and, if not perfectly controlled, there is a strong potential for drifts resulting in poor quality output or damages to its costly components.

In an extremely simplified form, the pyroprocess can be described as successive chemical transformations: limestone decarbonation, solid-state reactions, liquid phase sintering, and reorganization of the clinker microstructure through cooling. These chemical reactions occur in the preheater cyclones, the precalciner vessels, the rotary kiln, and the cooler.

In the previous part, it was shown that the raw mix design process is not as trivial as it seems as many parameters should be taken into account, more than only achieving an optimal target composition. Similarly, the goal of pyroprocessing is not only to achieve the requested clinker composition. It has to take into account numerous, complex, and interconnected parameters:

- Kiln feed flow rate
- Fuel and alternative fuel types (and their pre-processing)
- Successive chemical reactions at various temperatures and locations
- Presence of gas, liquid, and solid matters
- Equipment limitations (temperature limitations for instance)
- Fuel cost
- Gas emissions

The clinker quality is first of all given by the chemical composition and the physical parameters of the kiln feed that is designed according to the expected clinker crystallographic composition. However, pyroprocessing itself impacts the clinker, modifying its chemical composition, crystal phase ratio, crystallite sizes, and arrangement.

Such a complex process which involves many subprocesses, multiple interactions, technical and economic constraints, and essential feedback is perfectly suited to the use of ML tools for a part, even better, the whole of the process. A few specific application areas for ML tools are described below.

a. Pyro operation process control:
   Very recently, Petuum and the building materials company Cemex [22, 23] released some information after the implementation of an all-in-one "AI" package covering pre-heater, kiln and cooler with a goal to optimize energy savings and maximize the production rate.

   Petuum explained that AI tools can now predict, prescribe, and finally control processes in an autonomous way.
b. Cyclone blockage detection:
   From the introduction of the raw meal extraction to clinker cooling, the process has to run continuously and does not support any failures of any stage. Among critical issues is the cyclone blockage that requires one of the most dangerous maintenance operations to eliminate it. Cyclones, where the cold kiln feed is heated by the countercurrent hot kiln gas, are part of the preheater system. Complex chemical reactions occur in them between gas, solid, and liquid matters that may cause formation of buildups blocking the material flux. These blockages can be identified by monitoring temperature, pressure, and material flow rate.

Different approaches using ML tools are used to predict cyclones blockages. Thermal image analysis (Figure 2.8) is a good one to locate the points of temperature increase, but these points are not always indicators of build-up formation. On this example, the zone identified by the rectangle were supposed to be a potential build-up location but not confirmed by the real observation (but some defects were clearly identified). Process and quality data are also an obvious area where ML algorithms can be used. The main issue in that case is the lack of in-situ chemical data that is available only at the input (kiln feed analysis) and at some output (clinker mainly), involving specific data processing such as time series analysis. Some models co-developed by LH using random forest and logistic regression algorithms demonstrated that blockages are shortly (some minutes) preceded by specific patterns and are detected with success rates about 90%. The models also confirmed the relevance of certain features including the expected chlorine and sulfur concentration. Some improvements are expected by increasing data quality and data processing and by combining image analysis results.

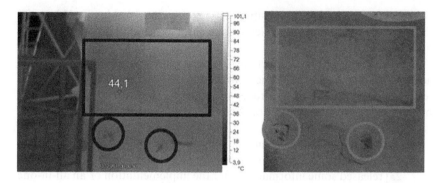

FIGURE 2.8   Left: IR picture of the bottom part of a cyclone (Temperatures are of the external surface but a perfect indicator of the internal ones). Right: Corresponding standard picture of the internal face of the same empty and cooled cyclone. (Pictures from H. Jansen, LafargeHolcim).

2.3.1.4   ML to Build a Free-Lime-in-Clinker Prediction Tool

Here is the description of the development of a ML prediction tool to estimate the free-lime content in clinker at the kiln output. This model, developed at LafargeHolcim Innovation Center (LHIC), is based on a specific dataset and strategy that have to be varied according to the location or the data availability.

The free lime is known to be closely related to the quality of cement, as high free lime could cause longer setting time or reduced strength of the cement. Today, the free-lime amount is measured in the laboratory with various sources of uncertainty due to the clinker sample being not representative and due to the measurement errors. In addition to that, the results are delayed due to the sampling and measurement time. Thus, predicting the amount of free lime in clinker production in real time and with low uncertainty can result in higher cement quality. The steps followed are:

a. Gathering Data

   Data sources: There are two main sources of data in the plant that could provide information about the free lime in clinker. One is process data gathered from sensors across different positions in the whole pipeline (e.g., fan speed, kiln temperature); the other is quality data, obtained via the analysis of the materials (e.g., kiln feed lime saturation, silica ratio). The objective is to predict the free lime percentage in the output clinker. Since the clinker is the final product of the process under consideration, all these process and quality parameters can be gathered before the clinker is produced, so the prediction can be done several minutes before we actually get the clinker. Although the prediction may not be early enough to change the free lime content, it could guide the technicians at the plant to act on the ones produced in the next hours.

Data frequency: Generally, the process and quality data are in time series format, but the time-step may be different among them. Process parameters can be gathered automatically from sensors without cost, so the time-step can be every minute or even every second. On the other hand, quality data usually need to be measured with more advanced and costlier techniques, so the time-step may be in hours. Similarly, the free lime percentage in the clinkers can sometimes be measured at the time-step of hours.

b. Exploratory Data Analysis

Time plot: To get the first impression about each parameter, we could create a time plot of 1 month for process data or 1 year for quality data. Time plots can show the evolution of the parameter over time, whether the data move smoothly or abruptly, periodically or randomly. Besides, a time plot of both a parameter and free lime could reveal the relationship between them. For example, it is not difficult to detect from the time plot that when the temperature at certain positions in the kiln rises, the free lime tends to be lower, and when the kiln feed lime saturation ratio (LSF) is high, the free lime also tends to be high.

Detecting events: At specific moments over the years, the plant made some permanent changes in the process or materials. If the change is significant enough, the corresponding parameter could be very different before and after the event (in the sense that the distribution of that parameter before and after is significantly different).

If a model is trained using data before the event and predicted on data after the event, then that input parameter could degrade the model quality. To detect such an event, we may plot the mean and standard deviation of each parameter monthly. If the mean and standard deviation of that parameter are very different before and after some particular moment, there is a high chance the plant has made some changes permanently impacting that parameter. If that is the case, we should be careful when training with such parameters.

Time shifting, Rolling average, and Feature engineering: The output clinker is made from materials from the kiln feed say 1 hour earlier (depending on the kiln) and is affected by the mid-kiln temperature 30 minutes earlier. Thus, it is necessary to shift the time series so that every parameter associated with a clinker sample is in line timestamp-wise. The amount of time to shift depends on the kiln and should be done with the guidance from the plant. The rolling average is to solve this issue by replacing values by the average of adequate period to reduce sensor noise or deviations and to smooth the time shift that is not constant. Rolling average is one in many techniques to create new time series from the original ones, which is called feature engineering and that should be done with the guidance from the plant.

Data Transformation and Normalization: After time shifting and feature engineering, the time series are converted into a single table where each row is a timestamp and each column is an original or newly created parameter (also called feature). After cleansing, each row now represents a clinker sample with corresponding process/quality parameters as input and the free lime percentage as output.

Some ML models (e.g., Neural Networks) are sensitive with input data values; input parameters with too large or too small values may lead to difficulty in calculating the derivatives to update the model. Hence, a general guideline is each input should be in range -1 to $+1$ or follows the standard normal distribution. One simple way to achieve this is normalization. Both feature engineering and data normalization can be seen as techniques to chew the data before feeding it to the model, so that the model can easily digest the data and produce better results rather than letting the model figure out by itself.

c. Training cycles

Feature Selection: Which parameters (also called features) are chosen as input is important. A good parameter is one that contains valuable information that the model can learn from to predict the output better (e.g., LSF or temperature, mentioned before). Adding a parameter that the model has nothing to learn from only causes confusion and difficulty for the model; the same observation can be made for adding too many parameters. To choose the right parameters, it is necessary to train with different sets of parameters and find out which works better. Feature engineering and feature selection are two very time consuming parts of the whole ML pipeline but have significant impact on the quality of the model.

Learning Algorithms: There are many established ML algorithms such as Linear Regression, Support Vector Regressor, etc., but with recent advances of ML and Deep Learning, two most prominent family of algorithms are Tree-Based Gradient Boosting methods (e.g., LightGBM or XGBoost) and Neural Networks (e.g., TensorFlow, Pytorch). Each of the two has certain strengths and weaknesses; which one works better depends on every problem and dataset. Hence, trying them both or combining the two is generally recommended (refer to the first part of the chapter).

Validation Scheme and cross validation: To estimate the quality of models, data is split into a train-set and a test-set. Since the data is very time-dependent (clinkers in consecutive hours are likely to have similar free lime) a time-based split is necessary. Typically, a 1 year dataset shall be split into a 10 months dataset for training and the last 2 months for testing (i.e., a 80/20 ratio), The drawback of this method, and particularly in the case of a continuous process, is that a two-month dataset is not long enough to confidently say that one model is better than the other. To have a better validation scheme, the "Cross Validation" technique is applied: data is divided into six groups, each of them is two consecutive months, and then six models are trained using the six combinations of test and train groups. The resulting output is the average of the six models with a reduced error.

Iterations: After each cycle of feature engineering, feature selection, and training, we can perform result analysis such as plotting the true versus predicted free lime and analyze the samples with high errors. We can also consult the plant to find the root causes of these hard samples and from that build a new set of feature engineering, feature selection, design a new model, and repeat the cycle until reaching a good accuracy.

Note that while aiming for accuracy, it is necessary to explain every step we do (e.g., why do we create or discard this feature; why do we set this hyperparameter like that) to avoid blindly overfit the model.

Deployment: Before deployment, we train a single model with the architecture and hyperparameters that we found the best on all the data and then test it on newly coming live data. We can continually update the model via retraining new data or change hyperparameters.

The free-lime predictions obtained by this model are more accurate (and, highly valuable at "real-time" (Figure 2.9). The graph shows clearly the right trends and value compared to the measurements with some clear benefits of the predicted values that can be obtained "on demand" (usually every five minutes at plant level), without bias due to sampling and at no cost. The predictions are intended to be input in the current HLC system of the kiln to replace the laboratory data.

Here are some studies reported on ML algorithm applications at various locations of this process [24–29]:

- "Multivariable nonlinear predictive control of a clinker sintering system at different working states by combining artificial neural network and autoregressive exogenous"
- "Imaging of Flames in Cement Kilns To Study the Influence of Different Fuel Types"
- "Identification and detection of the process fault in a cement rotary kiln by extreme learning machine and ant colony optimization"
- "A Temporal Neuro-Fuzzy System For Estimating Remaining Useful Life In Preheater Cement Cyclones"
- "Cyclone Track Prediction with Matrix Neural Networks"
- "Improvement of a cement rotary kiln performance using artificial neural network"

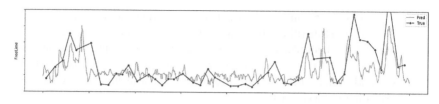

FIGURE 2.9   Free lime prediction curve and measurement curves over a week.

2.3.2   Cement Grinding, Property Evaluation, and Hydration

The output of the previous process is the clinker which is subjected to the next step of grinding with sulfate compounds (gypsum, hemihydrate, or anhydrite) and other additions (filler and mineral admixtures) to produce cement as the final product.

The clinkerization process requires a precise tuning that takes into account both the input (the kiln feed and the fuel) and the output material specifications (composition, granulometry, etc.), plus other considerations such as material availability, equipment limitations, or process costs.

The grinding process is similar because the quality of the output material (the cement) is a function of all the input materials and the grinding process settings that are themselves to be adapted to the input materials.

The main parameters that are known to impact the cement quality are:

- The clinker chemical composition, the crystal structure of the individual phases, and their microstructure that determine the clinker grindability
- The chemical and physical parameters of the added sulfate phases (gypsum, hemihydrate or anhydrite) and, if any, of the cement substitute material if any (limestone, fly ash, slag, etc.)
- The mill parameters set according to the grindability and the expected cement quality

ML algorithms are good candidates for several parts of this process such as predictive maintenance, clinker grindability prediction, cement quality prediction, mill set point prescription, and so on. Some examples of ML tools are discussed below.

2.3.2.1   Developing an ML Tool for Cement Compressive Strength and Setting Time Prediction

Here is an example of a ML tool developed by LH Innovation Center and implemented in one of the LH plants. This operating tool is in practice to deliver the 28-day compressive strength and the initial setting time predictions (respectively named "CS28" and "SETINI" in the following text) for all the cement types produced at that plant for dispatch and mill output.

The setting time and the strength acquisition of cement are two of the most important intrinsic values of cement. The initial setting time is the time that elapses from the moment water is added to dry cement powder until the paste ceases to be fluid and plastic. The hardening of cement is monitored by the measurements of the compressive strength of standard cement mortar samples, after various durations: typically 2, 7, and 28 days of hardening.

Several factors affect both the properties. The aluminate phase content of the clinker is one of them but cannot be corrected at that step of the cement manufacturing process. Among the other factors, the cement composition (amount of the sulfate phases, filler, and/or cementitious materials) and fineness are the most important ones, which precisely act as the levers available in the grinding process.

Having these two "predicted" values as fast as possible (and it is 28 days in advance for the compressive strength!) may help to better optimize the grinding process or prevent the dispatch of improper cement.

As mentioned in the first part of this chapter, the implementation of a ML new tool follows several unavoidable steps:

1. Business case and technical feasibility study:
   Obtaining the CS28 and SETINI predicted values is a way of reducing process costs by

   - optimizing the mill parameters (including the grinding aid consumption),
   - minimizing the over-quality of the cement,
   - reducing some time-consuming laboratory operation,
   - optimizing the clinker and mineral additives mix to impact the clinker factor, and
   - avoiding potential customer quality claims.

   In parallel, a quick feasibility study is carried out using

   - some off-line historical data,
   - standard ML algorithms (SVM, random forest, neural network), and
   - an a priori selection of basic features measured at the plant laboratory.

   The study demonstrated that both predictions could be obtained (and optimized) at acceptable uncertainties.

2. Availability (and quality) of data:
   As already mentioned, the cement plants have been fully equipped with several thousands of sensors for decades. These sensors produce a huge amount of data, from process data to measurements done in the plant

laboratory, thus giving a broad overview of all the processes in real time and any historical data if necessary.

In the present case, the relevant features are hourly measured and a historical dataset spanning over 3 years is available.

In addition to that, data can be extracted, processed, and sent to the network in an automated way as a standard readable file (like csv file), and this operation is easy to set.

3. Expertise availability and end user requests:

The third key point of success is the expertise sharing between the tool developers and the cement process experts. It allowed the developers to focus on the relevant features, optimize data cleansing, and set the adequate metric limits for the model validation.

The experts were also asked to list their requests concerning the tool quality, handling, and HMI (Human-Machine Interface). Here is a list of the main requests:

- No manual operations to connect and update the tool
- All visible data present at launch are the last data
- A single screen gathering relevant data
- Easy selection of the cement type and the predicted feature
- Calculation of the confidence level of the prediction
- Visualization of the predictions and measurements for a specific time range
- Visualization of the feature importance level
- Possibility to "predict" by varying one to three specific features manually

This exchange, to be done prior to ML development, is the best way to stay focused on the expectations and involve the local users.

4. Model building and validation

Various strategies and algorithms were tested regarding the objective of predicting a continuous value (the compressive strength and the setting time) from numerical input (all the measurements acquired in the laboratory).

Three ML algorithm types were identified: neural network (using the TensorFlow library), random forest (lightGBM), and SVM (sklearn).

The dataset is limited to a full year of data to minimize the impact of the long-term process variations and contains measurements of several cement types (low to high compressive strength Portland cements, and including limestone cement).

The relevant features that are used to build the model were rapidly identified, thanks to local expertise and knowledge. In addition to that, it was also decided to keep only "cement powder" sample measurement (to get rid of all the measurements obtained on cement paste or standard mortars samples). Therefore the features were limited to:

- the crystal phase concentrations refined by the Rietveld method (from the XRD signal)
- A part of chemical element concentration (XRF measurement) and the Loss of Ignition value
- The granulometry of the cement

The SVM was rapidly abandoned due to its low performance compared to the two others (on that specific dataset)

The random forest algorithm reported acceptable predictions but was dropped because of its poor efficiency to report continuous prediction values from large ranges of input values (see Figure 2.10).

Finally considering only a NN model, we decided to build one model per predicted feature. Due to the small number of input features (around 30) and the single output, we started from a simple architecture made of only two layers of less than 25 neurons each (this architecture had proven its efficiency on some preliminary studies).

Training and testing were done on a single dataset containing all the data from all the cement types. No improvement was observed when using a smaller database containing a single cement type. Another noticeable aspect is that the same architecture is used for CS28 and SETINI predictions. The model validations were done by challenging several values such as the typical MAE and RMSE but also some indicators of trends, requested by the users. Finally, the CS28 prediction values exhibit an uncertainty (comparing the predicted and measured values)

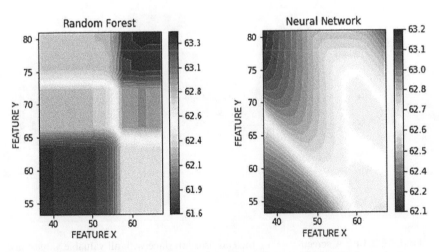

FIGURE 2.10 2D decision chart that predicted compressive strength value as a function of two features, calculated with the random forest model (left) and the neural network model, the latter displaying continuous prediction values.

higher but very close to the laboratory uncertainty. The SETINI prediction uncertainty is about 10% of the setting time measured values.

5. Automation and production rollout

As soon as the ML tool is validated, it becomes an "application" and is to be plugged to the existing infrastructure. From "fully independent tool" to an all-integrated one, various strategies can be applied at that moment of implementation (but should have been decided at the beginning of the project to facilitate that step and anticipate potential issues). The objective is to test the model on site, in real time, on "live" data to detect possible bugs in the data flow or visualization and check if the output is consistent with the expectations. An example of the HMI is shown Figure 2.11).

6. Monitoring and update

When implemented, tested, and validated, the application has to be monitored so that a drift or an unexpected output can be immediately attributed to a model failure or seen as a real effect. In the case of the 28-day compressive strength prediction, there is no way to compare it immediately with a measurement, which can be done only 28 days later but, like any other output, local experts should be able to look at some other indicators to check and validate (or not) the prediction. The power of ML is its ability of learning! As soon as the new data are validated (that is compulsory!), the model can be "retrained" taking into account the new samples, with the expectation of reducing the uncertainty of the prediction. In case of a model failure (unexplained drift,

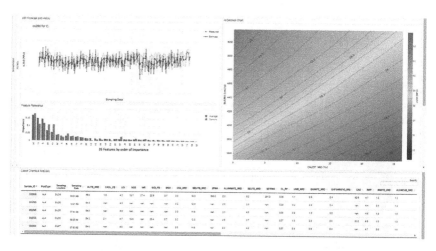

FIGURE 2.11   A screen-copy of the end user interface with all valuable values: last prediction values, chart showing the predicted and measured values, the feature importance diagram, a decision chart, and a table with the feature values of the last five samples.

error increase, or exotic predictions of the properties, a deeper analysis should be done with the experts of both sides to check the source of error.

2.3.2.2 AI in the Laboratory: Semi-automatic Classification of Cementitious Materials Using Scanning Electron Microscope Images

For many years, the microstructure of clinkers has been studied by petrographic microscopy. After adopting a specific and elaborate sample preparation technique, using a limited number of clinker nodules, the well-versed microscopists were able to identify the crystal phases and measure their shape, size, and distribution.

The complexity of the sample preparation, the required expertise (and its subjective judgment) and the time of analysis made this technique quite obsolete and not adequate to drive the clinkering process in a quick and efficient way.

Scanning electron microscopes (SEMs) are definitely not standard equipment of a cement plant, but the following example shows that its combination with in situ chemical analysis and some ML tools allows to perform analyses of much more complex cement related materials than clinker in an almost automated way. The following example will demonstrate the results of a powerful ML tool developed jointly by the LHIC and the GIPSA laboratory [30].

The challenge: From a piece of a hardened blended cement paste, evaluate:

- The chemical composition of the sample
- The ratio of anhydrous/hydrated compounds
- The ratio of the compounds
- Information on microstructure

Why such a request?: With growing pressure to reduce the CO_2 footprint of cement, blended cement appears to be a strong lever: Portland clinker is partially substituted with supplementary cementitious materials (SCMs) such as slag, fly-ash, and natural pozzolans. To get the mechanistic understanding of the blended cements, it is critical to get quantifiable information on the microstructure of the hardened cement paste (the degree of hydration of the cement, and more generally, the volume fraction of phases). The SEM seems to be the most appropriate equipment to extract these data.

Why is it challenging?: Presuming that the sample preparation for the SEM studies has been properly carried out, it is possible to perform backscattered electron (BSE) imaging, in which the SEM produces grey scale images with a contrast caused by the atomic number of all the chemical elements of the sample "hit" by the electronic beam. Several authors showed that different phases in cement paste could be identified from BSE images: clinker (the densest phase) is the whitest phase, pores (the lowest density phase) are close to black, and hydrates are of an intermediate intensity. Playing with the frequency distribution of the intensities and some stereological principles, the volume fraction of these phases can then be quantified. However, the applicability of this technique to complex blended cements is somewhat limited as SCMs such as fly-ash, slag, and natural pozzolans produce intermediate grey scale intensities that overlap in the same range as calcium-sulfate-hydrates and calcium sulfoaluminate hydrate phases. The separation of phases from grey scale intensity becomes virtually impossible. To overcome this limitation, the idea is to combine the SEM-BSE image with a multiple channel X-ray microanalysis map (SEM-EDS). Due to the very high amount of data and number of interactions, AI tools were developed and adopted. The Figure 2.12 is a representation of both images of a blended paste (cement + fly ash). The X-ray multichannel collection is made here of 14 images, the first one corresponding to the aluminum mapping.

Principle: The whole of a BSE-SEM image and its corresponding SEM-EDS elemental maps forms is a multispectral image, to be seen as a "cube" dataset, in which each pixel represents a vector of several elements (typically 10 to 20 major elements, depending on the chemistry of the sample).

The whole process is a two-step one: 1) apply a Support Vector Machine (SVM) algorithm, which exploits the spectral information at each pixel and 2) run a Markov Random Field (MRF) algorithm, which uses the spectral

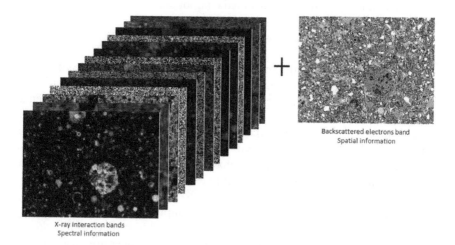

Backscattered electrons band
Spatial information

X-ray interaction bands
Spectral information

FIGURE 2.12 Representation of a multispectral SEM dataset (left: the multichannel chemical mapping; right: the corresponding BSE image).

information of the neighboring pixels, in order to assign the most probable phase to the pixel of interest. This process is close to the spectral-spatial classification approach that has shown excellent performance on complex hyperspectral satellite images of various terrains, with only limited human intervention.

The SVM approach requires two main inputs: a) the multispectral image and b) a set of "learning" points that are user-defined areas in the image (to be seen as some "signature" points of the sample). This step is clearly visible in Figure 2.13, showing the BSE and corresponding X-ray chemical mapping of an OPC/Fly Ash cement paste after 91 days of hydration. The "learning points for the SVM algorithm are color-marked on the BSE picture.

All the large picture widths shown in the Figure 2.13 to 2.17 represent 360µm.

At the end of this process, each pixel is mapped and assigned to a class together with its probability to be included (Figure 2.14). In this figure, Fly ash particles are shown in pink/purple; all other hydrated phases (Portlandite, CSH, AFt, AFm) and porosity are in different shades of green.

The Markov Random Field (MRF) regularization strategy is applied to overcome the "spatial" distribution issue (due to the fact that a single pixel is much smaller than the volume impacted by the electronic beam), making the assumption that it is generally likely to have neighboring pixels belonging to the same class. By this method, outlying or isolated points are "corrected" to their most likely class, and the resulting image is "smoothed" (Figure 2.15). Quantification of a phase is easily extracted from the latter image.

The same process can be applied on various complex systems such as the micronized Portland cement + fly ash + limestone system after 1 day of

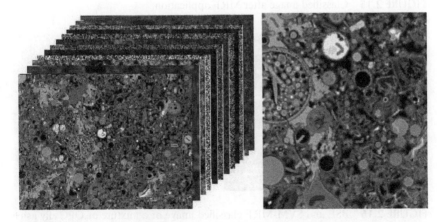

FIGURE 2.13 Stack of X-ray mapping and BSE image of a cement + fly ash paste after 91 days of hydration (left). The middle area is zoomed on the right with learning points marked with thick curved lines at the top, bottom and lower right areas in the picture.

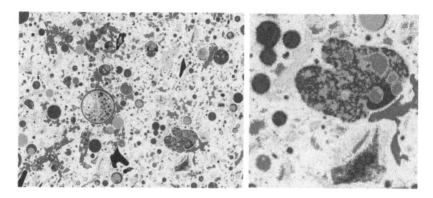

FIGURE 2.14　View of the field in Figure 2.13 after SVM classification. The lower right corner area of the left picture is zoomed on the right.

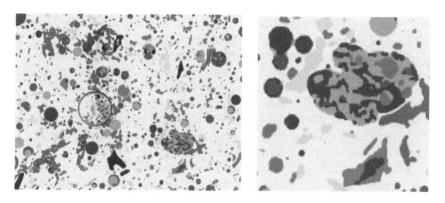

FIGURE 2.15　Classified image after MRF application.

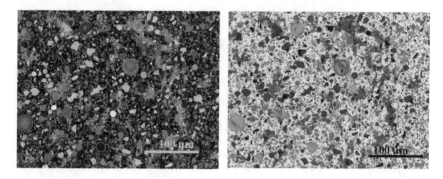

FIGURE 2.16　BSE and SVM-MRF classified image of a mixture of OPC+fly ash+ limestone, hydrated for 1 day. Fly ash, limestone and clinker grains are seen in different gray shades.

 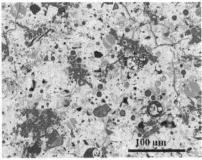

FIGURE 2.17 BSE and SVM-MRF classified image of a mixture of OPC+slag+fly ash after 91 days of hydration. Fly ash, slag and clinker grains are seen in different gray shades.

hydration (Figure 2.16) or Portland cement + slag + fly ash system after 91 days of hydration (Figure 2.17).

From the above illustrations, it is observed that a combination of ML tools applied on multispectral images (BSE- and EDS-SEM) of complex mixes has the capability of refining a very large number of data and all its interactions with a minimum input of human experience and knowledge.

2.3.3 ML in Field: An Example of Sound Analysis

Every running machine is emitting sounds or vibrations that are signature-like. These kinds of signals are very sensitive to all types of intrinsic variation (like the rotating speed or the ageing of its component) or extrinsic variations such as air temperature, pressure or material flow rate, which make them good indicators of process variations.

The mills are a perfect example at a cement plant. The level of material inside is usually monitored with the help of microphones, or some vibration sensors used for predictive maintenance applications (vibrations sensors are typically set on crushers, conveyors belts or elevators).

The following example illustrates the use of ML algorithms that take place in a simple sound recognition application that is able to classify sounds emitted by a mixture of aggregates, sand and water in a concrete mixer. After being "trained" with an adequate dataset, the application is able to predict, for instance, the content of water in the mixture.

Principle: The dataset: About 30 different mixtures of aggregates, sand and water at various ratios in a rotating mixer are "recorded" for 1 minute using a standard sound recording application of a smartphone and its integrated microphone, at a distance of about 1 meter.

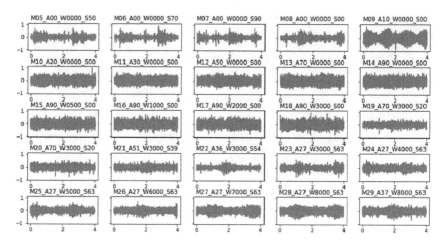

FIGURE 2.18   Waveforms of 25 samples of 4 seconds duration

The raw signals are typically represented by their "waveforms" (amplitude as a function of time). Figure 2.18 shows waveforms of 25 sound samples of 4 seconds duration, randomly extracted from the original 1 minute duration recordings (all recordings with various ratios of sand/water/aggregates).

These "raw" signals have to be processed to reduce their size and extract the valuable information. The output of that process is a series of only a few tens of scaled "mfccs" (mel frequency cepstrum coefficients) per sound samples, to be compared to several tens of thousands data for a 4 second sample (Figure 2.19).

Figure 2.19 shows the waveforms and the representations of mfccs (before scaling) of one sound sample. This 4 seconds sample is processed at a sample rate of 22,050 Hz with a result of a file length of 88,200 values. The "standard" mfccs process cuts the sample into small pieces of 512 data and distributes the calculated frequencies along the desired log distribution (here 40 values), making a matrix of $173 \times 40 = 6920$ values. Taking the average of all the frequencies, the sample is characterized by only 40 numbers. This feature extraction is of course to be adapted to the input quality and the application objectives.

After the extraction of the valuable features (the mfccs) of all the testing samples, typical ML algorithms are set and optimized to develop, depending on the objective, an application able to classify (clustering or classification) or predict the composition of a similar mix from a sound sample recorded under similar conditions (Figure 2.20).

Application: To give an idea of the accuracy and the potential of such applications, 300 hundred samples with a duration of half a second each were

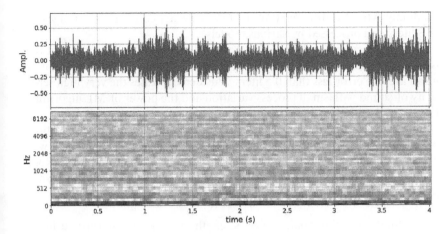

FIGURE 2.19 An example of a 4-second sample in the waveform format and its corresponding sound samples "mfccs" representation.

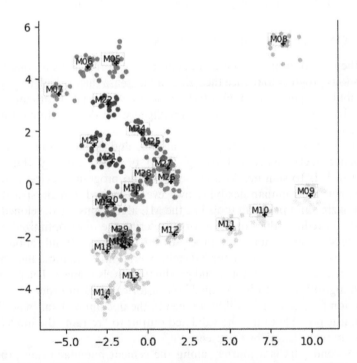

FIGURE 2.20 2D representation of all the samples showing the ability of the machine learning tool to cluster all the samples with a good precision.

randomly extracted from mixtures that have a water content in the range of 3 to 8 liters and processed by various models.

The classification tool was able to correctly recognize all samples but one (meaning a success of 99.7%), and the prediction model gave output with a mean average error less than 0.5 liters. This demonstrates the very high sensitivity of sound with small process variation and the power of ML algorithms to extract the added value of a dataset and provide accurate tools.

The three last examples were selected to show three different applications of ML algorithms that use measurement values, image, and sound data with a purpose of improving the process control, supporting fundamental R&D teams and/or reporting data to the maintenance teams. Certain other examples are easily found in the scientific publications [31–34]:

- "Analyzing the compressive strength of clinker mortars using approximate reasoning approaches - ANN vs MLR"
- "A machine-learning based solution for chatter prediction in heavy-duty milling machines",
- "Artificial intelligent milling system",
- "Determining cement ball mill dosage by artificial intelligence tools aimed at reducing energy consumption and environmental impact"

2.4 CONCLUDING OBSERVATIONS

Since the invention of the first digital computers in the 1940s followed by a tremendous progress to reduce their size and increase their calculation speed, their "ability to perform complex tasks commonly associated with intelligent beings" is the definition of "artificial Intelligence" as demonstrated. But are they really intelligent?

In the first part of this chapter, definitions and details about "Artificial Intelligence" (AI), "machine learning" (ML), or "deep learning" (DL) have been provided. To sum up, AI does not exist (no programs/devices/machine can compete with human flexibility over even a limited domain space), but some amazing and powerful tools like the ML algorithms are developed that are able to predict values or behaviors; prescribe the most optimized set of parameters; or even run some processes by themselves (think about self-driving cars). An overview of the ML tool operation, evaluation, and implementation at both the developer and production levels is given. The principle of such algorithms is to be "trained" on historical data in order to build a "prediction function" that will be applied to the upcoming data. In addition to that, it is possible to use the validated output to "retrain" the model and improve it: they "learn" by experience.

The second part is a journey along the cement manufacturing process, starting at the quarry, ending at the final product. Building an exhaustive list of all the possible locations and ML tool types is impossible (and would become obsolete and require to be updated in a very near future), so the choice was made to cite a few examples and describe others in detail, with the aim of covering the greatest number of algorithm and input data types:

Data types:
    measured values from the plant laboratory (material analysis…)
    continuous values from sensors (kiln temperature, mills rotating speed…)
    Pictures from drones, microscopes or digital cameras
    Sounds or vibrations signals
    Mix of various data types (cf. free lime prediction)
Output types:
    A prediction as a single number or as a "yes/no" value
    Prescription as a set of optimized parameters calculated to reach a target
    Self-driven machines (drones, massive dumpers or full process)
Algorithm types:
    Neural networks
    Clustering tools
    Random forest tools

The "full autonomous intelligent cement plant" does not exist but is no longer unthinkable. Before inaugurating it, there are plenty of operations that can be improved all along the manufacturing processes with the help of AI tools.

At the quarries, the optimization of the raw material logistic and the prediction of the limestone composition before blasts are good.

The optimization of the fuel mix and the raw meal in addition to a better kiln control will have some benefits at the pyroprocessing stage.

Improvement in the mix design of the cement with the help of all the quality characteristics of the clinker and the added material will allow to better fit the cement production amount with the customer requests.

But to be as efficient as possible and make the most of the tools of AI, it is necessary to look at the entire ecosystem. For instance, the output of the pyroprocessing is the clinker, the properties of which are given by the raw mix, the fuel mix, and the process parameters. Going backward, the raw mix property is a function of the limestone and the raw mill parameters, but is designed to produce a targeted clinker. Going forward, the cement characteristics are given not only by the clinker ones but also by the added material compositions and amounts and the cement mill process parameters.

Optimization of the full process cannot be done by only focusing on the final product quality. Parameters such as process costs (including the raw material, the maintenance and the equipment replacement costs), respects of regulations, and willingness to reduce the global footprint (CO_2 emission reduction) have to be taken into account.

This demonstrates that the cement manufacturing process is fully interconnected and complex with many sources of variances and a large amount of interactions of all types that make its global optimization extremely difficult to achieve without the help of ML tools of all kinds.

2.5 PERSPECTIVES

In the next decade, there is no doubt that we will face an increase of standard computer power, a huge development and easier access to ML algorithms, an improvement of the digitalization at all levels, and, hopefully, a better understanding and acceptance of what AI can bring us.

In parallel, the cement manufacturing will have (in fact, already has) some big challenges to solve. By considering only the CO_2 emissions to be reduced, this implies modifying or increasing the fine-tuning capacities of the entire process to make it much more reactive and efficient from the quarry to the production of cement. "Transforming cement manufacturing through application of AI techniques" is a promising route to achieve this.

The examples detailed in this chapter demonstrated the huge potential of the ML algorithms. It has been mentioned that combining data of all types is no more an issue and helps to extract valuable information but, like the light bulb that did not come from the continuous improvement of the technology of a candle, an efficient cement manufacturing process transformation can't be achieved by only improving the current software and sensors.

The next challenge and an essential step to build an "intelligent" cement plant would be to produce "new" data, either by using existing signals that are not fully exploited (think about the kiln thermal images, X-ray diffractograms, or sounds of the mill, etc.), or by testing new features that are not adapted to the conventional tools but could be of value to ML algorithms.

ACKNOWLEDGMENTS

The author would like to thank Tien Nam Le and Samuel Meulenyzer, both research engineers at LHIC for their contributions to this chapter ("Prediction of the clinker free-lime" and "Semi-automatic classification of cementitious materials using scanning electron microscope images" sections respectively).

Thanks also to Jesus Subero, Christophe Levy, Fabrice Pourcel, and Françoise Freeman for their contributions and advices all along the writing process.

REFERENCES

1. https://en.wikipedia.org/wiki/ENIAC
2. http://www.ibiblio.org/apollo/
3. https://www.fujitsu.com/global/about/innovation/fugaku/
4. https://www.britannica.com/technology/artificial-intelligence

5. Luc Julia, *L'intelligence artificielle n'existe pas*, FIRST, Paris, France, 2019.
6. https://deepmind.com/research/case-studies/alphago-the-story-so-far
7. *Ted Dunning (MapR Technology)*, conference at Hadoop Summit 2014, Amsterdam.
8. https://longmont.pastperfectonline.com/photo/A8D1EABC-74FF-4983-85EF-093043174876
9. http://globalcement.com/magazine, report-2017–2018.
10. http://pubs.usgs.gov/periodicals/mcs2020/mcs2020-cement.pdf
11. A.K. Chatterjee, *Cement production technology - principles and practice*, CRC Press, Boca Raton, FL, 2018.
12. https://delair.aero/
13. Tom Simonite, "Mining 24 hours a day with robots", *MIT Technology Review*, January 2017.
14. R. Trivedi et al., *Journal of Rock Mechanics and Geotechnical Engineering*, 6, 2014.
15. A. Parida et al., *Procedia Earth and Planetary Science*, 11, 2015.
16. Xiu-zhi Shi et al., *Transactions of Nonferrous Metals, Society of China*, 22(2), 2012.
17. Khan Muhammad et al., *Archives of Mining Sciences* 62(4), 2017.
18. Aysun Egrisogut Tiryaki et al., *Materiali in Tehnologije* 50(4), 2016.
19. Dimitris Tsamatsoulis, *The Canadian Journal of Chemical Engineering* 92(11), 2014.
20. Lian Guodong et al., *Second International Conference on Advances in Information Processing and Communication Technology IPCT* 2015.
21. Kausar Sultan Shah et al., *3rd Conference on Sustainability in Process Industry*, 2016.
22. Globalcement.com, published on 21 March 2019.
23. Worldcement.com, publication, January 2020.
24. Meiqi Wang et al., *Advances in Mechanical Engineering*, 12(1), 2020.
25. Morten Nedergaard Pedersen et al., *Energy & Fuels*, 31(10), 2017.
26. Ouahab Kadri et al., *Academic Journal of Manufacturing Engineering*, 15(2), 2017.
27. Rafik Mahdaoui et al., *International Journal of Reliability, Quality and Safety Engineering*, 26(3), 2019.
28. Yanfei Zhang et al., *International Joint Conference on Neural Networks (IJCNN)*, May 2018.
29. Aghdasinia, H. et al., *Journal of Ambient Intelligence and Humanized Computing*, 2020.
30. Samuel Meulenyzer et al., *14th Euroseminar on Microscopy Applied to Building Materials*, 2013.
31. Ahmet Beycioglu et al., *Computers and Concrete*, 15(1), 2015.
32. Ibone Oleaga et al., *Measurement*, 128, 2018
33. Farhan Alfin, millermagazine.com, March 2020.
34. Julio Rafael Gómez et al., *Ingeniería e Investigación*, 33(3), 2013.

3 Process Automation to Autonomous Process in Cement Manufacturing
Basics of Transformational Approach

Anjan Kumar Chatterjee

Conmat Technologies Private Limited, Kolkata, India

CONTENTS

3.1 Introduction 79
3.2 Automation to Autonomy in Manufacturing: Basics and
Approach 80
    3.2.1 Steps Toward Achieving Autonomous Operation 81
    3.2.2 Technological Imperatives for Transition 83
3.3 Expanding Concepts of Machine Learning 87
3.4 Current Process Control Infrastructure in Cement Plants 89
3.5 Advances in Process Control Strategy for
Cement Manufacturing 91
3.6 Future Considerations in Proliferating APC Systems in
Cement Manufacturing 95
3.7 Concluding Observations 97
References 98

3.1 INTRODUCTION

Tracing back the technological history of the cement industry, spread over almost two centuries, one may observe that the scale of production and the commensurate level of mechanization have all along been the primary drivers of its phenomenal growth. During 1970s and 1980s, automation was used as a means to maintain the plant operations in safe and continuous mode. During 1990s, the process control and information system of the then vintage were integrated with the plant operation, leading to higher productivity, better efficiency, lower manufacturing cost, and improved product quality. During the above period and thereafter, increasing emphasis was laid on the adoption of computer-integrated manufacturing with progressively

DOI: 10.1201/9781003106791-3

enhancing levels of computing functionality and information technology (IT). In late 1990s and in early years of the twenty-first century, the cement industry became more and more market-driven from its earlier orientation of being production-driven. Added to the above transformation are the push for reducing the greenhouse gas emissions and enhancing the resource sustainability. In practice, the above change drivers have led the industry to strive for the following:

- man and machine productivity of high order
- significant conservation of natural resources
- maximization of alternative fuels and raw materials
- high plant availability, coupled with energy efficiency
- production of environmental load-reducing cements

All along the way, the cement industry has been undertaking timely efforts to reshape itself with energy efficiency, resource conservation, and sustainability. In recent years, the efforts are directed toward adopting the path of digital transformation due to the fact that the introduction of digital technologies into operational technology (OT) at both the plant and enterprise levels offers the potential of achieving the above objectives to an extent that would not be possible otherwise. Under these circumstances a very pertinent question that deserves a careful look is what roles autonomous operations will play in the manufacturing plants. If they are significant, how distant is their implementation and what might be the impediments. This chapter is intended to review briefly the status of cement manufacturing in the evolving path of making automated processes into autonomous operations.

3.2 AUTOMATION TO AUTONOMY IN MANUFACTURING: BASICS AND APPROACH

Automation is not a human attribute, whereas autonomy is, but both are relevant paradigms of the present-day process industries with varying effects. Automation is a human activity for creation and application of technologies to produce and deliver goods and services with minimal human intervention. Automation is ubiquitous around us. We see it extensively in manufacturing, utilities, transportation, security, and several other areas. It performs a sequence of highly structured pre-framed tasks, each of which requires human overview and interaction. We also observe that even in highly automated systems, the human involvement, though relatively less, is highly crucial and unavoidable. In addition to the continued human involvement, automation suffers from diminishing returns and exhaustion of opportunities for increased levels of implementation. In this backdrop the feasibility of introducing autonomy in process industries seems to be opening up a new horizon.

Quite different from machine automation, the sense of autonomy, originating from human behavior, refers to the human desire and capability to define 'nomos' (law) that will guide future actions. For this to happen, it is essential that autonomy has to have an inherent link with intelligence, the origin, and manifestation of which are extremely difficult to decipher. Perhaps, the most prominent manifestation of intelligence is through 'posing dilemmas and solving problems' [1]. It is, therefore, pertinent to examine what autonomy and intelligence mean for machines, and when it is not possible to define autonomy and intelligence for humans in a broad consensual way. No doubt that, compared to human autonomy and intelligence, these attributes for machines and systems will run the risk of reductionism. In the context of machines, the term 'intelligent' is limited to computational systems using algorithms and data to address complex problems and situations, and such systems could be regarded as 'autonomous' in a given domain as long as they are capable of accomplishing their tasks despite environmental changes within the given domain [2]. This concept has given rise to the new paradigm of 'artificial intelligence' (AI), quite extensively used nowadays but not with universal understanding of its sense. For example, a new concept of 'extended intelligence' is being researched [3]. Instead of thinking about intelligence in a framework of human versus machines, it may be more appropriate to consider the smart machine systems as integrated entities with 'extended intelligence'. The human intelligence excels at processing sensory data, understanding abstract thoughts and free association, while the AI may excel at remembering, processing, predicting, and analyzing. Consequently the two forms of intelligence are likely to balance each other's weaknesses and to complement each other's strengths. While humans have difficulties of scale, computers can perform trillions of operations a second. Thus, the concept of extended intelligence is emerging as a new paradigm.

3.2.1 Steps Toward Achieving Autonomous Operation

Advancing the automated operations to autonomous state can only be conceived as an evolving multi-stage process, and the transition has to be planned and implemented in a calibrated manner. The path to reaching autonomy in process industries is generally considered to have five steps or levels, based on the steps widely perceived necessary for making self-driving or autonomous cars [4]. In targeting autonomous cars, it is conceived that the bottom four levels will need a driver, but with diminishing roles, as his functions will be progressively taken over by the autonomously operating systems. Only after reaching level 5 human intervention will be completely dispensed with. More or less in a similar pattern, the transition plan for manufacturing systems to autonomous operations may be tentatively contemplated as schematically shown in Figure 3.1. The figure shows that there is an important demarcation between the lower levels (0–2) and the upper levels (3–5). In the lower levels,

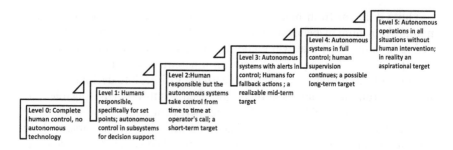

FIGURE 3.1   Tentative levels of process transitions from no autonomy to complete autonomy.

the autonomous systems are subservient to human functions and responsibilities, while in the upper level the human roles are taken progressively to a passive state. It may be pertinent to mention here that the legal dimensions of risks involved in institutionalizing autonomous systems in process industries are still not adequately addressed, though they are very crucial in accelerating the progress of autonomous operations in manufacturing.

While autonomy could be brought to any part and life-cycle phase of a plant, in the context of operations some sections might be given priority, depending on the situation and the major requirements. In [5], the authors have shown preference to select the following priority areas:

- field operations/maintenance: operations that are done at site such as valve manipulation, maintenance of instruments and actuators, equipment malfunctioning that cannot be handled from the control room, etc.
- control room operations: operations that are primarily done by operators using human–machine interfaces, such as the automated actuators and fixing set points for standard control loops
- planning and scheduling: production planning in accordance with customer orders and demands, as it has a significant impact on what needs to be done and how it should be done

Investments and relative priorities for making such subsystems autonomous would obviously depend on the current condition of the plant and the available technology infrastructure. In control room operations, the activities such as the start-up and shut-down, control under steady-state conditions, switchover provisions from autonomous controls in the event of process abnormalities, fault detection and correction, and emergency and safety measures are some specific sections of a manufacturing plant where the step-up from supervised operation to autonomous operation needs to be planned and implemented in a calibrated manner.

To achieve remote operations for autonomy several steps might be necessary. Some of the important steps are the following:

- converting manually operated equipment to fully automated machines
- implementing procedural automation for manual functions
- using resilient and redundant communication systems, controllers and critical equipment
- adopting intelligent sensors
- conducting remote monitoring, surveillance, and inspection via robots
- applying predictive and prescriptive maintenance using data analytics, machine learning and other AI-based techniques
- adopting AI and data analytics also for predicting process abnormalities and corrective measures
- using complex modeling techniques as necessary
- creating and using digital twins

The above points indicate the essentiality of moving toward a different technological paradigm, the dimensions of which are discussed below.

3.2.2 TECHNOLOGICAL IMPERATIVES FOR TRANSITION

Nowadays it is a widely held view that in manufacturing industries there is a progressive convergence of OT and IT. Enterprise resource planning and manufacturing execution systems were introduced several years back in industrial plants in order to boost production planning and optimization. These IT systems have become more effective in recent years due to enhanced availability of data and information about the involved physical processes, primarily with the aid of smart sensors and interconnection with OT, essentially consisting of PLC (Programmable Logic Controllers) and DSC (Distributed Control Systems). Added to these features is the advent of industrial internet-of-things (IIoT), which has transformed industrial plants into large-scale cyber-physical systems (CPS).

Simplistically speaking, such CPSs are underpinned by two technological pillars: AI and big data analytics. There is a plethora of literature on the development and application of both these disciplines, but for the specific and limited purpose of this discourse, we may look into the interrelation of AI and data analytics, yielding new technological possibilities in manufacturing (Figure 3.2). In recent years it has become evident that the development of big data technology depends on AI and the development of AI must rely on big data technology. Big data is defined as high-volume, high-velocity or high-variety information assets that cannot be processed by the traditional methods, and instead, demand cost-effective and innovative forms of

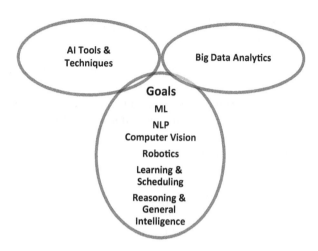

FIGURE 3.2   Major goals of AI and data analytics.

information processing to provide insights from the data. The insights generally refer to hidden patterns, unknown correlations, market trends, customer preferences, etc., which help in decision making, process automation, and so on. There is no defined volume for big data but generally the data volume in terabyte range is regarded as big data, which, for all practical purposes, must have intrinsic 'value' and 'veracity', in addition to the other attributes mentioned above.

The big database is the basic prerequisite for application of AI, a discipline that deals with the development of theories, methods, technologies, and systems that essentially simulate the human intelligence. One of the most relevant goals of AI application is to let machines perform some complex tasks that would have required intelligent humans to perform. For this goal to reach, the essential infrastructural requirements are reliable and versatile connectivity, cloud data services, and data analytics. The Seattle Report on database research has summarized the advancements made and further progress expected in the above related fields [6]. Some relevant aspects are discussed below.

A. **Cloud data services**: Use of various software development platforms, data processing, data storage, and servers on no-ownership model is already in practice. Variants of PaaS (Platform as a Service), SaaS (Software as a Service), and IaaS (Infrastructure as a Service) are well known. While PaaS allows the deployment of software and applications on the cloud, SaaS permits consumers the ability to access or use an application or service hosted by the cloud. Unlike these two variants, IaaS allows consumers to control and manage the operating systems, applications, network connection and storage without controlling the cloud themselves. Edge computing is an altogether different approach that allows computer resources and application services to be distributed along the communication path via decentralized computing infrastructure. Edge computing is ideal for high-latency concerns. Both cloud computing and edge computing have their pros and cons, which are not discussed here. However, it may be relevant to mention about 'fog computing', which, as a mediator between hardware and remote devices, regulates the type of information that should be sent to the server and the type that can be processed locally. In other words, it is an intelligent gateway that offloads clouds, enabling more efficient data storage and analytics.

Cloud computing has become mainstream and usage of cloud data systems has grown significantly in recent years. The industry now offers on-demand resources that provide high elasticity. For analytics, the industry has converged on a 'data-lake architecture', which uses elastic computer services to analyze data in cloud storage on demand. This architecture disaggregates compute and storage so that they can scale independently.

B. **Data science**: It focuses on the processes and systems that enable the extraction of knowledge or insights from data available in various formats such as the structured, semi-structured, or unstructured types. It provides the pipeline from the raw input data through data processing and wrangling, to data analysis, data visualization and finally the insights. The traditional environment of data science has been the extraction of insights from databases by applying SQL (Structured Query Language), OLAP (Online Analytical Processing), data mining techniques, and statistical software suites. The modern data science shows a shift to an environment that relies more on the rich ecosystem of open-source libraries for sophisticated analysis including machine learning techniques. Another significant departure is that the data scientists also work with data lakes that hold structured and unstructured data sets with varying sets of data quality, not necessarily restricting themselves to carefully curated data warehouses.

Another feature that deserves mention here is the use of NoSQL (Not Only SQL) for databases. Traditionally SQL has been the standard language for dealing with relational databases. It could be used to insert, search, update, and delete the database records. It can also do optimization and maintenance of database. The main advantage of NoSQL is its non-adherence to relational database concepts, although the ACID (Atomicity, Consistency, Isolation, Durability) features are not always guaranteed.

C. **Hardware and connectivity**: With the rise in compute-intensive workloads, a new generation of powerful accelerators leveraging FPGA (Field Programmable Gate Arrays), GPU (Graphics Processing

Units), and ASIC (Application Specific Integrated Circuits) are being used. These technologies are seen as viable approaches for training big models. The memory hierarchy continues to evolve with the advent of a new generation of SSD (Solid State Drives) and low-latency NVRAM (Non-Volatile Random Access Memory). Specialized interconnects as well as improvements in network bandwidth and latency are already in place. The advent of 5G with ample bandwidth is on the verge of reshaping the workload characteristics of data platforms.

D. **Scope of big data analytics**: there is a recent prediction that each user would create 1.7 megabytes of new data every second. This would mean that within a year there would be 44 trillion gigabytes of data accumulated in the world [7]. This volume of data needs to be analyzed for decision making, optimizing business performances, studying customer trends, and delivering better products and services. This is an indication of the growing demand and importance of data analytics and it has resulted in the development and application of a large number of tools, many of which are open-source type. The consequence of this large-scale development of tools is the impending challenge being faced by the user community to choose the right tools for the intended applications. Some of the tools with varying capabilities have gained certain levels of popularity. For example, R programming in data mining, Tableau Public and Python in data visualization, APACHE Spark for fast large-scale data processing, Hadoop in data storing, Talend in data integration and management, SAS for data manipulation, STORM for computational systems, MongoDB for processing frequently changing data sets, etc. Generally speaking, the tools that can operate on various platforms or can be assembled on different databases are preferred. It must be borne in mind that there are widely varying types of data sources such as Access, Excel, Microsoft SQL, Teradata, Oracle, IBM DB2, Ingres, MySQL, IBM SPSS, Dbase, etc., which need to be handled in data processing. It is also important to consider that the big data analytics is required to be performed for different objectives in view such as the following:

- descriptive that summarizes the past data into a form that can be easily read and converted into a report form
- diagnostic that is done to understand what caused a problem
- predictive that looks into historical and present data to make predictions for future trends
- prescriptive that prescribes solutions to a specific problem

In essence, the data analytics is the science of examining the colossal volume of raw data to reach certain conclusions and the focus is in drawing inferences.

3.3 EXPANDING CONCEPTS OF MACHINE LEARNING

As shown in Figure 3.2, Machine Learning (ML) as a discipline emerged out of the quest for AI and data analytics. It is a systematic approach to leveraging advanced algorithms and models to continually train data, test with additional data, and apply the most appropriate algorithm to a problem. The ML algorithms differ from other algorithms in a way that the data itself creates the model. The ML algorithms are most often written in Java, Python or R language and each of them includes libraries that suggest a variety of ML algorithms. The ML algorithms may have different features and purposes. The commonly encountered varieties are briefly presented in Table 3.1.

It is important to note that the application of ML is not a one-off action. The application development has to be conceived as a cyclic process as shown in Figure 3.3. The ML requires the right set of data that can be applied to a learning process. In order to use the techniques an organization does not have to have big data. However, big data can help improve the accuracy. Further, the link of AI with ML lies in three important dimensions: reasoning, natural language processing (NLP), and sequence planning. Reasoning refers to making inferences from data, NLP to the ability to train computers to understand both written text and human speech, and planning to construction of sequential acts for applying the outcomes.

Machine learning is generally classified into four categories as shown in Figure 3.4. The broad features of these categories are discussed below.

The categories shown in the above figure are not strictly hierarchical but can be thought of as algorithms of increasing complexity. The 'unsupervised learning' relates to large amounts of unlabeled data that is used as input in order to find structure in it, which could be a grouping of examples or a grouping of features or of some other criteria. The 'supervised learning'

TABLE 3.1
Common ML Algorithms

Serial No.	Type of Algorithm	Salient Feature
1	Rule-based	Relational rules to describe the data
2	Bayesian	Encoding prior beliefs about the model, independent of what the data states
3	Dimensionality reduction	Removing data that is not useful for analysis
4	Instance-based	Used to categorize new data points, based on similarities to training data
5	Regularization	For avoiding overfitting
6	Regression	Quantifying the strength of correlation between variables in a data set
7	Clustering	Bringing together objects more similar to each other than in other clusters
8	Decision tree	Branching structure to illustrate the varying likelihood of occurrence of different outcomes of a decision

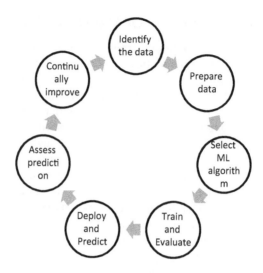

FIGURE 3.3   The cycle of ML application development.

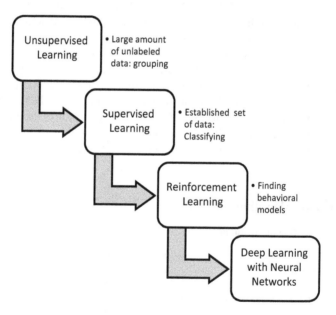

FIGURE 3.4   Categories of machine learning algorithms.

begins with an established set of data with a view to classifying it in a manner that can be applied to the analytics process. The algorithms include classification, regression, etc. The 'reinforcement learning' is a behavioral learning model and it differs from other types of supervised learning due to the fact that the system is not trained with the sample data set. The system learns through trial and error. Thus, a sequence of successful decisions will result in the process being 'reinforced' as the best solution to the problem at hand. One of the common applications is in robotics, in which the data is recalibrated on the basis of the robot's navigation failures. It is also good for control theory and operation research. Admittedly, the reinforcement learning has emerged as a powerful complement to supervised learning. Lastly, the 'deep learning' uses hierarchical neural networks to learn a combination of unsupervised and supervised learning. It is pertinent to mention here that neural networks may consist of thousands or even millions of simple processing nodes that are densely interconnected. With rising process complexity in IoT manufacturing applications, the number of hidden layers increases. With incorporation of multiple hidden layers in neural networks, in deep learning, the machine learns the data in an interactive manner. It is not only useful for learning patterns from unstructured data but also for image recognition and computer vision applications. Some of these areas have received unprecedented push for progress with deep neural networks.

The barrier to writing ML-based applications have been sharply lowered by widely available programming frameworks such as TensorFlow and PyTorch as well as FPGA, GPU and specialized hardware for use in private and public clouds [6]. The database connectivity has also greatly helped the ML users. The technological breakthroughs in AI and ML will pave the way for autonomous operations of industrial plants.

3.4   CURRENT PROCESS CONTROL INFRASTRUCTURE IN CEMENT PLANTS

For the last three decades, the structure of control systems in cement manufacturing plants is engineered with Distributed Control Systems (DCS), in which PLC are distributed throughout the process, each to control a certain section of the process. In recent years the entire system is further networked for communication and monitoring. In most plants, the DCS includes Supervisory Control and Data Acquisition (SCADA) software and modern PCs as human–machine interfaces. The production conditions in different production sections are monitored and the signals are transmitted to PLCs via input/output (I/O) devices and the control signals generated at PLCs are then returned through I/O devices to each of the execution devices. A simplified diagram of the systemic architecture of the control system is given in Figure 3.5. The Master Station refers to the servers and software responsible for communicating with the field equipment and also to human–machine interfacing software on workstations in the central control room. At the workstations, the process flow charts, operating conditions, and logic diagrams are displayed on the viewing screens with the help of CRT (Cathode Ray Tube). The start/stop and set points are accessible to the operators who can archive data and visualize data trends for process analysis and trouble shooting.

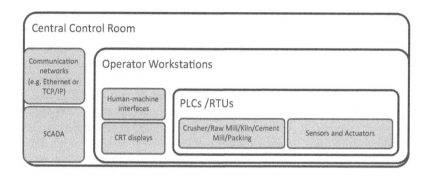

FIGURE 3.5 Structural configuration of the process control system.

It is pertinent here to observe that the functionality of PLCs has evolved over the years to include sequential relay control, motion control, process sequence control, and, as already mentioned, DCS and networking. The data handling, strategic processing power, and communication capabilities of some modern PLCs are approximately equivalent to desktop computers. A Programmable Automation Controller is a compact controller that combines the features and capabilities of a PC-based control system with that of a typical PLC. In most plants, the PLC performs PID (Proportional–Integration–Derivative) control functions, data compiling, and formatting; manages alarm logic; and executes alarm signals and interlocking. The PID control function has been an important component in the control philosophy (Figure 3.6). The controller input is the error signal from the comparison of the set point with the feedback process variable. Accordingly, the output of the system is adjusted by varying the coefficients (K_p, K_i, and K_d) of the proportional, integration and derivative functions for response and control. This process continues until near-zero error is obtained. For PID tuning and optimization, software packages are available, which collect data and make a mathematical model of the system. The PID control function can be represented by the following equation:

$$y(t) = K_p \times e(t) + K \int_i e(t)\,dt + K_d \times de(t)/dt \qquad (3.1)$$

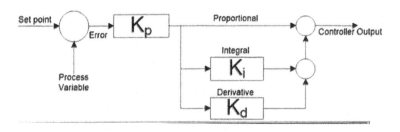

FIGURE 3.6 Schematic diagram of a PID control scheme.

where, e is the error, t is the time, and K with subscripts refer to proportional, integral, and derivative applications.

In order to meet the necessity of connecting to multiple industrial networks in the DCS, various discrete and process industrial networks, such as DeviceNet, PROFINET, ETHERNET/IP, Modbus TCP, HART, Foundation Fieldbus, are utilized. Ethernet is one of the most common protocols used today to interconnect general computing systems and servers. The advantage of Ethernet is that it has a variety of speeds available from 10 to 1000 Mbps and also the newly developed 10 GbE and can use many different forms of cable. The industrial networks are built from a combination of these networks and at least one real-time network or fieldbus to connect devices and process systems. These networks are nested deep within the enterprise infrastructure so that some layers of protection against external threats are provided.

A modern production line is engineered with the process control structure and communication networks as outlined above. From the beginning of the present century, the control strategy has been influenced by certain basic forms of AI, which have been discussed below.

3.5 ADVANCES IN PROCESS CONTROL STRATEGY FOR CEMENT MANUFACTURING

From the perspectives of process control, the advanced strategy that first caught up with the cement industry more than two decades back was the installation of 'Expert Systems' for different process sections and more particularly for pyro processing. The fuzzy logic, based on the 'IF-THEN' rules, has been applied in designing the controllers [8]. The basic architecture of a fuzzy logic controller is shown in Figure 3.7 [9].

The fuzzy logic controllers appeared to be workable as the knowledge base for the cement manufacturing process was generally available and the control engineering accepted the principle of 'good enough' solution for the

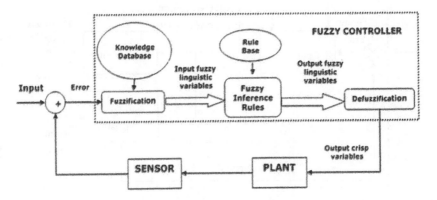

FIGURE 3.7 Basic architecture of fuzzy logic controller [9].

complexity and variability of the process. Conversion of the crisp values into fuzzy values, application of fuzzy rule base of the operational domains, simulation of human decisions by performing approximate reasoning, and finally conversion of the fuzzy values into crisp values through defuzzification could be achieved to a reasonable extent. Notwithstanding the initial success, the proliferation of the system was affected by the lack of requisite database, inadequacy of human expertise, and problems of updating the rules.

Consequently thereafter, the Model Predictive Controller (MPC) technology was developed with greater promise to handle challenges associated with the control of the most critical pyro processing section. The MPC allows the controller to receive information about the current operating conditions of the process, and then it uses a model to predict process response to a sequence of future moves in manipulated inputs over the prediction horizon. To begin with, the plant models could be represented by conventional mathematical approaches like transfer function, state space, and ARMA (Auto-Regression Moving Average). However, the control performance of MPC fundamentally depends on the accuracy of the model describing the plant behavior and other controller parameters. It has also the ability to handle multiple-input–multiple-output (MIMO) process with time delays, which made it suitable for the cement process. For a better appreciation of the controller, a kiln model with four state vectors (x) is shown in Figure 3.8 [10]. The feed (U1) and fuel (U2) flow rates were considered as control inputs. The output matrix (y) was defined by two of the possible measurements namely torque (Y1) and the burning zone temperature (BZT) (Y2). Coefficients A, B, and C corresponded to the state input and output matrix that governed the dynamics of the kiln. The state-space model of the kiln described in Equations (3.2) and (3.3) was used to predict the future behavior of the kiln.

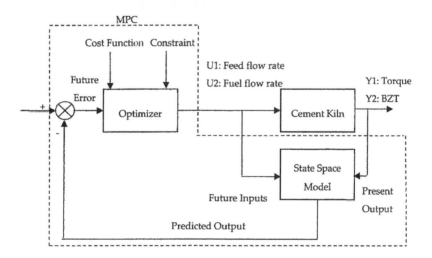

FIGURE 3.8 Block diagram of a MPC for a cement kiln process [10].

$$x(n+1) = Ax(n) + Bu(n) \qquad (3.2)$$

$$y(n) = Cx(n) \qquad (3.3)$$

Kalman filter-based state observer was used to predict the states of the kiln. The filter used the state observer parameters (A,C) and the noise covariance matrices to determine the Kalman gain, which was applied to fuse the prediction and measurement to assess state of the kiln accurately. Then, the estimated states were employed to determine the future error, and a cost function was formulated using this error along with the rate change of inputs. An online optimizer was used to estimate the control input that could minimize the cost function. Finally, the control input was implemented on the kiln to regulate both the torque and BZT. Practical limitations of the control variables and plant behavior were fed as constraints in minimizing the cost function as illustrated in Figure 3.8. The MPC was designed to work on the principles of the receding horizon approach as shown in Figure 3.9 [10].

The future behavior of the kiln was estimated using the state-space model and past outputs. The prediction was made for a finite number of samples described by the prediction horizon (P). The prediction horizon had to be optimally determined and the choice of control horizon (M) depended on the prediction horizon and the close-loop response of the kiln. Further, the optimal tuning of the MPC weights was done using genetic algorithm with interactive decision tree.

An important lesson from the above study was that MPC weights are predominantly specific to plant dynamics and the nature of disturbances the plant encounters. The weights, in fact, play a vital role in determining the optimal control of the process. The tuning procedure of MPC continues to

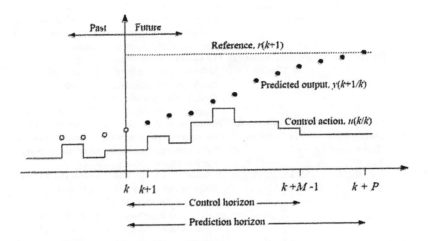

FIGURE 3.9   The receding horizon operation of MPC [10].

depend on operator-driven stochastic optimization technique. Several other approaches have been tried by researchers to overcome the deficiency including the adoption of nonlinear MPC optimization.

Considering the overall trends of field experience of both the 'Expert Systems' and MPC-based 'Production Optimizers' that were put into practice, it can be observed that they needed a high level of adaptation to intrinsically changing operating conditions of cement plants. In most cases, the performance and robustness of these control systems have not proved satisfactory. Further developments have, therefore, taken place, resulting in the introduction of Advanced Process Control (APC) systems in plant operations. The APC systems are aimed at moving the process control from isolated discreet functions to one in which the processes are connected via IoT technologies. They are more extensively data-driven. Several APC systems having different bases and features are being offered commercially. A few examples are presented in Table 3.2.

TABLE 3.2

Examples of a Few Commercially Available APC Systems

Serial No.	Technology Provider	Trade Name of APC System	Basic Features as Reflected in the Corresponding Trade Literature
1	ABB Group	ABB Ability	MPC-based, data analytics covering raw materials and the unit processes, sustained monitoring of predefined KPIs,
2	aixergee GmbH	aixPert Optimizer on aixProM platform	Big data analytics, statistical and AI algorithms combined with deterministic models like process and real-time CFD simulation, parallel operation of digital twins, aimed at total plant optimization.
3	FLSmidth & Co. A/S	ECS/Process Expert	MPC-based with symbolic expert systems and fuzzy logic and non-symbolic machine learning and deep learning technologies to cover all unit operations and business functions.
4	Petuum, Inc.	Petuum Industrial AI AI-pilot	Data analytics covering control systems, historians, unstructured and structured databases, machine data and images for predictions, optimal prescriptions, and supervised steering of the end-to-end processes
5	Powitec Intelligent Technologies GmbH	PIT Navigator & PIT Stabilizer	Nonlinear model based process simulations with the help of neural networks coupled with optic flame structure and use of automatic step-and-ramp of manipulated variables.
6	Siemens AG	SIMATIC PCS7/ CEMAT	Modular with Kiln Control System (KCS) and Mill Control system (MCS), a standard library to address the majority of closed-loop control tasks, knowledge base approach using the current plant data, digital twin for virtual commissioning and operator training

TABLE 3.3
Parametric Changes under AI-Based Optimization

Variable Parameters	Kiln 1	Kiln 2
Clinker output, t/d	3000	6000
Increase in secondary air temperature in AI-based optimization, °C	55	47
Increase in tertiary air temperature in AI-based operation, °C	8	44
Increase in clinker temperature, °C	2	7
Excess of energy recovered ($\times 10^7$ kJ/d)	9.6	27

It may be relevant to mention that the APCs are new in the field and have so far been adopted in a limited number of plants. Wherever applied, the experience has been positive in achieving higher benefits in energy conservation, improved productivity, reduction in emissions levels, larger use of alternative fuels, longer equipment lifetimes, and other yardsticks, compared to the results obtained through traditional measures. As an example, a specific mention is made here of a particular study on the application of AI technology to optimize the clinker cooler operation [11]. The energy recovery for two coolers of different capacities were measured indirectly by the two key variables: secondary air temperature and tertiary air temperature and the differences in the parameters as obtained under routine operation and AI-based optimization are shown in Table 3.3, which also includes an estimation of energy savings computed by a thermodynamic analysis. Some increase in the clinker temperatures remained a subject of further studies with possible tuning with weight factors. It was also reported by the authors that the model performance was tested at the plant by randomly switching on and off the 'autopilot' system and the mean values of the secondary air temperature and tertiary air temperature differed by 96°C and 80°C, respectively, but with high standard deviations.

3.6 FUTURE CONSIDERATIONS IN PROLIFERATING APC SYSTEMS IN CEMENT MANUFACTURING

The most important requirement for implementing APCs is the access to **historical data** from the plant operation. DCS store short-term information required for different analyses. The data volume is not large enough to train a model that will be able to learn the dynamics of the process with the objectives of automation and optimization. Hence, it is important to have 'data historians' in place. A data historian is a program that records the data of processes running in a computer system. The data from numerous sensors, intelligent electronic devices, DCS, PLC, laboratories, and even the manually entered data are collected by data historians. Most cement plants generally archive inactive data or past operational data of no current relevance. It will be more relevant to upgrade the system in many situations to the configuration of data historian, in which the data is time-staggered and cate-gorized in quickly machine-readable format. In this context, the option of open-source time-series database is worth mentioning. As already mentioned earlier, it has been essentially developed for non-relational data or NoSQL format. The difficulties encountered in data historians to link the data to a context or to suffer from limitations of scale and analysis can be overcome in the time-series database with ease of integration and data handling.

In addition to the database, it is important to be able to do **streaming of data** between the plant site and the cloud-based service within the frequency and latency requirements for which a proper architecture has to be designed and implemented. Broadly speaking, the basic **elements of the AI platform** consist of the following:

- Application Programming Interface (API) covering speech recogni-tion, recognizing objects from vision, language choice for identifying entities from the free format texts, and translating texts
- machine learning libraries such as TensorFlow, Pytorch, Keras, Scikit.Learn
- machine learning tools such as Notebooks, AI pipelines
- cloud infrastructure including GPU, data storage, IoT, etc.

There are several important considerations for selecting the AI platform. The most important feature is whether the cloud platform should be a single pre-ferred one or hybrid in composition. Since the cloud providers have different focuses on their APIs, it is possible to mix and match the APIs from multiple cloud platforms to suit the needs of the customer. The hybrid approach is often seen to achieve the best combinations. The data ownership and privacy including the data storage location is another critical feature. Machine learn-ing uses training and test data sets. The availability of pre-trained models is certainly an advantage, but the data compliance by the cloud provider must be ensured. ML tooling is important when working with data sets. Each plat-form provides a Notebook environment for experimentation. It is important to know if the Notebook environments are effectively integrated with an ease of access to ML libraries and infrastructure of the cloud providers. The API availability and ML capabilities, including NLP, recommendation engine, pattern recognition, and prediction, are extremely significant in the evalua-tion process because the APIs provide the foundation to build applications by the client. Finally, the life-cycle cost of the application must be computed, which should include the costs relating to data migration and data transfers. The cost calculator can be an advantage for the cloud platform.

Extending **deep learning and reinforcement learning** to process control problems is a recent endeavor to make the application of ML more effective in this field. The classical controller design involves careful analysis of the process dynamics, development of an abstract mathematical model, and finally, derivation of a control law that meets certain design criteria. It has already been explained earlier that unlike the classical approach, in reinforcement learning the closed-loop controllers learn by interacting with the process and effect incremental improvement of the control behavior. The same learning principle can be applied to a wide range of processes that include linear and nonlinear systems, deterministic and stochastic systems, single I/O systems, and multi-input/multi-output systems. A particular study on the application of deep reinforcement learning to process control has been reported in [12]. It has been shown that the controller set-up followed the typical reinforcement learning set-up, whereby an agent (controller) interacted with the environment (process) through control actions and received a reward in discrete time steps. Deep neural networks served as function approximators and were used to learn the control policies. Once trained, the leaned network acquired a policy that mapped the system output to control actions. Though the policies were not explicitly specified, the deep neural networks were able to learn policies that were different from the traditional controllers. Machine learning has already made its debut in the cement APCs. The application needs to be intensified to deep reinforcement learning to handle the complexities of the cement production process.

3.7 CONCLUDING OBSERVATIONS

The advent of 'Industry 4.0' strategy in the large manufacturing industries, including cement, has compelled them to look for ways to become operationally hyper-efficient and production-wise highly adaptable to the changing conditions and environments. The path laid out already to achieve the above goals is that of comprehensive digitalization leading to autonomous operation with minimal or no human intervention. Technologically, the destination of autonomous operation cannot be reached without the application of AI in to-day's common parlance or 'extended intelligence' as being propagated through a modified concept of balancing and complementing human and AI.

The path for implementation of artificial or extended intelligence obviously includes the interconnection of machines and automation devices with advanced IT tools and techniques based on IIoT system. The prerequisite for this pathway is to acquire and process large volumes of data and information from the plant and many other open sources by big data analytics via cloud services. While a fully autonomous plant, particularly in the cement sector, is still a distant reality, it is time now to design and install intelligent machines that operate autonomously by collecting and analyzing digital data from the shop floor and maintain themselves with predictive analytics. The natural follow-up step thereafter is the interconnection of different autonomous machines and related processes in production workflows. This step will ultimately lead to autonomous plants comprising end-to-end autonomous production processes.

In this roadmap there are several issues of data security and business viability on one hand and socioeconomic aspects on the other, which cannot be overlooked or set aside. Nowadays, there are acceptable technological solutions available for data security. The business viability is essentially an enterprise management issue and can be examined appropriately at various stages of implementation. The socioeconomic problems are more crucial as it may involve redundancy or alternative deployment of workforce as well as extensive retraining and reskilling of manpower. Notwithstanding the barriers, the future lies in realizing the objective of operating autonomous plants.

REFERENCES

1. K. Karachalios, J. Ito, Human intelligence and autonomy in the era of extended intelligence, *CXI Essay*, Available at https://globalcxi.org
2. https://standards.ieee.org/industry-connections/ec/autonomous-systems.html
3. https://whatis.techtarget.com/definition/extended-intelligence-EI
4. Taxonomy and definitions for terms related to driving automation systems for on-road motor vehicles, Standard J 3016_201806, Rev 2018-06-15, SAE International, 2018.
5. T. Gamer, M. Hoernicke, B. Kloepper, R. Bouer, A.J. Isaksson, The autonomous industrial plant – future of process engineering, operations and maintenance, *IFAC PaperOnLine*, 52-1, 2019, 454–460.
6. D. Abadi, A. Milamani, D. Anderson, P. Bailis et al, The Seattle report on database research, *SIGMOD Record*, 48(4), Dec 2019.
7. https://hackr.io/blog/top-data-analytics-tools
8. M.G. Sellito, E. Baluguni, R. Gamberini, B. Rimini, A fuzzy logic control application to the cement industry, *IFAC PaperOnLine*, 51-11, 2018, 1542–1547.
9. https://tutorialspoint.com/fuzzy_logic/Fuzzy_logic_control_system_htm
10. V. Ramasamy, R.K. Siddharthan, R. Kannan, G. Muralidharan, Optimal tuning of model predictive controller weights using genetic algorithm with interactive decision tree for industrial cement kiln process, *Processes*, 7(12), 2019, 938.
11. P. Acharyya, S.D. Rosario, R. Flor, R. Joshi, D. Li, R. Linares, H. Zhang, *Autopilot of cement plants for reduction of fuel consumption and emissions*, *Proceedings of the 36th International Conference on Machine Learning*, Long Beach, California, PMLR 97, 2019.
12. S.P.K. Spielberg, R.B. Gopaluni, P.D. Loewen, *Deep reinforcement learning approaches for process control*, *6th International Symposium on Advanced Control of Industrial Processes (AdCONIP)*, May 28–31, 2017, Taipei, Taiwan.

4 Electrical Systems for Sustainable Production in Cement Plants
A Perspective View

Peddanna Shirumalla
ERCOM Engineers Private Limited, New Delhi, India

Anjan Kumar Chatterjee
Conmat Technologies Private Limited, Kolkata, India

CONTENTS

4.1 Introduction 100
4.2 Electrical Installations – Backdrop of Standards and
Regulations...101
4.3 Power Supply and Receiving System...103
    4.3.1 Incoming Voltage Considerations for Power Grid
        Supply 103
    4.3.2 Approach for Determining the Power Requirements and
        Transformer Capacity...104
    4.3.3 Selection of Transformers...106
4.4 Sourcing of Power 107
    4.4.1 Power from Waste Heat Recovery Systems 108
    4.4.2 Power from Renewables 111
4.5 Power Distribution System 113
    4.5.1 Power System Design Considerations 114
    4.5.2 Load Centre Substations 114
    4.5.3 Voltage Selection for Power Distribution 115
    4.5.4 MV Switchboards...116
    4.5.5 LV Distribution Transformers 117
    4.5.6 Main LV Distribution Boards..117
    4.5.7 Intelligent Motor Control Centres....................................117
    4.5.8 Variable Frequency Drives..118
        4.5.8.1 LV Drives 118
        4.5.8.2 MV Drives...119
        4.5.8.3 Energy Losses Due to Harmonics....................119

DOI: 10.1201/9781003106791-4

        4.5.9   Earthing/Grounding 120
        4.5.10   Battery with Charger 120
        4.5.11   Power Factor Improvement 120
        4.5.12   Plant Lighting 120
        4.5.13   Power, Control, and Instrumentation Cables 121
        4.5.14   Power Supply from an Emergency Generator 121
4.6   Energy Efficient Motors and Drives 121
        4.6.1   Types of Motors 122
        4.6.2   Protective Enclosure of Motors 124
4.7   Electrical Energy Conservation 124
4.8   Control, Automation and Information System for
       Power Distribution 127
        4.8.1   Automation of Electrical Distribution System 129
4.9   Process Signal Communication, Integration and Automation 130
        4.9.1   Communication Protocols 130
        4.9.2   Industrial Wireless Communication 132
4.10   Advanced Process Control and Emergence of AI Techniques 132
        4.10.1   AI Applications to Electrical Systems 134
4.11   Internet of Things and Data Processing Infrastructure 136
4.12   Concluding Observations 137
References 138

4.1   INTRODUCTION

Production of Portland cements is one of the most energy-intensive industrial processes with a huge scale of operation. The global cement production has crossed 4.0 billion ton per year and the energy consumption in the industry is more than 11EJ. Further, the energy-related expenses in the cement sector, mostly on fossil fuels and electricity, account for 30% to 40% of the production cost. It has also been observed that, while the pyro process of a cement manufacturing facility is dependent on the simultaneous use of both the thermal and electric forms of energy, all the process steps in material preparation prior to pyro section and the post-clinker cement-making steps, including the entire chain of material handling systems, rely critically on supply, distribution, and efficient utilization of electricity. In fact, the productivity and efficiency of an integrated cement-manufacturing facility are overwhelmingly dependent on the reliability and stability of the power distribution design and installation.

There are grave concerns in the world about the emissions of greenhouse gases, and particularly the carbon dioxide from the cement sector. To a significant extent, these emissions are related to the use of thermal power, supplied from the grid and generated from the captive facilities, in addition to the quantity emitted from the manufacturing process itself. Hence, energy conservation and use of renewable forms of energy are the important avenues being explored further by the cement sector all over the world.

The modern cement manufacturing units are heavily instrumented and automated. In installing the modern plants, therefore, a critical responsibility lies with the electrical, instrumentation, and automation engineers to design and install systems, which will run the plant successfully with sustained production levels and the most optimum product quality, ensuring that the production process remains commercially viable for years to come. For reliable and sustainable production, all equipment needs to be specified properly. It is important to adopt standardization to the extent possible and to keep spares inventory as minimum as possible.

In the last two decades, researchers and engineers have explored how different tools and techniques of artificial intelligence can be applied to cement manufacturing in general and electrical engineering, instrumentation, and automation in particular. Enormous progress in computers and computational science has opened up new horizons of productivity and efficiency. Expert systems and fuzzy logic controls have already made their appearance in cement manufacturing. Newer opportunities with the application of cloud computing, model predictive control, artificial neural network, machine learning, deep learning, etc. are being unfolded.

This chapter is intended to review the essentials of the electrical systems in cement plants, advances made in achieving energy efficiency and conservation, and the future directions of development. The chapter is not meant to serve as a manual for design and operation of the electrical systems in cement production.

4.2 ELECTRICAL INSTALLATIONS – BACKDROP OF STANDARDS AND REGULATIONS

The construction and the characteristic features as well as the performance and tests carried out on electrical installations of three arbitrarily defined ranges, low voltage (LV < up to 1.0 KV), medium voltage (MV 1.0 to 35 KV), and high voltage (HV > 35 KV), along with the equipment used, are subjected to national and international standards and regulations. Similarly, the operation, maintenance, network reconfiguration, safety precautions, and even the protection equipment related to such installations are regulated by national and international standards and by the national laws. The electrical systems being inherently hazardous, the regulatory framework for safety is specifically important and relevant. In most countries, the electrical installations comply with more than one set of regulations issued by the national authorities or by the recognized professional bodies. Although these regulations are based on national standards, they are primarily derived from the codes and standards of the International Electro-Technical Commission (IEC), which are formulated through sharing of the global experience by experts from different countries. It is probably not an exaggeration to state that the IEC standards 60364

(Electrical installations for buildings), 61440 (Secondary cells and batteries), 60479 series (effects of current on human beings and livestock), and 61201 (Guidance on use of conventional touch voltage limits) are often considered as the fundamentals of most electrical installations [1]. The endeavour is to achieve widespread harmony in the standardization of equipment for power generation, transmission, distribution, and utilization. In fact, such harmonization is expected to extend to even insulating materials, winding wire, and measuring and process control instruments. The practice-oriented details of electrical codes and standards are comprehensively dealt with in [2].

It is important to note that, despite the efforts of standardization, the international practices differ. For example, two principal properties of power supply, voltage, and frequency vary between regions. A voltage of nominally 230 V and a frequency of 50 Hz are used in Europe, most of Africa, most of Asia, and much of South America and Australia. In North America the most common combination is 120 V and a frequency of 60 Hz [2]. Other voltages also exist. The transmission level voltages are generally 110 kV and above. In many instances, the voltages are 220 kV or 132 kV. The lower transmission voltages such as 66 kV and 33 kV, though used in some regions, are regarded as sub-transmission voltages. Voltages less than 33 kV are used for power distribution purposes. Transmission voltages above 765 kV, used in some regions, are considered extra-high voltage [2].

Because of the relevance and importance of harmonization in standards, in addition to IEC, several organizations, nationally and internationally, are engaged in this mission. Specific mention may be made of International Standards Organization (ISO), European Committee for Electro-Technical Standardization (CENELAC), American National Standards Institution (ANSI), American Society for Testing and Materials (ASTM), National Electric Manufacturers Association (NEMA), Institution of Electrical and Electronics Engineers (IEEE), etc. In practice the electrical codes and standards are examined in conjunction with other national regulations pertaining to factory acts, pollution controls, health, and safety provisions. It is therefore obvious that the design, construction, and operation of electrical systems cannot proceed without referring to multiple standards and regulations. While designing the power distribution system, a prior approval from the local statutory authorities is essential. After erection is completed in accordance with the approved drawings, a final approval from the local statutory authorities is also a prerequisite for energizing the supply to the premises to operate the plant.

Like many other countries, the electrical systems standardization in India is the responsibility of the Bureau of Indian Standards (BIS) through its Electro-Technical Division. Hence, the electrical power generation, transmission, distribution, utilization equipment, and materials of construction are governed by the relevant standards as applicable. This chapter is based on the Indian standards and practices, unless any specific reference has been made to other standards.

4.3 POWER SUPPLY AND RECEIVING SYSTEM

Generally speaking, the power supply to an integrated cement manufacturing unit is tapped from the national or regional power grids. The power supply system is a large installation as it must support an extensive supply network, starting with the substation of the grid where power is received and ending with individual drives and points of usage. The state grid, typically of 220 kV linking several generating substations is schematically shown in Figure 4.1 with 132 kV supply to plants having more than 5 MVA requirement and 33 kV supply to plants with requirement of less than below 5 MVA [3].

4.3.1 Incoming Voltage Considerations for Power Grid Supply

Although the incoming voltage for plants with power up to 5 MVA has been shown as 33 kV in Figure 4.1 for illustration purposes, in practice the supply voltage at 33 kV is considered adequate in many situations for power up to 15 MVA. For power up to 30 MVA, the incoming voltage of 66 KV is considered necessary and for still higher power up to 60 and 100 MVA, the incoming

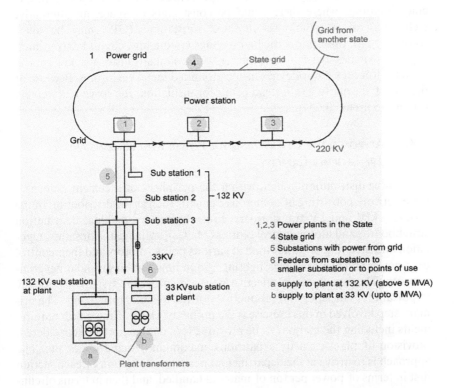

FIGURE 4.1 Schematic diagram of power supply from the state grid to a cement plant [3].

voltages of 132 and 220 KV, respectively, are the practical norms. It is evident, therefore, that the receiving voltage for supply from overhead power grid depends on the quantum of power required. For overhead lines, current rating is a limitation. Hence for higher power requirement, incoming voltage is increased to reduce the current. As stated earlier, these voltage standardizations vary from country to country and due to power frequency of 50 or 60 HZ adopted in different countries. It is also relevant to mention that if the incoming power is received through underground power cables and not overhead lines, multiple power cables at different voltages may be laid in parallel to meet the total power requirement at any receiving voltage.

For reliability of the incoming power source, the incoming line is provided with a dedicated feeder from main grid substation to ensure minimum disturbance in the power supply voltage and frequency fluctuations. For reasons of economy and for ensuring lower line losses, the grid power is taken from the nearest grid substation. Incoming grid power is normally received in one corner of the plant layout for safety reasons and for saving of factory land. In this context, it is relevant to note that the overhead high-voltage grid substations are conventionally air-insulated type (AIS) and need more space. In specific situations, where there could be constraints of space, an alternative technology of installing gas-insulated substations (GIS) may be taken recourse to. In this option, the high-voltage conductors, circuit breaker interrupters, switches, current transformer, and voltage transformer are kept in SF-6 (Sulfur Hexafluoride) gas inside grounded metal enclosures. Because of the use of the above gas, instead of air for insulation, the space requirement is reduced up to 70% [4].

4.3.2 Approach for Determining the Power Requirements and Transformer Capacity

The power distribution system within the periphery of a cement plant is a large network consisting of numerous elements such as the distribution transformers, MV and LV transformers, control panels, individual distribution switchboards and motor control centres (MCCs), switchgears for safety, regulation and metering of power used at various points, motors and their control gear, power and control cables, lighting, grounding (earthing), and other components. The layout of an effective power distribution system is critically important for efficiency and productivity of the manufacturing unit. The primary step involved in this exercise is the proper working of the power requirements including the margins to be provided for, load factor to be considered, provision of plant capacity expansions, maximum demand issues, etc. The approach is to arrive at the departmental power consumption of each section first in terms of power per ton of material handled, and then in terms of clinker and cement produced by using appropriate conversion factors. The sectional power consumption estimates are based on listing all drives with their ratings and listing power drawn by them at the rated capacity. From the

TABLE 4.1

Typical Power Consumption Estimates for a Whole Plant [3]

Sections	Unit/t of Material, kWh	Material Conversion Factor	Unit/t of Clinker, kWh	Unit/t of Cement, kWh
Quarrying	0.5	2.0	1.0	0.96
Crushing	1.5	1.5	3.25	2.16
Raw grinding	16.0	1.55	24.8	23.85
Blending	2.0	1.55	3.1	2.98
Kiln feed, kiln, calciner	0.5	1.65	0.825	0.79
Preheater, Cooler	25.0	1.0	23.0	24.0
Coal mill	30.0	0.15	4.5	4.33
Cement mill	35.0	1.0	–	35.0
Packing	2.0	1.0	–	2.0
Water supply, lighting of factory and colony, other utilities, and losses	–	–	–	5.0

Total: 101.07
(Say, 101 kWh/t of OPC)

aggregation of all the sectional consumptions, the total power requirement for the plant is estimated. A typical illustration is furnished in Table 4.1 [3].

Table 4.1 illustrates the power requirement, based on production of Ordinary Portland Cement (OPC). However, globally there is a major shift towards blended cements with additions of pozzolanic, cementitious, and filler materials having different grinding behaviour. Hence, in the modern plants the product mix turns out to be a major consideration in working out the power requirements. It is pertinent to note that among various cement substitute materials, the metallurgical slags are harder to grind and, consequently, production of cements containing the granulated blast furnace slag as a major constituent becomes an overriding consideration in power consumption. Further, at the project stage the details required for estimating the power requirements, as shown in Table 4.1, may not be available. Hence, a more practical approach is to work out the power requirement on certain assumptions, mostly provided by the original equipment manufacturers. Typically, for a 3000 t/d cement plant, the power demand for the clinkering section is worked out as follows:

$$\text{Clinker production assumed} = 3000 \text{ t/d}$$

$$\text{Maximum specific energy consumption} = 60 \text{ kWh/t of cement}$$

$$\text{Peak load factor} = 1.2$$

$$\text{Average Power Factor} (\text{minimum}) = 0.95$$

$$\text{Running hours per day} = 24$$

$$\text{Power demand} = 3000 \times 60 \times 1.2$$

$$24 \times 0.95 \times 1000$$

$$= 9.47 \, \text{MVA}$$

The power demand for cement grinding will vary, depending on the product mix as mentioned above. However, for the normal or OPC it is estimated on the basis of the gypsum addition factor of, say, 0.05, maximum specific power consumption of 40 kWh/t of cement, and mill running hours of 20 per day, the peak load factor and power factor remaining the same as for the clinkering section. The power demand for OPC grinding works out to be 7.95 MVA. If the product is a blended Portland cement containing granulated blast furnace slag, the power demand for grinding is mostly taken as 1.2 times that of OPC, which in the present example will work out to be 9.54 MVA. For working out the requirements of power to be drawn, it is desirable to add 5%–10% on the estimated power demand of about 19 MVA for the total clinkering and grinding process in the present instance. Furthermore, the substation capacity should allow for margins for additions and expansions that are not always foreseen. A margin of 25% in the capacity is preferred in practice [3].

In selecting transformers, it is possible to install a single transformer of the requisite capacity, but generally it is logical to select two identical rating transformers in parallel rather than one to improve power reliability. However, where there are other alternative power sources operating in parallel with the grid, a second grid transformer may not be necessary. Power transformers are generally loaded up to about 80%. If the total load is estimated after providing for all margins at 26 MVA, then in effect the transformer rating may be taken as 32 MVA.

4.3.3 SELECTION OF TRANSFORMERS

Main power transformers may be ONAN (oil natural air natural) cooled or ONAN/ONAF (natural air cooled up to about 75% rating and force cooled up to about remaining 25% rating). These power grid transformers are complete with on-load tap changer (OLTC) to automatically adjust the incoming voltage by suitable number of taps to keep the secondary voltage constant. Normally voltage fluctuation persists for about one minute or so, then only voltage is adjusted by OLTC to avoid frequent operations. In case of power transformers for captive power generation, there is no need of OLTC as voltage and frequency are well regulated in the generator itself.

Delta winding connection is preferred for power transformers so that the third harmonic currents can flow, ensuring near sinusoidal wave form on the secondary side of the transformer. However in case of higher primary voltage of 66 KV and above, due to higher insulation costs involved for delta connection, star connection is adopted. In such cases, it is important to ensure that the flux density in the transformer is limited to linear range of magnetization curve to avoid harmonics. If it cannot be avoided, tertiary delta winding on no-load or sometimes for some load is necessary so that the third harmonic currents flow in this delta winding to get sinusoidal wave form on secondary side of the transformer. The distribution transformers are also designed with similar logic.

4.4   SOURCING OF POWER

Normally, the large cement plants are provided with multiple sources of power for energy economy and uninterrupted operations. Typically, the power sources are of four broad categories:

- national or regional grid power
- captive thermal power (diesel, coal, or gas fired)
- waste heat recovery systems generated power
- renewable power.

Apart from the above, open access power, purchase by bidding, is also available in some countries. Grid power tariff varies within the day, during peak time, non-peak time, and normal time. Plant operations need to be planned to take advantage of lower tariffs available during non-peak time.

Although the grid power is the commonest form of power supply to the large and integrated cement manufacturing units, the general practice is to keep a provision of at least 40% of their power requirements through captive sources. An emergency diesel generating (DG) set of small capacity, such as 1 to 1.5 MW or as required to cater for emergency loads envisaged from the process consideration, is planned for the system. Sometimes the emergency DG sets are used to run entire kiln section during the outage of the grid power supply, when the DG set rating has to be higher. While a large number of captive power plants (CPP) in the cement industry in the past were diesel-based generating (DG) sets of varying low-range capacities, several factories recently have opted for coal-fired CPP, generally of 18–25 MW capacity. Since the grid power and captive power have to work simultaneously in a manufacturing unit, they are synchronized, when working in parallel, for their voltage and frequency. Alternatively, the captive power is used only to supply power to specific sections, such as the kiln or a major mill, when there is a shortfall of power or power cut from the grid power source. The trend, however, is to have large CPPs and WHRS (Waste Heat Recovery Systems), synchronized with the grid power. WHRS power is the cheapest of all sources of power.

4.4.1 Power from Waste Heat Recovery Systems

In cement manufacturing by the dry process, nearly 40% of the total heat input comes out as waste heat from the exhaust gases of the preheater and the clinker cooler. The heat content in the preheater exhaust gases emitted at 300°-400°C ranges from 750 to 1050 MJ per ton of clinker and similarly, the heat content in the hot air coming out from the clinker grate cooler at 200°–300°C is about 330–540 MJ per ton of clinker. A traditional practice has been to use a part of it for drying raw materials and coal, which leaves a large quantity of heat that can be used for power generation with the help of a heat recovery boiler and a turbine generator system. There are three technological alternatives, depending on the working fluid used: Steam Rankine Cycle (SRC), Organic Rankine Cycle (ORC), and Kalina Cycle (KC).

The steam cycle is the commonest and the most relied WHR system in the world. In fact, it was reported way back in 2012 that out of 865 installations of WHR systems in the world, 854 units were based on steam cycle, 9 units were ORC based and only 2 units were of KC technology [5]. The steam based WHR systems can recover waste heat from sources having temperatures above 260°C and consists of a suspension preheater (SP) boiler, an air quenching cooler boiler (AQC), and the turbine generator as shown pictorially in Figure 4.2 [6]. It is understood that the WHR system is designed to be connected to the internal grid of the plant and is expected to yield double benefits of cost economy and CO_2 emissions reduction. China is recognized as the industry leader globally for the installation of WHR systems (Figure 4.3) [5].

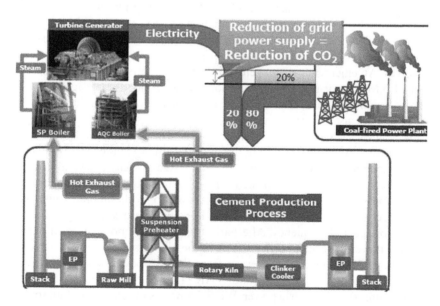

FIGURE 4.2 Simplified depiction of a WHR system with steam cycle [6].

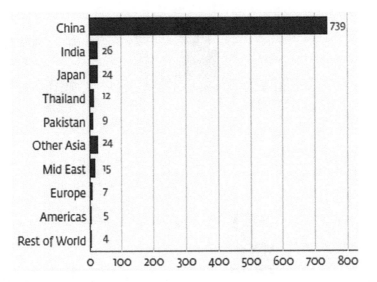

FIGURE 4.3 Illustration of country-wise installations of WHR systems in 2012 [5].

The operational experience is that 22–36 kWh per ton of clinker can be generated from the system. Further, if the kiln operation is modified with a view to increasing the exhaust temperatures of the preheater gases and the cooler air, an output up to 45 kWh per ton of clinker may be achieved. Implementation of such modified systems would obviously depend on the viability of the process modifications.

The development of ORC based WHR system was aimed at utilizing the waste heat sources with temperatures lower than 150°C. An ORC system typically uses a high molecular-mass organic working fluid such as butane or pentane that has a lower boiling point, high vapour pressure, and higher mass flow than water. The ORC systems are widely used to generate power from the geothermal and solar heat and also from biomass combustion. Systems of ratings lower than 2 to 3 MW are mostly ORC based.

It is important to note that although the ORC installations in the global cement industry have not proliferated much over the last two decades, there is a visible revival of interest in the technology in recent years. Some of the major cement producing countries are exploring more intently the choice of the working fluids and their effects on performance and viability of the ORC-based WHR systems [7, 8]. The distinctive feature of an ORC based WHR system in cement manufacturing is the two stage transfer of heat as shown in Figure 4.4 [10]. In the first stage, heat is transferred from the waste gases to an intermediate heat transfer fluid, a diathermic oil, which maintains temperature at a stable value. In the second stage, the heat transfer is carried out from the diathermic oil to the organic working fluid and electricity is generated by the ORC unit.

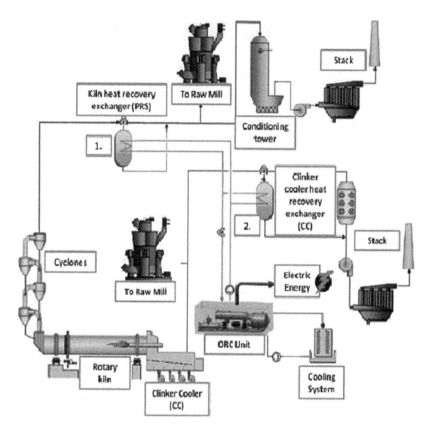

FIGURE 4.4   Basic features of a WHR system based on ORC [9].

The third technological option is the KC developed by Alexander Kalina [11]. In essence, it is a thermodynamic process for converting thermal energy into electricity. It uses a solution of two fluids with different boiling points as its working fluid. Since the solution boils over a range of temperature, more of heat can be extracted from the source than with a pure working fluid used in the other two technologies described above. The same phenomenon takes place in the condensing end as well and consequently the KC provides higher efficiency, comparable to a combined cycle but with less complexity. With proper proportioning of the two components of the solution, the boiling point can be adjusted to suit the heat input temperature. Although water and ammonia are the widely used in the KC, but other combinations are feasible. A 3.6 MW electricity generating unit, installed at the Kashima Steel Works of Sumitomo Metal Industries of Japan in 1999, is the longest running commercial application of the technology. Although the first generation systems were intended for waste heat recovery from sources with temperatures in the range of 120°C–180°C, the second generation design is applicable for low and relatively higher temperature sources. As already mentioned, the installations of

WHR systems with this technology are very few in the cement industry, but the potential is high as the cement manufacturing process has numerous low-temperature waste heat sources.

4.4.2 Power from Renewables

It is reported that globally renewable energy sources comprised 13.5% of the total energy supply in 2018. Since 1990 they have grown at an average annual rate of 2.0%, which is marginally higher than the growth rate of the total energy supply in the world. Growth has been specifically high for solar PV and wind power, which grew at average annual rates of 36.5% and 23.0%, respectively (Figure 4.5) [9].

Specially, in the context of global electricity generation, it is important to note that the renewables accounted for 25.2% in 2018 and occupied the second position after coal. Of all the different renewables for power generation, the significant growth of the solar photovoltaic (PV) system has happened due to its technological maturity and supportive policies extended in most countries. Competitiveness of solar PV is noticed in various parts of the world.

The cement industry is an important industrial user of solar PV technology. However, the applications are limited to lighting of plant buildings, residential colonies, water heating, and similar relatively low temperature requirements. For the objective of using the solar power for applications involving higher temperatures, a significant technological progress has been the development of concentrated solar power (CSP). The CSP technology can be integrated with existing steam cycle based power generating plants at different stages in the process such as feed water heating, steam generation, and similar requirements. A concept proving study of the solar steam gasification project for low-grade carbonaceous feedstock to produce a high-quality

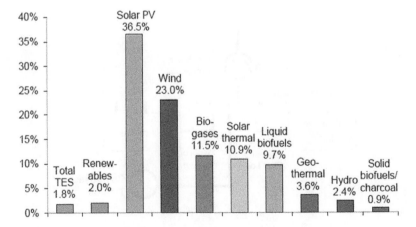

FIGURE 4.5 Average annual growth rates of different renewable energy sources in the world over the last three decades [9].

synthetic gas has also been reported [11]. Another EU backed pilot project SOLPART (solar reactor for industrial production of reactive particulates) is on the anvil, in which a 50 kW solar reactor is continuously operated to heat a fluidized-bed system at 950°C [12].

A technology start-up company is exploring the use of improved CSP in industrial processes above 1000°C [12]. It uses a closed-loop control system to improve the accuracy of concentrating the solar heat energy by aligning an array of mirrors by using computer vision software to reflect sunlight towards a single target. The target is to achieve temperature up to 1500°C for processes like clinker making. It is evident from all these developments that the solar energy is ahead of most other renewables for sourcing power.

In this context, the next important renewable source is the wind power. It is known that a wind power plant consists of multiple individual wind turbines, separated by an electrical collector system. Since the wind plants are generally connected to the utility transmission grid, the needs of the power system are governed by the fluctuations occurring at the point of interconnection with the host system. In a modern wind plant having doubly fed asynchronous generator (DFAG) based wind turbine, a supervising plant controller is used to provide closed-loop voltage regulation for the entire plant, which creates effectively a hierarchical control system. Notwithstanding the advances in the design of wind plants, integration of such plants with cement production units is fraught with difficulties of voltage fluctuation under varying wind conditions and lack of reactive power control and voltage regulation in the absence of wind. It appears that a solution to this problem lies in using hybrid systems consisting of modern wind plants and the latest generation of advanced gas turbines. A schematic diagram of the hybrid system is shown in Figure 4.6. The adoption of such a technology will allow

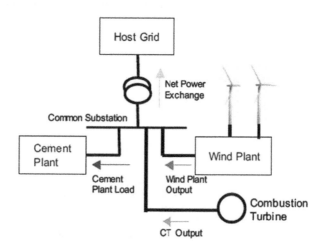

FIGURE 4.6 Integration of a hybrid captive wind power plant with a cement production unit [13].

seamless integration of a cement production unit, the hybrid captive power plant and the regional transmission grid [13].

4.5 POWER DISTRIBUTION SYSTEM

From the foregoing discussions it is evident that the internal grid in a cement plant is fed with power received from the transmission grid and also from the captive generation and WHR systems. When working in parallel, they are all synchronized. Some sections of the plant will have only LV supply, and others will have both MV and LV supplies. Three different alternatives in practice are shown schematically in Figure 4.7, Figure 4.8, and Figure 4.9 [3].

In Figure 4.7 the option of stepping down voltage from the transmission grid voltage for the total plant at one place and leading high-tension line from it to respective section is indicated. If the captive power plant is large enough, there may be two main rings in parallel and double bus will not be required. Figure 4.8 shows the option of locating 132/6.6 kV transformers near the load

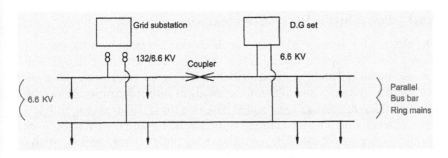

FIGURE 4.7 Power distribution arrangement with supply from the transmission grid and from the captive DG source in parallel [3].

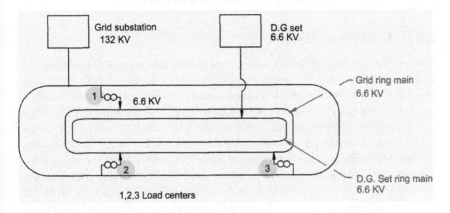

FIGURE 4.8 Transformers of 132 kV and 6.6 kV located near load centres [3].

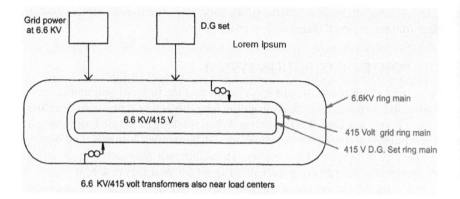

FIGURE 4.9   Distribution of 415volt transformers near load centres [3].

centres, while Figure 4.9 presents the distribution option 415 volt transformers near the load centres.

4.5.1 POWER SYSTEM DESIGN CONSIDERATIONS

The electrical system is designed to achieve overall economy on the basis of energy efficiency, system reliability, ease of maintenance, and safety of personnel. Radial system of power distribution is adopted as it is more economical, simpler, and safer and allows for convenient future expansion. The fault levels at different voltages are decided based on the calculations for fault levels, considering mainly impedances of transformers, fault contribution from parallel power sources, motors, etc. The respective medium and low voltage switchgear are rated for these fault levels for safe operation when faults occur. The proposed typical power single line diagram of a large cement plant at main substation level with grid incoming supply, in parallel to captive power, WHRS power, and emergency DG power is shown in Figure 4.10.

4.5.2 LOAD CENTRE SUBSTATIONS

To keep the distribution losses low, power at medium voltage is taken nearer to the plant process department substations and distributed with minimum lengths of cables. In case of LV loads, the currents will be larger and distribution losses (mainly I^2R losses) are very large. Hence, to minimize LV distribution losses, MCCs are located close to the plant and they feed LV loads through minimum length of cables. All feeders of MV and LV switchboards are provided with necessary numerical protective relays, digital meters, control switches, indication lamps, etc. as required. A graded relay coordination from downstream to upstream is ensured for the entire power system for selective isolation in case of faults with minimum interruption of power

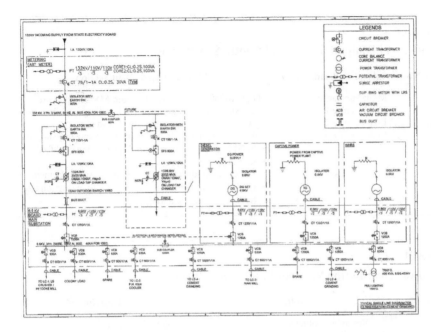

FIGURE 4.10 Typical power distribution single line diagram at main substation level.

before commissioning the electrical system. Electrical equipment for power distribution viz. MV switchboards, distribution transformers, LV switchboard, MCCs, capacitor banks, etc. are housed in respective load centre substations as indicated in Table 4.2.

4.5.3 Voltage Selection for Power Distribution

Considering the economy in overall capital cost of electrical equipment, the various voltages for power distribution are selected beforehand and standardized for a given project as illustrated in Table 4.3. Distribution voltages are either 11 or 6.6 KV or their equivalent voltages. The voltage of 11 kV is preferred for large plants in order to restrict current ratings of bus bars, to limit the fault current ratings for MV circuit breakers, to reduce the power cable sizes, and to reduce the number of cables in parallel for a given power rating. The voltage of 6.6 kV is selected for small- to medium-size plants.

Although the low voltage motors are cheaper than medium voltage motors, low voltages have problems of high currents, associated with high-voltage dips during start-up, causing tripping or disturbing other motors on the bus. Further, more number of cables are required to be laid and maintained. Thus, a limit for low voltage motors is set around 200 kW rating for fixed speed motors (Table 4.3).

TABLE 4.2
Electrical Equipment for Load Centre Substations

S. No	Substation/Load Center	Description of Plant Section
01	Main Substation	High-voltage switchyard and main MV switchboard
02	Load Centre 1	Limestone crusher and material handling and Transport
03	Load Centre 2	Raw material grinding, transport, and storage
04	Load Centre 3	Preheater, kiln, cooler, and clinker transport to silo, coal handling, crushing, and grinding and fine coal feeding to kiln and calciner
05	Load Centre 4	Clinker and additive handling and transport, Cement grinding, transport to cement silos, silo extraction and transport, packing and dispatch

TABLE 4.3
Power Distribution Voltages in a Cement Plant

Sl. No.	Voltages	Description
1	Receiving Voltages	33/66/132/220 kV
2	Distribution Voltages	
	a. Medium voltage (MV)	11 or 6.6 kV; 11 kV preferred for medium and large size plants
	b. Low Voltage (LV)	415 V
3	Motor terminal voltages	
	a. 200 kW and below for fixed speed and up to 350 kW variable frequency drives	Low voltage. DOL starter below 90 kW and soft starter 90 kW and above for fixed speed motors.
	b. Above 350 kW to 1000 kW, variable frequency drives	690 V/11 or 6.6 kV to reduce harmonic load, where required.
	c. Above 200 kW fixed speed and above 1000 kW variable frequency drives	11 or 6.6 kV
4	Control supply	
	a. AC control circuit	220V/110 V, etc., single phase through a control transformer.
	b. DC control circuit for electrical switchboards	110 V DC from battery with float-cum-boost charger
5	Welding set and Heaters	Low voltage
6	Plant and township lighting	230 V, single phase, or comparable voltage as adopted in a country

4.5.4 MV Switchboards

All MV switchgears should be of high speed, draw out motor operated vacuum circuit breaker/vacuum contactor. Each feeder should be equipped with necessary digital metering and protection relays with digital signals. The status signals, kW and kWh signals of all breaker feeders are to be taken to programmable process controller through a single data cable after collecting all signals of the switchboard through looping.

4.5.5 LV Distribution Transformers

Step-down transformers of unitized design for indoor substations are normally considered. These transformers are ONAN cooled and have off-load tap changer. Hermetically sealed transformers are also available, which have the advantage of no oil maintenance as oil in transformer is totally disconnected with external air, which brings moisture in to the oil and deteriorates oil quality, needing frequent oil quality checking and filtration, when required. The transformer capacities have been worked out based on expected loads plus 30% spare capacity. Standard transformer ratings of 1000, 1600, 2000, and 2500 kVA are normally considered as per the local regulations of the country. Distribution transformers are designed for suitable K factor to take care of LV harmonic loads of variable frequency drives on the transformer. The fault digital signals of these transformers are taken to control system for monitoring in the control room for power distribution.

4.5.6 Main LV Distribution Boards

The LV boards considered are with fully draw out motor operated air circuit breakers with necessary safety switches to facilitate safe operation and easy maintenance. These are located at load centre substations and receive power through bus duct/cables from distribution transformers. Bus ducts for higher currents are preferred to avoid many parallel cables, which may not share currents equally due to unequal cable lengths, which occur during installation. However, bus ducts are troublesome and take more time for maintenance.

4.5.7 Intelligent Motor Control Centres

The intelligent motor control centres (iMCCs) consist of advanced microprocessor based input and output modules for control of each low voltage motor, connected to automation system through Profibus cable. They can be of draw out or non-draw out modules as per customer choice, in compartmentalized design for indoor installation. The advantages of iMCCs over conventional type MCC are the following:

* power-related measured quantities obtained as digital signals
* less control wiring and cabling
* single module programmed as direct online (DOL), reversible DOL, and reversible actuator type feeder, etc.
* all digital signals of motor starters connected to programmable process controller directly through single data cable up to maximum no. of nodes allowable.
* higher availability due to faster troubleshooting.

4.5.8 Variable Frequency Drives

As motor speed depends on power frequency, the speed of the motor can be changed by varying the power frequency through a variable frequency drive using a converter for DC output and inverter to invert DC input to AC output with variable frequency. But to take care of torque of the load, which is same at any speed, voltage to frequency (V/F) ratio is maintained constant to get same flux which produces the torque. These are used as soft starter to reduce starting current of motors and also for saving energy with speed control of the motors and to control speed to optimize the process. Variable speed drive loads in modern cement plants have gone up to around 60% of the total load of the plant to achieve energy efficiency. But these drives add to the harmonic problems on the power system within the plant and affecting other customers on the grid. Though the harmonics at the point of common coupling (PCC) may be within the permissible limits as per IEEE-519 to protect other customers on the grid, these harmonics may cause some harm to the internal electrical equipment leading to unexpected breakdowns. When such problems arise, harmonics present in the internal system are studied to mitigate the harmonic effects. Some solutions to mitigate these harmonic problems are discussed below.

4.5.8.1 LV Drives

Drives up to 350 KW are normally considered with six pulse drive. These along with large chokes in series will produce current harmonic distortion up to about 40%. However, as the total load of LV drives is normally less than 20% of the total load on the distribution transformer, it will not affect the total system. However, if the total load of LV drives exceeds 20% of the transformer load, alternative methods to reduce harmonics, especially for large drives or for all drives, are adopted as mentioned below.

a. Active front end (AFE) drives are used which operate in four quadrants and limit the harmonics to less than 5% apart from improving power factor to near unity. But these are expensive and are rarely used.

b. Six pulse rectifiers with addition of passive filter reduce harmonics to less than 10%.

c. Six pulse rectifiers with addition of active filters will reduce harmonics to less than 5% to meet IEEE-519 requirements. Drives with IGBT also reduce the harmonics to less than 5%. But these drives are expensive.

d. It is likely that the seventh harmonic may create resonance condition with capacitor banks and capacitors may be damaged, and it is necessary to use suitable reactor in series with capacitors to limit high currents.

e. The LV drives are connected to LV distribution transformers. Since they are not normally specified to be converter duty transformers, they get heated up with harmonics, causing temperature rise problems, and huge fans are used at site to cool the transformer. Distribution transformers act as low pass filter for harmonics and the lower level harmonics will move towards the MV system. For higher harmonics, distribution transformers act as a high impedance and so high harmonics are dropped. These harmonics move up to the MV system with reduced amplitude, and hence, less effects are observed for these harmonics. Distribution transformers having non-linear loads such as variable speed drives should be rated for suitable K factor and suitably specified in order to take care of heating effects of harmonics, generated by these non-linear loads.

f. Drives of rating above 350 kW and up to 1000 kW are connected to 690 V or any nearby voltage, which causes current harmonic distortion of about 9.5% with 12 pulse configurations through a three-winding transformer. In some plants, 690V drives have been used up to 2000 kW rating, which is not recommended as numerous cables will be hanging on the motor and may cause frequent tripping problems due to high currents. If this 12-pulse load is within 20% of the total load of the plant, it may be within the permissible margin, but in the field it is often more than 20%. Harmonics of 9.5% produced will affect the motors, transformers, and capacitors, etc. It is therefore desirable to shift 12 pulse drives to higher pulse drives like 36 pulse drives, where the harmonics are less than 4%. But the initial cost is much higher for 36 pulse drives as compared to 12 pulse drives. Despite the higher cost considerations, in the long run, 36 pulse drives are more economical and may turn out to be viable.

4.5.8.2 MV Drives

Drives above 1000 KW with 36 pulse configurations are normally connected to 6.6 or 11 KV depending upon the distribution voltage selected. These drives will have harmonics less than 4% within permissible limits of IEEE-519. Further the load of these drives in large plants is much higher up to about 40% of the total load of the plant. Thus, these drives help in mitigating the effects of the harmonics on the power system.

4.5.8.3 Energy Losses Due to Harmonics

Harmonics also cause additional energy losses. It is estimated that harmonics contribute to energy losses to the extent of about 15 to 20% of the total distribution losses. By mitigating the effects of harmonics by several methods, as discussed above, it is estimated that energy can be saved up to about 0.5 to 0.7 unit per ton of cement. In addition to energy saving, the harmonics factor demands attention for saving the life of many costly equipment such as capacitors, distribution transformers, motors, and others, which may burn out and would need replacement.

4.5.9 Earthing/Grounding

The MV system is resistance earthed through neutral of the respective power transformer to limit the earth fault current to about 100 amps or so as per the local regulations to reduce damages to motors during earth fault. However, some countries go for insulated neutral system as per local regulations. The LV system is normally solidly earthed through neutral of the respective distribution transformer. Electrical power distribution system is effectively earthed with the help of galvanized iron/copper earth continuity conductor of suitable cross-section, following the local regulations. Earth pit with galvanized iron/copper pipe electrodes are provided around all the load centres and process buildings to keep earth resistance below one ohm or so. Separate earthing systems are considered for power equipment, lightning protection of the buildings and neutral of the distribution transformers and all earth pits are interconnected in accordance with the local regulations. For automation equipment, separate earthing system is provided as recommended by the suppliers.

4.5.10 Battery with Charger

A control voltage of 110 V DC is essential for trip coils of MV/LV switchgears and for all indication lamps on the switchboards as at the time of tripping due to fault or after tripping, the normal AC power supply is not available. This is done with the provision of batteries of proper quality and capacity. The Ni-Cd batteries are often selected. The battery charger is generally a thyristor type with automatic and manual float-cum-boost charging facility. Some emergency lighting loads of the substation building may be connected on the batteries.

4.5.11 Power Factor Improvement

To maintain a high overall plant power factor to reduce the demand charges, static power factor improving capacitors of suitable KVAR ratings and voltage grade must be considered. Improved power factor also reduces the distribution losses and voltage drop due to reduced current with higher power factor.

Normally suitable size capacitors are considered for direct connection across stator terminals of MV motors. Further, for compensation of LV loads, multi-step automatic controlled capacitor banks are considered for connection to main LV boards. The capacitor banks are sized to maintain the overall power factor to about 0.98 lag.

4.5.12 Plant Lighting

A very efficient illumination scheme with energy efficient LED lamps for all electrical rooms, load centre buildings, control rooms, plant buildings, storage areas, and roads inside the plant area are considered with the following aims in view:

- boosting up productivity
- minimizing accidents
- facilitating maintenance round the clock
- providing the comfort of daylight to the operating personnel.

Plant lighting is designed to get the illuminations lux levels in accordance with the standards of the industry. Being a specialized area, the task is normally assigned to the specialists for complete design and supply of lighting equipment.

4.5.13 POWER, CONTROL, AND INSTRUMENTATION CABLES

The power cables are required to distribute the MV/LV power from power source to the drive motors and other electrical loads through MV/LV switchgears. The power cables have to be designed suitable for the fault levels and the current carrying capacity, limiting the voltage drops within the permissible limits to be defined. Selection of the cable sizes for different loads is mostly done with the help of software. Control cables are unscreened LV copper cables, laid for carrying control signals only. Instrument signal cables are special screened LV cables. The cables are selected for the ambient conditions of the plant site, considering deratings for grouping and laying method of cables. In cold countries with freezing temperatures, specially designed cables are necessary. The data cables such as the twisted pair type, concentric type or fibre optic type are normally supplied by the automation system providers in accordance with their own specifications.

4.5.14 POWER SUPPLY FROM AN EMERGENCY GENERATOR

The emergency generator power supply is connected to the MV switchboard in the central control room through draw out type vacuum circuit breaker (VCB) having suitable electrical interlocking system with normal grid power supply incomer to prevent parallel operation between normal and emergency power supplies. From this MV switchboard the emergency supply can be taken to all emergency loads on medium voltage and on low voltage (LV) through the distribution transformers.

4.6 ENERGY EFFICIENT MOTORS AND DRIVES

In a cement plant, there are numerous motors and drives to move fans, to rotate kilns, to transport materials and, most importantly, to grind both raw materials and clinker. Generally speaking, in a single-kiln plant, about

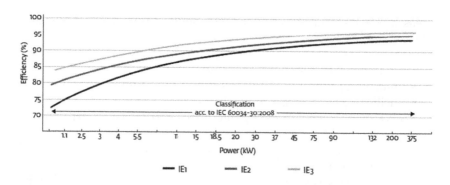

FIGURE 4.11 Efficiency of different motor classes [5].

700–900 electric motors may be used, their ratings varying from a few kW to MW. The typical efficiencies of electric motor classes, as defined in IEC 60034-30:2008 are shown in Figure 4.11 [5]. Over 70% of the electrical energy is consumed by motors. Generally, the MV motors have higher efficiency up to around 95% due to lower currents involved. Efficiency of low voltage motors is not high due to higher current (I) with higher copper losses (I^2R) involved.

Normally IE3 motors are used in the industry. A still higher efficiency class motors, IE 4, are also available and can be considered, where the cost-benefit is favourable. These higher efficiency motors, however, have higher starting currents due to low resistance of windings and cause more voltage dips during start-up, which need to be taken care by soft starters in higher rating motors.

4.6.1 TYPES OF MOTORS

The types of motors used in cement industry are squirrel cage induction motors in general and slip ring induction motors, where higher starting torque is required, such as for loads of crushers, grinding mills, etc. For all LV motors, a starting torque of 200% of full load torque is specified so that it can be used for all types of load applications. For very low speed applications, geared motors are used. Where variable speed is required, a squirrel cage motor is specified suitable for variable speed application. For kiln application, squirrel cage motor with starting toque of 250% of full load torque, suitable for variable speed application is specified. The motors in use in the main machinery are briefly described below:

a. **Crushers:**

   Mainly slip ring induction motors with rotor starters are used to get high starting torque. In case of smaller rating with LV drive, squirrel cage motors with high starting torque with special design of rotor with deep rotor bars can also be used if starting torque requirement of the crusher for starting on load is met.

b. **Ball mills:**
Mainly slip ring induction motors with rotor starters to increase the starting torque by inserting resistance in rotor circuit with or without brush lifting arrangement are used Normally the rotor winding is shorted in rotor starter through a contractor after the motor starting is over and in this case the brushes on the slip rings remain in circuit and require brush inspections and replacement at some intervals and thus more maintenance prone. In case of brush lifting arrangement, the slip rings have a brush-lifting device and a sliding contact bar, allowing the slip rings to be short-circuited inside the motor itself and brushes are lifted, after motor starting is over through external rotor starter. Thus, brush maintenance problem is very much reduced.

c. **Gearless drives for ball mills:**
Gearless drives for ball mills are also available with rotor poles mounted on the mill itself and stator in two pieces or so, floor mounted and mounted over the rotor poles, with air gap. The field on 3 the rotor is supplied from a DC source through slip rings. The stator is supplied through cyclo-converter with variable voltage and variable frequency (lower than power frequency), so that it can be operated at variable speeds also. The advantage of this drive is that it eliminates the gear boxes with power losses and lubrication requirements. However, it is a complicated drive as it has to be built on to the mill along with additional items like DC supply, slip rings, cyclo-converters, as compared to simpler slip ring motor with rotor starter with gear box to reduce the speed

d. **Vertical roller mill**
Mainly slip ring induction motors with rotor starters are used to get high starting torque required for starting on load.

e. **Hydraulic roller press**
Normally slip ring motors with rotor starter for starting are used. In case of cement grinding, where different blended products are produced in the same mill, variable speed frequency drives are used to run at different speeds for different products as necessary.

f. **Rotary kiln**
AC variable speed drive with squirrel cage motor, which gives high starting torque of 250% of full load torque required for kiln starting duty is employed to control the kiln throughput by adjusting the motor speed,

which is a process requirement. Earlier DC drives were used when reliable AC drives were not available, and these are since replaced with more energy efficient AC drives.

g. **Process fans**

For process fans, starting torque requirement is not critical, however higher starting torque is preferred to accelerate heavy mass of the fan faster and squirrel cage induction motor meets the requirement. However, for energy saving purpose, AC variable speed frequency drives are invariably used, which also provides higher starting torque. For fan applications, power is proportional to cube of the speed and hence by reducing the speed as per process requirement, power drawn by motor is drastically reduced and thus there is potential for energy saving, though motor efficiency is reduced at lower speeds.

h. **Geared motors**

For slow speed applications such as damper actuators, motors with inbuilt gear to run at slow speed are used and these are normally supplied by the machinery supplier.

4.6.2 Protective Enclosure of Motors

Due to dusty conditions and mostly outdoor locations in the plant, all LV motors are recommended to be housed in IP55 enclosure suitable for outdoor installation. Though some of the motors located indoors can have IP54 enclosure also, for standardization and inter-changeability, all motors are specified to be in IP55 enclosure only.

MV motors are available in totally enclosed fan cooled (TEFC), closed air and air cooled (CACA), closed air and water cooled (CACW) and totally enclosed tube ventilated (TETV) housing. In smaller ratings, TEFC or TETV are available. For larger ratings, CACA type motors are available. For very large ratings, CACW type motors are available as water cooling is more effective, but water cooled motors are not preferred as there are chances of damage to the motor windings due to accidental leakage of water, apart from continuous supply of water and pumping arrangement needed for such systems.

4.7 ELECTRICAL ENERGY CONSERVATION

Energy consumption for various process departments varies from country to country, and from plant to plant. A typical relative consumption of electrical energy in a cement plant is shown in Table 4.4. It is advisable to prepare similar tables for individual plants for reference, monitoring, and improvement.

It is evident from the table that the opportunity for energy conservation pervades through the entire process and involves the main machinery,

TABLE 4.4

Electrical Energy Consumption Pattern in Different Sections of the Cement Manufacturing Process

Sl. No	Energy Consumption (%) in Process Departments	Subsections of Each Process Department	Energy Consumption (%) of Sub Sections
1	Raw material preparation – 5.5	Crushing	3.0
		Transportation and conveying	1.7
		Pre-homogenization	0.8
2	Raw mill – 22.5	Raw material handling	0.5
		Homogenization	9.0
		Raw mill gas handling	1.0
		Raw mill grinding	12.0
3	Kiln – 21	Kiln gas handling	14.0
		Kiln material handling	1.0
		Clinker cooler drives	1.5
		Coal handling and grinding	4.5
4	Cement grinding – 42.0	Cement mill material transport	2.0
		Cement mill gas handling	6.0
		Cement mill grinding	34.0
5	Cement packing and dispatch – 1.5	Cement packing	1.1
		Cement loading and dispatch	0.4
6	Services – 7.5	Air compressors	4.0
		Plant lighting	1.3
		HVAC and other	2.2

auxiliaries, and both the raw and in-process materials. The process optimization aspects and use of energy-saving hardware including the replacement of ball mills by low-energy milling systems are not discussed here. Other important measures for conserving the electrical energy, over and above the use of high-efficiency motors, which has already been discussed earlier, are briefly highlighted below:

a. Energy metering and load profiling

A very important step in energy management is to meter the incoming utilities and divide it to different electricity allocation centres, as illustratively shown in Table 4.4. There might be preference in many plants for installing main incoming utility meters on generators, plant substations, kilns, MV motors, and other major power consuming centres. This measure helps in load profiling exercises, which chart the energy consumption patterns in order to identify the peak demand periods, correlate electricity consumption with the facility activities, and production in real time. The utility conservation plans may emerge realistically from such load profiling exercises.

b. Power quality monitoring

The quality of power, as discussed earlier, depends on harmonics, voltage excursions, and sometimes distribution system events including problems on the energy grid. It is important to measure and record trends and alarms on the quality parameters.

c. Use of variable frequency drives (VFD)

For the benefits explained earlier, the VFDs are used quite extensively in cement plants. The application is particularly beneficial for induced-draft and forced-draft fans, kiln drives, mill drives, material handling systems, centrifugal pumps and fans and compressor controls. In this context, it may be relevant to mention that fans and fan drives constitute a significant energy consuming feature of a cement production process. The relation of power consumption with volume flow under different control mechanisms is shown in Figure 4.12 [5]. Normally the fan flow is controlled with the help of dampers, which is an energy-inefficient system. By installing a variable speed drive for the fan motor, the fan flow can be controlled by motor speed control and the damper may remain fully open or removed from installation. Hence, variable speed drives are provided for all fans operating with damper controls and the technology is highly viable.

d. Optimization of compressed air systems

The compressed air systems constitute an essential but energy consuming auxiliary in cement manufacturing. Over 80% of electricity consumed in these systems is wasted as heat. As much as 20%–30% of the compressor output is often leaked through couplings, pressure regulators, condensate traps, shut-off valves and pipe joints. Minimizing such leakages below 10%, reducing the pressure drop from the compressors

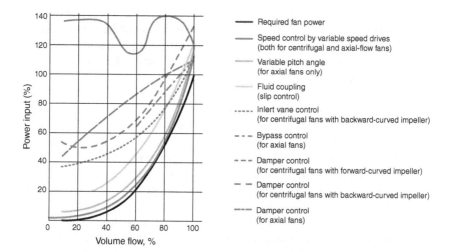

FIGURE 4.12   Control of fan power with different mechanisms [5].

to points of use below 10%, eliminating the unnecessary and inappropriate use of compressed air are some of the important measures taken in the plant to save energy. A more effective step is to apply controls for compressed air. Variable speed motors are used for air compressors to save energy by eliminating unload operation, which consumes about 15 to 35% of full load power. Further, by running the compressors at lower speed with variable speed drive, substantial energy can be saved.

e. Correction of voltage imbalance

A voltage imbalance causes imbalance in current, which results in torque pulsations, increased vibration and mechanical stress, increased losses, and motor overheating. Occurrence of the voltage imbalance can reduce the life of a motor's winding insulation. Motor losses increase rapidly when voltage imbalance exceeds 1%. Hence, voltage monitoring needs to be done regularly and any imbalance should be expeditiously corrected.

f. Power factor correction

Power factor at both high and low voltage levels should be improved to around 0.98 lag to reduce the distribution losses and voltage drops in the system. Improvement in power factor may be effected by using VFDs, synchronous motors, installing capacitor banks for induction motors, etc.

g. Lighting loads

In the past, less efficient lights such as fluorescent, metal halide or sodium vapour lamps were provided. If the existing lights are replaced with energy efficient LED lamps, the power saved is estimated at about 1.5 kWh/t. of cement. The investment returns are seen to be attractive as well.

h. Air conditioner loads

The total load of air conditioners installed for electrical substations and control rooms is substantial. The latest models of air conditioners are energy efficient, as they run at lower speed with inverter drives and may save up to 63% of power as claimed by the suppliers. Thus, the replacement of old air conditioners by the latest energy efficient ones with variable speed drives may save up to about 2 units/ton of cement. The practice of increasing the temperature settings also brings down the energy consumption.

4.8 CONTROL, AUTOMATION AND INFORMATION SYSTEM FOR POWER DISTRIBUTION

Normally the process control system and electrical control and information system are set up in proximity in central control room (CCR) for better coordination. All main incoming and outgoing feeders to the power and distribution transformers can be switched on from electrical control system, whereas all motors are switched on and controlled from process control system in CCR in the sequence of interlocks.

The current and potential transformers act as the field sensors. As usually done, the transformers are monitored for pressure build-up inside the tank, oil and winding temperature, and oil level. For circuit breakers, sensing signals may come from gas pressure in case of SF6 breakers, and from the number of operations undergone. The modern circuit breakers are provided with contacts or sensors to indicate their 'on-off' state as well as fault and power information from digital multi-function meters provided on each feeder.

The control system for power distribution and energy monitoring includes communication networks, Modbus/Profibus, remote terminal units (RTU), sub master, master control stations, operator workstations, supervisory control, and data acquisition system (SCADA) and communication networks and systems in substations with an open standard protocol (IEC 61850). A typical configuration for control system for power distribution is shown in Figure 4.13. The communication network is designed to effect data transfer

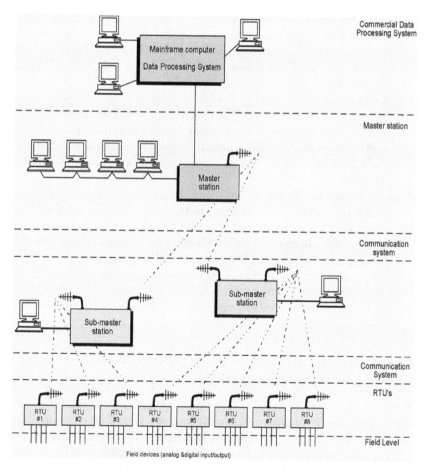

FIGURE 4.13.   A typical configuration for control system for power distribution.

among the central host computer servers and also the field data through RTUs. The medium of transfer can be data cables, radio waves, etc. The operator workstations are the computer terminals consisting of standard HMI (Human Machine Interface) software and are networked with a central host computer. These workstations are operator terminals that request and send the information to RTUs to monitor and control the remote field parameters.

4.8.1 AUTOMATION OF ELECTRICAL DISTRIBUTION SYSTEM

The SCADA system performs automatic monitoring, protecting, and controlling different equipment in distribution systems with the use of intelligent electronic devices in RTUs, equivalent of programmable logic controller (PLC). It restores the power service during fault condition and maintains the desired operating conditions. It improves the reliability of supply by reducing duration of outages and gives the cost-effective operation of distribution system. Its major functions can be categorized into substation control, feeder control, and load control as described below.

a. Substation control

In substation automation system, SCADA performs the operations like bus voltage control, bus load balancing, overload control, transformer fault protection, bus fault protection, etc. It continuously monitors the status of different equipment in the substation and accordingly sends control signals to the remote-control equipment. Also, it collects the historical data of the substation and generates the alarms in the event of electrical accidents or faults. The Input-output (IO) modules of RTUs connected to the substation equipment gather the field data, including status of switches, circuit breakers, transformers, capacitors and batteries, voltage, and current magnitudes, etc. and transfers to remote master unit via network interface modules. The central control or master unit receives and logs the information, displays on human machine interface (HMI) and generate the control actions based on received data. This central controller is also responsible for generating trend analysis, centralized alarming, and reporting. The data historian, workstations, and central servers are connected by local area network (LAN) at the central control room. A wide area network (WAN) connection with standard protocol communication is used to transfer the information between field sites and central control room. Substation control eventually improves the reliability of the power system and minimizes the downtime with high speed transfer of measurements and control commands.

b. Feeder control

It includes feeder voltage control and feeder automatic switching. Feeder voltage control performs voltage regulation and capacitor switching, detection of faults, identifying fault location, isolating fault, and restoration of service. In addition, it facilitates historical data collection of feeder parameters and their status for remote energy management at the central monitoring station.

c. Load control

Smart meters with a communication unit extract the energy consumption information and make it available to a central control room as well as the local data storage unit. At the central control room, the control unit automatically retrieves, stores, and converts all meter data. Modems or communication devices at each meter provide secure two-way communication between remote sites and central control and monitoring room.

Energy consumption data is collected at feeder control and load control levels and taken to central control level for preparing the energy consumption report for each process section separately and disseminated in the department for review and for taking appropriate measures to improve the energy performance.

4.9   PROCESS SIGNAL COMMUNICATION, INTEGRATION AND AUTOMATION

Collection, communication, and integration of the field process signals from sub-control panels, variable speed drive panels, iMCCs, electrical switchboard panels, local IO racks, etc. are routed to local or central control room as shown in Figure 4.14. Inside each local substation, all Profibus signals are looped together and routed to local process controller through a common single cable, limited to the number of nodes permitted. In the case of Profibus cables, longer than 100 m, from the field to the local process controller, as the signal strength weakens over distance, a multi-core fibre optic cable is used to get the signal information correctly and to avoid several repeaters in between. Thus, all signals from the field are taken to respective programmable process controller for onward transmission to servers and operating stations in central control room.

4.9.1   Communication Protocols

The serial communication protocols are Modbus or Profibus or both protocols with their further variations. Developed several decades back by MODICAN for PLC in industrial applications, Modbus is an open protocol now, which is easy to implement and use. It refers to one of the three variants: Modbus ASCII, Modbus RTU, and Modbus TCP/IP. Modbus RTU is a popular master–slave protocol with three different physical media for point-to-point and multi-point communication. Profibus (Process Fieldbus), like Modbus, is also a master–slave protocol. Profibus has certain protocol features that let specific versions of it operate in multi-master mode on RS 485,

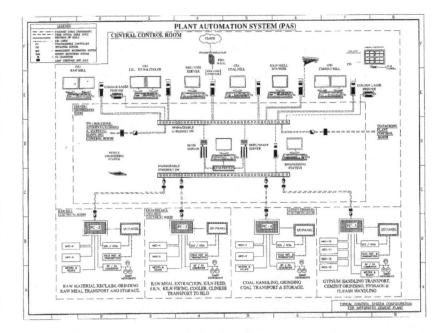

FIGURE 4.14 Typical configuration of a plant automation system.

whereas Modbus has only a single master. Profibus DP (Decentralized Peripherals) is used in machine automation and other discrete signal applications. The main physical layer for Profibus DP is RS485. Profibus PA (Process Automation) is used in process control applications. It connects to PLC by Profibus-DP through a segment coupler, which allows intrinsic connection to electrically isolate PA devices from DP network. A schematic diagram of Profibus master–slave interaction is shown in Figure 4.15 and the broad comparison of Profibus and Modbus is presented in Table 4.5.

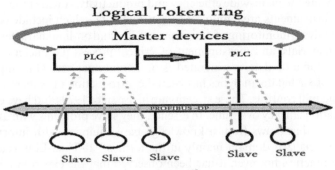

FIGURE 4.15 Profibus with multi master/slave interactions.

TABLE 4.5

Comparison of Modbus and Profibus Protocols

Parameter	Modbus	Profibus
Data rate	Not specified	Profibus PA-31.25 kbps: Blue sheathed screened cable; Profibus DP-9.6 kbps/5 to 12 mbps; violet sheathed
Communication type	Master/slave	Master/slave
Media support	Twisted pair	Twisted pair
Maximum number of nodes	32 slaves max (4000 feet); With repeaters 247 slaves	Profibus PA: 256 per network Profibus DP: 127 per network
Physical layer	RS 232 for point to point, RS 485 or RS or RS 422 for multi-point	RS 485

4.9.2 INDUSTRIAL WIRELESS COMMUNICATION

Wireless technologies have become progressively widespread in industrial automation applications. The advantage of wireless networks is the elimination of expensive cable infrastructure. In addition to locational flexibility, the wireless technologies offer real-time communication for applications such as in SCADA and RTU. A high bandwidth for video transmission, used for remote security, is another advantage. Various frequencies can be used for wireless transmission.

4.10 ADVANCED PROCESS CONTROL AND EMERGENCE OF AI TECHNIQUES

The development of Advanced Process Control (APC) for cement manufacturing has been dealt with in Chapter 3 of this book, Some of the points relevant to this chapter are briefly recapitulated here.

It is well understood that integrated process control systems are required for improving the plant-wide efficiency and productivity. A wide range of sensors and instruments are employed to generate data, which include continuous gas analysis, monitoring of emissions, off-line and on-line characterization of bulk materials, and measurement of critical process parameters. A substantial volume of data is generated, stored, and analysed. One important outcome of such data analytics has been the introduction of 'expert systems' in kiln and mill controls. These systems are in use since 1970s and have evolved in application efficacy over time. In essence, they solve problems with an inference engine that draws from a knowledge base equipped with information about a specialized domain mainly in the form of 'If-Then-Else' rules [14]. Through practice, however, it has been observed that the expert systems are highly domain-specific, strongly dependent on human expertise and have limitations for optimizing the entire process.

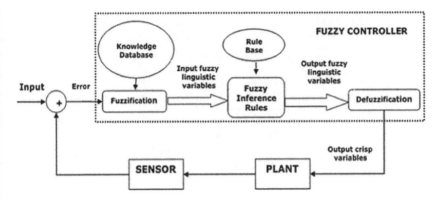

FIGURE 4.16 Major components of the fuzzy logic controller [15].

A step forward is the adoption of the fuzzy logic control system, which makes it possible to create rules for how machines respond to inputs that account for continuity of possible conditions rather than responding in the straightforward binary manner. A schematic diagram of a fuzzy logic controller is presented in Figure 4.16 [15].

As shown in the figure, the fuzzifier converts the crisp input values into fuzzy values. The inference engine simulates the human decision making by performing approximate reasoning with the help of rule base, and the defuzzifier converts fuzzy values into crisp values to be operated upon by the plant. However, there are several limitations of the fuzzy control systems, which primarily include the requirement of an optimum database, high input of human expertise, prompt updating of rule base, and lack of capacity to handle multivariable processes.

For meeting the requirements of complex process control, the technology of 'Model Predictive Control (MPC)' has been developed with multivariate control algorithm [16]. This technology allows the controller to receive information about the current conditions of the process, then uses a model to predict process response to a sequence of future moves in manipulated inputs over a specified timeline, known as 'prediction horizon'. There is explicit consideration of constraints in MPC and it predicts the change in the dependent variable of the modelled system that will be caused by changes in the independent variables. A schematic structure of MPC is shown in Figure 4.17.

Since, however, the input historical data forms the base of MPC, and since the past control moves determine deviations for forecasting the predictive moves, there are application limitations of MPC as well, though the magnitude of limitations is considered less than those of expert systems and fuzzy logic controllers. Hence, more intense applications of artificial intelligence (AI) tools and techniques are being developed [17] under Industry 4.0, which include:

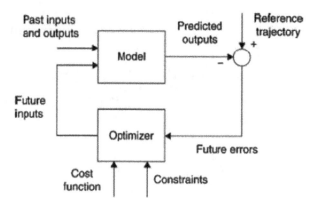

FIGURE 4.17   Basic structure of MPC [16].

- machine learning (ML)
- artificial neural networks (ANN)
- deep learning (DL).

It is important to note that the development of the above techniques has become possible due to the wider practice of data analytics and cloud computing. Machine learning is known to include a broad range of algorithms and statistical models that make it possible for systems to find patterns, draw inferences and learn to perform tasks without specific instructions. ML is generally adopted for addressing the knowledge acquisition bottleneck in implementing expert systems as knowledge has to be extracted automatically from the data base. In a simplified way, ANN can be seen as specific types of ML systems that consist of artificial synapses designed to imitate the structure and function of the human brain. The network observes and learns as the synapses transmit data to one another, processing information as it passes through multiple layers. The ANN approach involves a non-linear mapping between input and output. Deep learning, also known as deep structured learning or hierarchical learning, is the application of ANN to learning tasks that contain more than one hidden layer. The DL architectures have been applied to different applications including computer vision.

4.10.1   AI Applications to Electrical Systems

Generally speaking, the advantage of AI can be realized in shorter term from implementing the fault diagnosis techniques. Based on the basic principles of traditional expert systems, a fault diagnosis expert system, based on rule-based reasoning and case-based recognition may be constructed as shown in Figure 4.18 [18]. Such improved expert systems are considered suitable for fault diagnosis of transformers, generators, and motors.

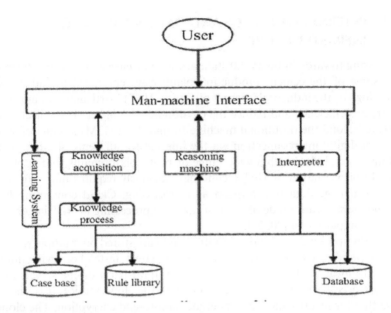

FIGURE 4.18 Basic structure of an improved fault diagnosis expert system [18].

Presently, many of the promising AI techniques are focused on power systems. For example, algorithms have been created, which are capable of identifying malfunctions in transmission and distribution infrastructure, based on images collected by drones [19]. Initiatives have also been taken to forecast how weather conditions will affect the solar and wind power generation, necessitating adjustments to meet power demand. Expert systems have also been perfected to reduce the work load of human operators by taking on tasks of routine maintenance, data processing, training, and schedule optimization.

The adoption of ML has been especially valuable for expanding the horizons of signal processing. When sound, images, and other inputs are transmitted, the ML algorithms make it possible to model signals, detect meaningful patterns, develop useful inferences, and make highly precise adjustments to signal output. The signal processing techniques can also be used to improve the data fed into the learning system. Further, by cutting out noise, the performance of IoT devices and other AI-enabled systems can be made cleaner.

It is strongly perceived that the automating tasks through ML and DL models, such as ANN or decision trees, will result in systems that might make decisions more accurately than what humans can do. However, robust and customized network architecture and billions of training examples will be necessary to optimize the performance of DL algorithms. Once an algorithm is trained, it would be essential to continue processing on progressively expanding volume of data generated by sensors, which brings in the essentiality of data analytics and cloud computing.

4.11 INTERNET OF THINGS AND DATA PROCESSING INFRASTRUCTURE

Progressing towards Industry 4.0 standard with extensive digital transformation, some of the cement production plants have embraced the advanced technology of the industrial internet of things (IIoT), particularly to improve the process efficiency and reduce the environmental impact. The objective is to move beyond the traditional machine-to-machine (M2M) interaction and to adopt digital interconnection via the internet for all items of equipment and machines, linked with a wide range of protocols, domains, and applications. Implementation of IIoT technology obviously requires more efficient ways to manage data transmission and processing. Cloud computing has become the standard mode of data storage and processing for IoT in general and IIoT in particular [20, 21].

Broadly speaking, in cloud computing, data are stored on multiple servers and can be accessed online from any device. In effect, instead of saving information to local hard drives, users store it on third-party online servers and can have access with legitimate accounts with the cloud service. It is understood that for security the data undergoes end-to-end encryption. The cloud architecture, however, consists of large data centres, which are mostly distant from the client devices. The cloud infrastructure, called community cloud, can be for a group of organizations with shared interest and service requirements. It can be privately held by specific organizations. It can be owned by a cloud service provider as a public cloud. It can be a hybrid one, consisting of several clouds with the capability to allow data and applications to move from one cloud to another. Further, different service models are employed in cloud It can be a Platform as a Service (PaaS), where users can deploy their software and applications. Another service model is Software as a Service (SaaS), in which clients have to purchase the ability to access or use an applications or service hosted by the cloud. The third option is to use cloud as Infrastructure as a Service (IaaS), in which users can control and manage the operating systems, applications, network connection, and storage without controlling the clouds as such. All these alternatives are before the cement industry for embracing the cloud solutions.

There is an option for cement plants to partially use cloud services for complex computations as and when required and run the plant from the central control room in the traditional manner. Some cement projects in India currently under implementation are going in for need-based partial use of cloud services in order to avoid interruptions in plant operations due to failure of internet connectivity. The complete switchover to the cloud will depend on the assured internet services.

It is relevant to mention here that in practice cloud computing is more suitable for projects and organizations which deal with massive data storage. Another important aspect is that cloud connectivity is via internet only. Because of remoteness and data filtering there is an element of latency as well in cloud solutions. Considering these aspects and also the ever-increasing number of connected devices, a modified form has been developed and adopted, which is known as 'fog computing', a term coined by CISCO. It defines a mix of traditional centralized data storage systems and cloud. Computing is performed at local networks, although servers are decentralized. The data therefore can be accessed offline because some portions of it are stored locally, whereas in cloud computing all the intelligence and computing are performed on remote servers. Thus, fog computing works as a mediator between hardware involved and remote servers used. Once implemented, it acts as a gateway and regulates which information should be sent to the server and which can be processed locally. Fog computing is a more complicated system as it constitutes an additional layer in the data processing and storage system, but it saves lag time and bandwidth. Obviously, the fog architecture is distributed, it consists of millions of small nodes located as close to client devices as possible. And the connectivity is through various protocols and standards.

There is another approach to data processing, which is called 'edge computing'. In this alternative, data is processed directly on devices without sending it to other nodes or data centres. Edge computing processes data away from centralized storage, keeping information on the local parts of the network, called edge devices. Although edge and fog computing are less known than cloud, they may facilitate cement manufacturing due to its intrinsic feature of IIoT. These networks solve many issues that cannot be solved by cloud computing services and adapt the decentralized data storage to particular needs. Some companies focus on edge computing, whereas others adopt fog computing as main data storage due to high speed and increased availability.

4.12 CONCLUDING OBSERVATIONS

The sustainable production framework for cement manufacturing has many critical dimensions, and use of electric power is one of them due to its essentiality, cost implications, and environmental impacts. For the modern large integrated cement plants, the power utility network is an extensive one, starting from the receiving substation of the transmission grid and ending up with power distribution to all drives and points of usage. Designing the power distribution system is a complex and specialized job that demands deep understanding of prevalent codes and standards as well as the regulatory policies and legal provisions of the country concerned. Broadly speaking, the power system design considerations include, inter alia, load centre substations, MV and LV distribution, lighting infrastructure, emergency provisions, and safety engineering aspects.

An important feature of cement manufacturing plants is the parallel use of power from multiple sources, which at the least include the grid power, captive power, and power from waste heat recovery systems. The grid power itself might be thermal, nuclear, and hydro-electric, which is not an area of concern for the cement plants, as long as the quality and consistency of receiving power are maintained. But the cement producers have to deal with captive power generation, which is predominantly thermal. In addition, in recent years there are increasing trends towards WHRS and renewable power from the energy conservation and climate change considerations. The perceptible directions of development in this regard are the recovery of heat from low-temperature waste gas streams for power generation and PV solar power. Since the solar power does not result in operating temperatures high enough for process use even with the present state of CSP technology, there is some emphasis on technological development for making the solar power suitable for process use. Simultaneously, research studies are in hand to make wind power stable and steady for captive use, in addition to the present mode of using the wind power in the grid.

Adoption of various conservation measures for the use of electrical energy is an ongoing focus area of the cement industry. Advancements in installing low-energy milling systems, such as vertical and horizontal roller mills and hydraulic roll presses in different configurations, are widely known. Installation of low pressure-drop cyclones in preheaters and optimization of comminution and pyro processes are extensively practised. Many other conservation measures such as using high-efficiency motors, drives, and fans, and taking recourse to variable frequency or speed drives particularly for process and cooler fans, compressors, etc. are the common practices in the industry. Cement plants have also adopted efficient control and automation systems with process control going hand-in-hand with electrical control and information management.

In recent years, the industry has adopted Advanced Process Control systems, based on Expert Systems and Fuzzy Logic Controls, and moved further into the realm of Model Predictive Control but most often the systems are in loops and not for the entire process chain. At the same time, it must be understood that the cement manufacturing process is not steady state; it is nonlinear, and it consists of interconnected items of equipment and machines. Hence, it is becoming increasingly evident that the tools and techniques of artificial Intelligence are needed to improve the plant productivity, efficiency, and environmental impacts beyond what has been achieved. Progress in data science and cloud computing has laid the way towards application of AI tools. Technologies based on IIoT, ML, DL, and ANN are being extensively explored for process control. The application of AI in power engineering and signal processing deserves a special mention. Potentials for sustainable cement production for survival and growth in the future therefore exist for cement manufacturers today with the implementation of these latest technologies.

REFERENCES

1. http://electrical-installation.org/enwiki/Electrical_regulations_and_standards
2. Robert J. Alonzo, *Electrical codes, standards, recommended practices and regulations*, e-book ISBN 9780815620467, Elsevier, 2010, 1–512.

3. https://electrical-engineering-portal.com/technical-articles/electrical-systems-cement-plants

4. https://electrical-engineering-portal.com/gas-insulated-substation-gis

5. IFC, Improving thermal and electrical energy efficiency at cement plants – international best practices, Available at https://ifc.org/Elect_Effic_Cement_05-23.pdf

6. Gen Takahashi, *Power generation by waste heat recovery in cement industry, Seminar on the Joint Crediting Mechanism (JCM) Project Implementation in Indonesia*, July 2017, Available at http://www.jfe-eng.com.jp.en/

7. L.F. Moreira and F.R.P. Arriera, Thermal and economic assessment of organic Rankine cycles for waste heat recovery in cement plants, *Renewable and Sustainable Energy Reviews*, Vol. 114, October 2019, 109315, Available at http://doi.org/10.1016/j.rser.2019.109315.

8. Omar Aboelwafa, Tamer S. Ahmed, Ahmed F. Soliman, Ibrahim H. Ismail, Organic Rankine cycle and steam Rankine cycle for waste heat recovery in a cement plant in Egypt: a comparative case study, *Water, Energy, Food and Environment Journal*, No. 1, 2020, 19–42.

9. IEA Statistics Report on Renewable Information Overview 2020.

10. Daniele Formi, Dario Di Santo, Francesco Campara, *Innovative system for electricity generation from waste heat recovery, Conference Paper*, ECEEE 2014, available at https://www.researchgate.net.publication/323999046.

11. https://lafargeholcim.com/solar-energy-cement-manufacturing

12. https://globalcement.com/news/item/10119-solar-power-cement-production

13. N. N. Miller, Dilip Guru, Kara Clark, Wind generator applications for cement industry, Conference paper uploaded on 15 December 2015, available at https://www.researchgate.net/publication/4342818.

14. https://tutorialspoint.com/artificial_intelligence/artificial_intelligence_expert_system

15. https://tutorialspoint.com/fuzzy_logic/fuzzy_logic_control.htm

16. https://sciencedirect.com/topics/engineering/model-preddictive-control

17. Weiyu Wang and Heng Sian, Artificial intelligence, machine learning, automation, robotics, future of work and future of humanity: a review and research agenda, *Journal of Database Management*, 30(1), January 2015, 61–79.

18. Han Feng, *The application of artificial intelligence in electrical automation control, IOP Conference, Journal of Physics: Conference Series*, 1087, 2018, 062008.

19. https://online.egr.msu.edu/ai-machine-learning-electrical-computer-engineering-applications

20. https://phoenixnap.com/blog/category/cloud

21. https://www.digiteum.com/cloud-fog-edge-computing-iot

5 Data-Driven Thermal Energy Management Including Alternative Fuels and Raw Materials Use for Sustainable Cement Manufacturing

Ashok Dembla
Humboldt Wedag India Private Limited, New Delhi, India

Matthias Mersmann
KHD Humboldt Wedag International AG, Cologne, Germany

CONTENTS

5.1 Introduction 142
5.2 Description of the Thermal Process 143
5.3 Sustainability in Cement Production through the Use of
    Alternative Resources 145
5.4 Alternative Raw Materials for Pyroprocessing 146
    5.4.1 Naturally Occurring Alternative Raw Materials 146
    5.4.2 Industrial Waste as Alternative Raw Materials 147
5.5 Alternative Fuels for Pyroprocessing 150
5.6 Storing, Dosing, and Conveying of Alternative Fuels 152
5.7 Operational Considerations in Using Alternative Fuels..................153
5.8 Adapting the Plant and Equipment to AF Combustion 154
    5.8.1 Criteria for Selecting Firing Locations...............................155
    5.8.2 Design Features of Pyroprocess Equipment 156
    5.8.3 Rotary Drum Reactor for Burning Coarser Fuels..............158
    5.8.4 NOx Control Technologies 158
    5.8.5 Process Instruments...160
5.9 Conventional Approaches for Process Optimization......................162
    5.9.1 Fuzzy Logic Control Philosophy 162
    5.9.2 Model Predictive Control 164

DOI: 10.1201/9781003106791-5

141

        5.9.3   Limitations of Conventional Automation Systems 165
5.10   Implementation Plan for Industry 4.0 Tools in Cement Plants 167
5.11   Integrated Robotic Laboratory for Quality Control 170
5.12   Advanced Process Control Systems Based on Artificial
         Intelligence 172
5.13   AI-Based APC for Thermal Process 176
        5.13.1 Kiln Control Module..181
        5.13.2 Calciner Module with Alternative Fuels Controller..........182
        5.13.3 Cooler Control Module..191
5.14   Evaluation and Implementation of Advanced Process
         Control System..192
5.15   Collaborative Operation in Data-driven Ecosystem.....................194
5.16   Concluding Observations 195
References...197

5.1 INTRODUCTION

Currently the cement industry is confronted with two mutually exclusive goals. On one hand, there is a huge pressure to increase profit and margins, while on the other hand there is considerable public interest on a sustainable and environment friendly use of natural resources. The fast and optimal reactions to continuously changing conditions are crucial for survival.

The major gains of substituting conventional fuels and raw materials by alternative fuels and raw materials (AFR) in the production of cement clinker are two-fold: conservation of natural resources and reduction of greenhouse gas emissions. The alternative fuels (AFs) that can be used are waste oils, mixtures of non-recycled plastics and paper, used tires, biomass wastes, municipal solid waste, and even wastewater sludge to substitute the fossil fuels. Similarly conventional raw materials can be partially replaced by alternative raw materials like wastes, tertiary materials, and by-products from the other industries. However, the introduction of the AFR may influence emissions, cement product quality, process stability, and process efficiency, if the process and fuels are not correctly selected and optimized.

AFR have different physical and chemical properties as compared to the conventional fuels and raw materials, respectively. Using these materials is quite challenging in terms of handling, preparation, and controlling the process. Recent developments in the cement process equipment technology in calciner and burner have encouraged cement companies worldwide to use the AFR to a large extent.

Selection of proper equipment for pyroprocess is the first step in enhancing the efficiency of thermal process, but one cannot forget the needs of process control system for stable operation, consistent quality control, and optimum thermal energy. The conventional process control system is not sufficient to handle the challenging and changing complex conditions of the thermal process due to the use of AFR. These conventional process control systems use rules-based control concept and have implemented fixed knowledge-based rules. This implies that, when a situation other than the pre-set problems would occur, the system will have limitations and will not be able to control or assist further. Hence, there is a need for advanced process control systems (Industry 4.0) which is based on the self-learning techniques for data-driven thermal energy management, including AFR that are used for sustainable cement manufacturing.

The concept of Industry 4.0 is based on integrated data-driven automation and optimization concept to bring the plants to their optimal economic performance within the degrees of freedom left by the technological, environmental, and contractual constraints. Data integration management system is a pre-requisite for the successful implementation of an advanced process control system. Further, in order to be successful, it must work in real-time data environment and be fed with consistent and correct information at all times.

The future advanced process control systems must be based on modeling and Artificial Intelligence (AI). These techniques need huge real-time and historical data to model the relationships of different variables with each other for predicting and optimizing the process. The wider the scope of production and the deeper the control level, the bigger will be the database requirement. Based on an arbitrary comparison with autonomous driving cars, it appears that the autonomous operation of a cement plant may require a volume of data which will be 1000 times more than the data usually acquired and handled for the traditional control strategies. Further, it is highly probable that the increased volume of data for autonomous operation will come from digital twins, rather than from conventional sensors, as the digital twins will form a system of soft sensors capable of capturing instantaneously 3D-sensitive data from critical positions in the manufacturing process. The big data, so collected, will need proper preparation and analysis to convert raw data into useful information, knowledge, understanding, and wisdom for prediction and decision making: (Figure 5.1) [1]. The specifics of the pyroprocess and the thermal energy use are discussed in this chapter.

5.2   DESCRIPTION OF THE THERMAL PROCESS

The thermal Process of cement manufacturing is divided into four parts: Preheating, Calcination, Clinker Burning, and Clinker Cooling as shown in Figure 5.2. The powdered raw mix is fed into the top-stage cyclone of the preheater. The hot gases from the kiln enter into the preheater from bottom. Then the powdered raw mix slides down through cyclones and comes in contact with hot air which travels from bottom to top. In the preheater, temperature of raw mix rises to 900°C to 1000°C and nearly 90% calcination (removal of CO_2 from $CaCO_3$) takes place before entering into the kiln.

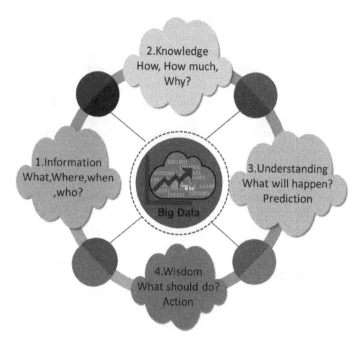

FIGURE 5.1   Turning raw data into wisdom [1].

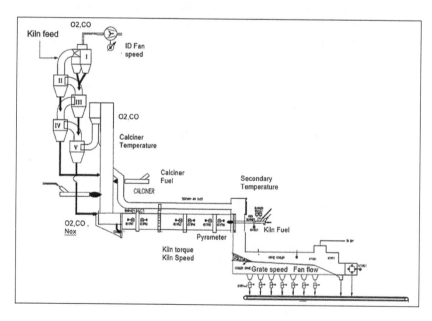

FIGURE 5.2   Flow diagram of the thermal process in cement manufacturing.

Then the powdered raw mix enters the kiln from one end and the burner is situated at the opposite end. The rotary kiln rotates at the speed of 1–5 revolution per minute (RPM) as per the requirement of the process. The raw mix in the kiln melts first into liquid form and then transforms into nodules due to the effect of the rotation of the kiln.

There are two zones inside the kiln, namely calcining zone and burning zone. The zone where raw mix enters into the kiln is called calcining zone, where temperature would be 950°C–1000°C. Burning zone starts after this zone where temperature would be 1350°C–1450°C. The hot clinker discharged from the kiln is cooled down quickly by air quenching with the help of efficient coolers. The temperature of clinker is brought down to 80°C–90°C from 1350°C. Fast cooling is very essential to get good quality clinker. If cooling is not quick, the compound stability in clinker will be adversely affected resulting in lower strength of cement after grinding.

5.3 SUSTAINABILITY IN CEMENT PRODUCTION THROUGH THE USE OF ALTERNATIVE RESOURCES

Since the production of cement is resource-intensive and the resources are primarily natural and non-renewable, the massively increasing growth trend of the cement industry can only be sustained by using AFR. Broadly speaking, AFR comprise industrial, agricultural, or household wastes and by-products, which can be used as a source of energy or as a substitute for raw materials and cement additives. The alternative raw materials such as metallurgical slag and coal ash are extensively used in producing blended cements, while alternative or waste fuels are steadily substituting the fossil fuels in the thermal process. The AFs are generally cheaper than the fossil fuels because most of the AFs are generated from wastes which would otherwise be discarded. Use of these AFs becomes feasible only with minimal pre-processing costs. The significant advantage of AF substitution is the preservation of non-renewable energy sources and the reduction of waste disposal sites. Since the AFs are generally characterized by different combustion behavior, the process needs to be adapted to their use. The extent, to which fossil fuels are replaced by AFs, known as thermal substitution rate (TSR), can be increased gradually to balance the process effects.

AFs, when coprocessed in cement kilns, significantly reduce the CO_2 emissions. Although the level of AF used in the world was only 6% in 2016, the average substitution in the EU countries touched 43% in 2018. In order to attain the sustainability goals, the global cement industry must reach 18% substitution of coal and petroleum coke by AF by 2030 [2]. A collaborative effort is required by the cement industry, equipment supplier, industrial automation supplier, research partners, and the policy makers to implement the best practices in the state-of-the-art technologies.

5.4　ALTERNATIVE RAW MATERIALS FOR PYROPROCESSING

In the cement industry, the alternative raw materials in use are broadly of two categories:

- Naturally occurring but generally not usable
- Industrial waste materials

5.4.1　Naturally Occurring Alternative Raw Materials

Limestone is the principal raw material for cement manufacturing, but in most deposits it is associated with some unwanted materials having no commercial value. Exploitation of the deposits involves handling of such unwanted materials occurring either as an overburden or as internal layers and patches. The mining activities including beneficiation of raw material add to the volume of mine rejects and tailings. Transportation and dumping of such waste in huge quantities lead to economic and environmental problems. The solution lies in utilizing such materials as much as possible. The overburden clay, which has to be removed and dumped for exposing the limestone band, may find use as a component of raw mix, where the limestone is deficient in SiO_2.

The composition of low-grade and marginal-grade limestone is shown in Table 5.1. These grades are not readily usable due to limitations of the basic chemistry in the cement manufacturing process. Hence, the following approaches are adopted or explored to utilize such materials:

- Production scheduling and blending of marginal-grade limestone with high-grade limestone to get the desired quality for raw mix
- Selective mining, manual sorting, screening/wobbling of the run-of-mine limestone for upgrading the quality
- Adoption of froth flotation technology for limestone beneficiation, which is already in practice in a few plants
- Introducing dry beneficiation technologies such as air classification, differential grinding and sieving, electrical separation, photometric sorting, etc., depending on the limestone characteristics such as

TABLE 5.1
Compositional Grading of Limestone

	Chemical Parameter	
Grade	CaO%	SiO_2%
Cement-grade limestone	44–52	6–12
Marginal-grade limestone	40–44	12–16
Low-grade limestone	36–40	16–20

differential hardness, electrical conductivity, optical reflectance, and other properties
- Exploring the potential of bacterial leaching of limestone

A general survey of cement plants, facing problems in limestone quality from their captive mines, shows that the common practices are screening, removal of clay fraction, blending with high-grade limestone, selective mining, and froth flotation. From the viability considerations the adoption of dry benefication technologies are still exploratory, although some of these technologies are well proven in other industries. The bacterial leaching, although in practice in metal mining, is yet to appear in the cement industry.

5.4.2   INDUSTRIAL WASTE AS ALTERNATIVE RAW MATERIALS

Waste materials from thermal power plants, fertilizer plants, aluminum industry, steel industry, lead-zinc industry, and other waste-generating manufacturing processes can be utilized effectively and economically in producing cement and building materials. At the same time, this strategy helps to keep the environment clean. Use of some important industrial waste streams is briefly described below.

A. **Fly ash from thermal power plant**
   In several countries, and more particularly in China and India, electricity is produced from a large number of coal-fired thermal power plants, which generate a huge quantity of a waste material called fly ash. Its storage and disposal cause grave environmental concerns. Although it is used extensively as a pozzolanic substance in making blended cement, it can also serve as an effective component of kiln feed for the manufacture of cement clinker. Fly ash contains significant amounts of Al_2O_3 and SiO_2 and has thus been used as a partial replacement of natural raw materials like clay or shale in the kiln feed. In some cases the Fe_2O_3 content in the fly ash may help in its use as a corrective material.
   In the dry process, the fly ash is either premixed with the raw kiln feed or introduced directly into the burning zone. The level of addition is dependent upon the fly ash composition. The use of high-carbon fly ash in raw kiln feed has the additional benefit of saving fuel. As a general rule, fly ash with higher carbon content should be introduced in the high-temperature zones rather than feeding it along with the raw material in order to avoid the emission of CO at the preheater outlet.

B. **Carbonate sludge from fertilizer plants**
   The carbonate sludge from the fertilizer industry has calcium carbonate as the major constituent along with varying amounts of free lime. The dry sludge in the form of fine powder has a high potential to be used as a partial substitute of limestone in clinker making and also as a component of masonry cement.

C. Red mud from the aluminum industry

Red mud, a residual waste obtained from aluminum industry, consists mainly of iron oxide, aluminum oxide, titanium oxide, and sodium oxide. The high alkali content in red mud often hinders its industrial use. Nevertheless the material is being tried for different applications. So far as the cement production is concerned, it has been used as a corrective material in raw mix preparation, particularly as a source of iron oxide and alumina. It has also been seen to have some mineralizing property in clinker making to reduce the free lime content or to bring down the burning temperature.

D. Slag from the iron and steel industry

Integrated steel plants have various unit operations to produce metallic iron, crude steel, and refined steel of different quality specifications. These plants are also equipped with captive power generation. As a result, the steel plants generate substantial quantities of waste from each of its operations. Blast furnace slag, steel slag, and fly ash constitute 75%–80% of the total solid wastes generated in a steel plant.

The typical chemical composition of grounded granulated blast furnace slag (GGBS) is given in Table 5.2. While GGBS is a widely used cement substitute material for making slag blended cement, it can be used in clinker making in small replacement of limestone, as it has a high content of CaO in it. Since this lime does not require any heat for decarbonation in the clinker making process, the specific heat requirement with the usage of GGBS may remain lower by 20%–30% as compared to normal operation with no replacement of limestone with slag. There are references of plants which are operating at 600 kCal/kg of clinker specific heat consumption for five-stage preheater systems, which is about 130–150 kCal/kg of clinker lower than conventional operation with limestone.

From Table 5.2 it is also evident that the use of GGBS as an alternative source of CaO is limited by its high SiO_2 content. In practice, the use of GGBS can be in the range of 20%–35% for energy saving purposes.

TABLE 5.2

Chemical Composition of GGBS

Composition	% Components Grounded Granulated Blast Furnace Slag (GGBS)
SiO_2	33–34
FeO	1–2
Al_2O_3	21–23
CaO	34–35
MgO	8–9
SO_3	0.2–0.5

TABLE 5.3

Chemical Composition of LD Slag

Constituents	Range%
SiO_2	10–13
FeO	22–26
Al_2O_3	1–2
CaO	40–50
MgO	8–12
MnO	2–4

The steel slag from the LD process is another alternative raw material for clinker making. The chemical composition of the LD slag is given in Table 5.3. Because of its high lime content, it can be used as a sweetener to compensate the lime content in low-grade limestone. The high proportions of FeO, MgO, and MnO in the slag are the limiting parameters for its use in clinker making.

E. **Ferro-alloy and non-ferrous metallurgical slags**

The ferro-alloy slags are those originating from the processes of manufacturing ferro-chrome, ferro-manganese, and ferro-silicon alloys, and the non-ferrous metallurgical slags are obtained from smelting of zinc and copper. Presently, these slags are used either as fillers or as road making materials. The copper slag is reported to have pozzolanic properties and can be used in making Portland pozzolana cement.

The chemical composition of copper slag is given in Table 5.4, which shows that it has very high content of iron oxide. It can therefore be considered as a potential substitute for iron ore, which is used in many

TABLE 5.4

Chemical Composition of Copper Slag

Constituents	%
L.O.I.	0
SiO_2	28.26
Al_2O_3	3.69
Fe_2O_3	57.74
TiO_2	0.22
CaO	2.70
MgO	0.24
Total S as SO_3	0.18
K_2O	0.41
Na_2O	1.31
Cl^-	1.160
P_2O_5	0.19

plants as a corrective material to an extent of 2%–3%. However, the high chloride content in this slag is a limitation for its use.

A few other varieties of non-ferrous metallurgical slags having potential to be used in the thermal process are highlighted below:

- Chromium and ferro-titanium slags: used in making high-alumina cement so far but can be used as a mineralizer in Portland cement clinker making
- Electro-thermal phosphorous slag: approximately 8.0 tons of slag of essentially calcium silicate composition is produced for every ton of phosphorus and the present application developments are more for building materials than for clinker making; some opportunities in cement grinding are being explored.
- Aluminum slag: Compositionally a dicalcium silicate rich slag, which may find use in making high-belite Portland cement.
- Magnesium slag: obtained from extraction of magnesium by decarbonation of dolomite and subsequent selenium reduction. The present use is for blended cements and making aggregates.
- Nickel slag: obtained from the metal extraction from the nickel ore; it has been found to be more pozzolanic than fly ash and other pozzolanic materials but it affects the early strength of the blended cement produced with 45% clinker replacement, although the late-age strength is maintained.
- Lead-zinc slag: obtained from the zinc metallurgical industry the slag is dumped without any effective industrial use, but it has some potential as a mineralizing and fluxing substance.

Although there are a large number of industrial waste that can be considered as potential sources for alternative raw materials, the practical use will be governed by the following norms:

- Quality parameters in terms of chloride, sulfur, alkali, and magnesia contents so as to comply with the cement specifications
- Continuous availability in required quantity
- Viable processing cost

Increase in the use of alternative raw materials will help not only to lower down the manufacturing cost but also to contribute to environmental sustainability and ecological balance.

5.5 ALTERNATIVE FUELS FOR PYROPROCESSING

The selection criteria for AFs are dependent on the prevailing circumstances of the operating plants, which in turn are defined by the process, product quality, and emissions.

AFs may either be of a single grade of segregated waste from a dedicated industrial production line or a mixture of various waste materials, mostly from the municipal waste collection. The dedicated industrial waste offers the advantage of purity and homogeneity but requires proximity of its availability at the cement plants. Otherwise, the cost benefit of using the AF will be nullified by the logistic cost. Sources for the municipal waste are often widely spread so that this type of waste can be supplied to many cement plants. The disadvantage of municipal waste is that its consistency in their composition cannot be guaranteed, given the heterogeneity of its composition and its usually open air storage. There is a need for controlling the chemical composition of the AF which is to be produced from the municipal waste by pre-processing in order to meet regulatory requirements for environmental protection. The cement producers must evaluate the following properties before using specific AFs [3]:

- Economic viability and regular availability
- Physical properties (scrap size, density, homogeneity, moisture content)
- Physical state of the fuel (solid, liquid, gaseous)
- Content of volatile and circulating elements (Na, K, Cl, S)
- Toxicity (organic compounds, heavy metals)
- Composition and content of ash and content of volatiles, calorific value
- PCB, cadmium (Cd), thallium (Tl), and mercury (Hg) contents
- Grinding properties
- Proper proportioning technology
- Gas emissions
- Negative effects on the cement quality and its compatibility with the environment

Industrial and municipal waste is available in almost all the cement-producing countries. As a general guide, any waste with a net calorific value or low heat value (LHV) of more than 10 MJ/kg fuel can be considered as an AF. Typical LHV ranges of different AFs are indicated Table 5.5.

TABLE 5.5
LHV and Moisture of Alternative Fuels

Description	LHV(MJ/Kg)	Moisture (%)	Remarks
Waste oil, refinery waste	30 to 40		
Waste tires incl. steel and inerts	28 to 32		High grade
Mixed plastic waste material	20 to 28	10	
RDF (from domestic source)	17 to 21	10	Medium grade
Dried wood, bark	15 to 17	10	
Grain straw	12 to 14	20	
Rice straw	11 to 12	20	Low grade
Dried sewage sludge	10 to 12	10	

The types of available AFs are mentioned below [3]:

- Gaseous: coke oven gas, refinery waste gas, pyrolysis gas, landfill gas, etc.
- Liquid: low-chlorine spent solvents, lubricating oils, vegetable oils, fats, distillation residues, hydraulic oils, insulating oils, etc.
- Pulverized, granulated, or fine-crushed solids: waste wood, saw-dust, wood shavings, dried sewage sludge, finely shredded used tires, residues from food production, etc.
- Coarse crushed solids: plastic waste, discarded wood strips, re-agglomerated organic matter, etc.
- Lumpy solids: whole scrap tires, plastic bales, plastic bags, and plastic drums, etc.
- Refuse-derived fuel of specified dimensions

Different AFs need different handling, preparation, and dosing and have different impacts on the process and quality.

5.6 STORING, DOSING, AND CONVEYING OF ALTERNATIVE FUELS

Specially designed installations are required to control the use of waste from its arrival at the plant to its entry into the kiln. These installations include storage systems, dosing systems, and conveyance systems to the kiln. Designs vary depending on the characteristics of the AF used and the space available at each plant. The manner in which the product is received and stored on its arrival and the way in which it is put in the kiln will therefore have to be specifically adapted to the plant [3].

Mechanical conveying of fuels to the kiln mainly depends on the particle size of the product and sometimes on the kind of material. Particle sizes of over 30 mm may cause problems in pneumatic conveying, and therefore, mechanical conveying is taken recourse to for particle sizes up to approximately 100 mm. If the mode of conveyance is mechanical, the point of fuel feeding to the kiln system normally changes to the calciner or the preheater. However, mechanical conveying entails far higher investment cost than pneumatic conveying, as it has to raise material several meters from the point of receiving to the point of delivery. To avoid false air entering into the process, a double or triple gate is used at the material's entrance to the calciner. When the dosing system is not in operation, the gate remains completely closed. The mechanical conveyance comprises different equipment depending on the specific needs of the plant and may include belt conveyors, worm screws, apron conveyors, chain redlers, bucket elevators, etc. [3].

5.7 OPERATIONAL CONSIDERATIONS IN USING ALTERNATIVE FUELS

Cement kilns offer nearly ideal possibilities for burning most of the AFs. The mineral or ash portion is incorporated into the clinker phase. Toxic organic harmful matter is destroyed under the high temperatures. Further, when the cement produced with AFs are used in the form of mortar or concrete, the heavy metals are incorporated into the crystal structure of the hydrated phases. Despite such benefits of using AFs, there are process limitations in actual usage, which need to be addressed. The AFs have different characteristics compared to the conventional fuels, both physically and chemically. For example, parameters such as the particle size, moisture content, ash content, volatile matter, and calorific value are different. In addition, inhomogeneous physical and chemical properties make the handling of AFs difficult. The important characteristics of some fuels are shown in Table 5.6, in which the positive, negative, and indifferent characteristics have been indicated [4]. It should also be understood that there are no fuels that will have no shortcomings.

High moisture or LHV of an AF may increase its burnout time, which may result in higher amount of exhaust gases. If the production line has no margin in ID fan capacity or the gas velocity is high at some critical points (e.g. in the kiln inlet chamber), then an increased use of AFs will reduce the kiln throughput. The fuel properties may influence the cement production and clinker properties in many other ways [2]. The important effects are highlighted below:

a. **Effects on energy balance:**
   - Increase of specific heat consumption (as most AFs contain higher moisture and lower HV)

TABLE 5.6
Characteristics of Some Solid Fuels

Description	Coal	Pet Coke	Anthracite	Tire chips	Fluff-RDF	Animal Meal	Sewage Sludge
Heat Value	A	A	A	A	B	B	C
Preparation	A	B	B	C	C	B	B
Handling	A	A	A	C	C	B	B
Chlorine Input	A	A	A	B	C	C	B
Sulfur Input	A	C	A	B	A	A	B
NOx Reduction Potential	B	C	C	A	B	A	B
Fuel Costs	C	B	B	A	A	A	A

A = positive, B = indifferent, C = negative

- Increased waste gas volumes (higher fuel moisture, fuel chemical composition, higher excess air demand, more fuel to maintain hot sintering zone)
- Higher amount of primary air (transport air) and leakage air; decrease of recuperation air from clinker cooler
- Increased heat losses by radiation (shifting temperature profile of kiln toward kiln inlet chamber)
- High amounts of alkalis, sulfur, and/or chlorides may affect kiln operation. Gas bypass may be required if these quantities cannot be absorbed in clinker. Exhaust gas volumes will increase and energy consumption will go up.

b. **Effect on plant operation stability**
  - High demands on fuel dosing equipment for continuous fuel feed.
  - The large fuel particles may not have sufficient time for complete burnout and unburned fuel may therefore fall into the clinker bed and continue combustion causing local reducing conditions, which may result in unstable operation of the calciner and the kiln.
  - Formation of build-ups in case of Cl- and S-rich AFs in the area of kiln inlet, riser duct, and bottommost cyclone. More manual cleaning efforts or a provision of a bypass system may be necessary.

c. **Effect on clinker quality**
  - Reducing conditions as mentioned above
  - Cooling down of the sintering zone
  - Possible enrichment of harmful elements in clinker, e.g. MgO, P_2O_5, etc.
  - Possible changes in the clinker composition due to ash absorption and reducing conditions

d. **Effect on emissions**
  - Avoidance or minimization of NOx formation
  - CO formation in case of inadequate calciner technology or unsuitable AFs
  - Reduced emission of CO_2

5.8 ADAPTING THE PLANT AND EQUIPMENT TO AF COMBUSTION

From the above discourse on the effects of AFs on energy balance, operational stability, clinker quality, and emissions, it is obvious that it is not easy to use AFR in a cement plant without deeper analysis and knowledge, as the process impacts can cause big problems. High substitution rates of AFs can only be achieved if the process and machinery are tuned and adapted to the new requirements. However, the use of AFs can be economically viable – and it usually is – when the process is optimally adapted. In the case of new installations the modifications required can be considered in the system design so that no production risk occurs. More than the rotary kiln end, the calciner, which operates at lower temperatures (typically at 850°C-1200°C), is better suited for burning AFs [5]. Different calciner designs are available from different suppliers; some of them are more suitable for burning large amounts of AFs. Residence time, temperature distributions, oxygen availability, turbulence levels, and velocity distributions are the key parameters for designing a good calciner. Increased use of lumpy AFs may have a negative impact on the internal cycles of sulfur, alkalis, and chlorine in the kiln system. By optimum calciner designs, such negative impacts can be avoided or considerably reduced.

Experience has shown, however, that it is difficult to obtain complete combustion of low-volatile fuels in calciner. Hence, the high-volatile AFs with relatively low calorific value are preferred for the calciner. The low-volatile fuels including petcoke, low-volatile bituminous coal, anthracite, etc. are more effectively burnt in the rotary kiln for which proven technologies are available.

The important process requirements for increasing the utilization of the AFs are summarized below:

- Determination of optimum location of firing
  - Main Burner
  - Inlet Chamber
  - Calciner
- Modification and adjustment of burner
- Modification and adjustment of calciner
- Adjustment of fuel split between calciner and main burner
- Adjustment of tertiary air flow for proper oxygen level
- Adjustment of raw mixt for ash absorption and combustion properties of AF
- Selection of suitable refractories
- Control of volatile circuits (SO_3, Cl, alkalis in hot meal)

5.8.1 Criteria for Selecting Firing Locations

The firing location of AFs is selected mainly on the basis of the particle size, moisture content, calorific value, volatiles cycle, and quality impact. Principally, the main burner flame provides good conditions for the use of AFs due to its high temperature, as it supports safe ignition of AFs with poor ignition behavior. However, the sintering process in the kiln also requires high flame temperature to ensure the product quality of the clinker. This limits the rate of deployment of AFs with low calorific value. The main problem with the use of AFs at the main burner is in the insufficient residence time of the

AF during the flight phase [5]. However, the effective utilization of AF in the main burner will depend on the specific fuel type and quality, co-firing ratio, burner design and settings, process stability related to the possible fuel spillage, etc.

The AF must be processed properly to obtain proper quality for suitable burning in the main kiln burner. Mostly, the pre-processed Refuse Derived Fuel (RDF) is used for this purpose. Other main burner fuels comprise waste oils, chemical solvents, animal meat and bone meal, sewage sludge, waste wood, and agricultural wastes, e.g. rice husks, straw, nut shells, etc.

Use of low-grade AFs such as waste coal, tires, sewage sludge, and biomass fuels (such as wood products, agricultural wastes, etc.) is more effective for calciners.

5.8.2 DESIGN FEATURES OF PYROPROCESS EQUIPMENT

Keeping in view the imperative need for enhanced usage of AFs, several innovative design features have been incorporated in the pyro section in particular. These developments are briefly described below.

a. **Preheater Cyclones**

Since burning of AFs result in high specific waste gas volumes, which in turn cause higher pressure drop in the preheater cyclones and increased electrical and thermal energy consumption or reduced production. The solution of these problems is found in the use of low pressure-drop cyclones.

b. **Calciner Design**

The main criteria for the calciner design are complete burnout of the calciner fuel without emission of organic and toxic residues, enough flexibility to burn a wide range of AFs, minimizing emissions of NOx and CO, and good controllability. For some of the AFs that are not sufficiently prepared and ground, the calciner path needs to be extended for more retention time to ensure complete burnout. For more difficult-to-burn fuels it may be necessary to further add a special module of combustion chamber to the extended riser duct. These alternatives are shown in Figure 5.3. The functional details of different parts of a calciner are shown in Figure 5.4 [4].

In the extended calciner, a hot spot burner creates a zone of high temperature (1000°C–1200°C) and the burnout happens in the extended portion. For very coarse AFs, a special module of an ignition chamber is integrated with the pure air side of the calciner. Fuels can be fed into the hot core zone of the combustion chamber at temperatures higher than 1200°C. The partly burnt-out fuels are fed directly to the calciner, where they are lifted up with the gas stream for complete burnt out. A small amount of fuel particles may fall down in the inlet chamber and burn out in the kiln [3]. The

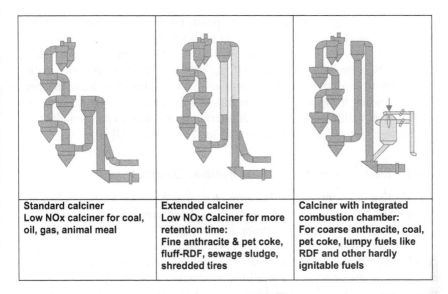

| Standard calciner
Low NOx calciner for coal, oil, gas, animal meal | Extended calciner
Low NOx Calciner for more retention time:
Fine anthracite & pet coke, fluff-RDF, sewage sludge, shredded tires | Calciner with integrated combustion chamber:
For coarse anthracite, coal, pet coke, lumpy fuels like RDF and other hardly ignitable fuels |

FIGURE 5.3 Changes in the design of calciner to suit the fuel types [4].

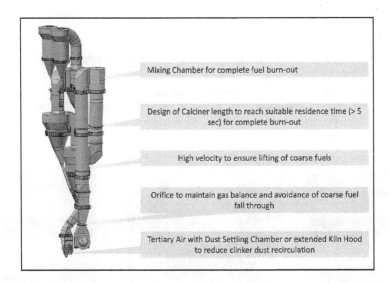

FIGURE 5.4 Functional details of the latest calciner for burning of alternative fuels [4].

combustion chamber system is also energy efficient as it operates on recuperated process heat.

It is relevant to mention here that, although the NOx generation is related to the content of volatile matter and nitrogen in the fuel, the calciner design and atmosphere for burning significantly influence the NOx emissions.

Generally speaking, an inline calciner operated with a high-temperature oxidation zone yields lower NOx emission than a separate-line calciner. It is also observed that with regular designs of calciners, while it might be possible to meet the NOx emission limit of 600 mg/Nm³ at 10% O_2 with high-volatile fuels, without secondary reduction measures, it is unlikely to achieve this performance when firing a low-volatile fuel like the petroleum coke. Hence, in order to deal with various streams of alternative and waste fuels, a multistage combustion system has become popular for NOx control [3].

5.8.3   ROTARY DRUM REACTOR FOR BURNING COARSER FUELS

An effective development for burning of coarse AFs like waste whole tires, crude RDF, etc. is the rotary drum reactor. The reactor is arranged above the kiln, located in between the tertiary air duct and the kiln riser duct. Fuel is transported through the reactor mechanically with sufficient residence time. A rotary combustion reactor constantly revolves and tumbles AFs inside it to ensure sufficient retention time at high temperatures (up to 1200°C) for complete burnout. The schematic details of a rotary drum reactor are shown in Figure 5.5.

5.8.4   NOx CONTROL TECHNOLOGIES

In cement kilns, NOx emissions occur by two primary mechanisms (Figure 5.6):

- Oxidation of the molecular nitrogen present in the combustion air which is termed as thermal NOx formation

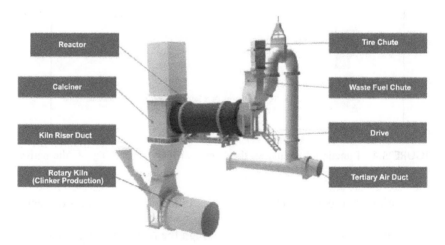

FIGURE 5.5   A schematic diagram of a rotary drum reactor.

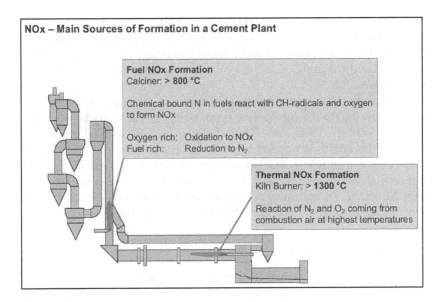

NOx – Main Sources of Formation in a Cement Plant

Fuel NOx Formation
Calciner: > 800 °C

Chemical bound N in fuels react with CH-radicals and oxygen to form NOx

Oxygen rich: Oxidation to NOx
Fuel rich: Reduction to N_2

Thermal NOx Formation
Kiln Burner: > 1300 °C

Reaction of N_2 and O_2 coming from combustion air at highest temperatures

FIGURE 5.6 Primary mechanisms of NOx formation in the pyroprocess.

- Oxidation of the nitrogen compounds present in the fuel, which is termed as fuel NOx formation

Sometimes the raw material feed to the kiln may also contain nitrogen compounds, which may result in the formation of NOx, similar to the fuel NOx formation. Because of the high temperatures involved in clinker burning, the thermal NOx formation happens to be the dominant mechanism. The term NOx includes both NO and NO_2 species, although NO_2 normally accounts for less than 10% of the NOx emissions from a cement kiln exhaust stack [6].

There are four primary types of cement kilns used in the industry: wet kilns, long dry kilns, kilns with a preheater, and kilns with a preheater/calciner. The wet and long dry kilns and some preheater kilns will have only one fuel combustion zone, whereas the modern preheater and calciner kiln designs have two or three fuel combustion zones: kiln burning zone, riser duct, and calciner. Because the typical temperatures present in the combustion zones are different, the factors affecting NOx formation are also somewhat different in different kiln types. In addition to the specific NOx formation mechanisms, the energy efficiency of the cement-making process is also important as it determines the amount of heat input needed to produce a unit quantity of cement. A high thermal efficiency would lead to less consumption of heat and fuel and would generally produce lower NOx emissions [6].

As NOx formation is directly related to fuel combustion, reduction in the amount of fuel burned per unit amount of clinker produced will reduce NOx emissions per unit clinker. Attempts to improve energy efficiency of the process by avoiding excessive clinker burning and utilizing waste heat effectively for preheating combustion air, coal, and raw mix is likely to reduce NOx emissions. The newer preheater and calciner kiln designs provide very efficient preheating and calcining of the raw mix with intimate gas-solids contact in cyclone towers [6].

To reduce the NOx formation, low NOx primary techniques can be applied to the burner and to the combustion systems used in the rotary kilns and calciners, respectively. These low NOx techniques need low capital investments and entail no additional operating cost, and hence are more preferable. If these primary combustion techniques cannot adequately reduce the NOx emissions to the levels necessary for complying with increasingly stringent rules, the selective non-catalytic reaction (SNCR) systems can be installed to meet the most stringent regulations for NOx emissions [6].

5.8.5 Process Instruments

Application of various process instruments is an integral part of the thermal process to obtain data and information from the field of operation. This is an indispensable step for advance process control (APC) and predictive condition monitoring. Some of the special process instruments are described below:

a. **Process Camera (Video and Thermography)**
   Process cameras provide high-quality images from the kiln or cooler. The kiln operator visualizes the condition of kiln and cooler from the images and video output for operational controls. An illustration of the kiln and cooler camera along with thermography is presented in Figure 5.7, which also shows the prediction of free lime based on online spectroscopy and image processing techniques.

b. **Kiln Shell Scanner**
   The operating principle of the kiln shell scanner is based on the contactless measurement and monitoring of kiln shell temperatures by a thermal imaging camera. The kiln shell scanner captures the real-time temperature profile of a rotary kiln surface. A photograph of the

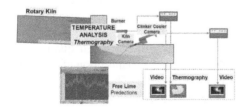

FIGURE 5.7   Kiln and cooler camera and free lime prediction.

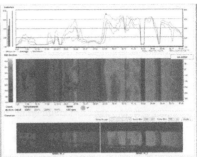

FIGURE 5.8 Kiln shell scanner and thermal images.

camera along with the 2-D and 3-D views of the thermal image is shown in Figure 5.8.

The camera captures the infrared radiation of kiln shell that is converted into an electrical signal, which finally results in the calculated temperature profile across the kiln surface. The temperature map thus generated and further processed with the aid of software is finally presented as a dynamic display in colored images. Multiple benefits can be derived from the kiln shell scanning such as the following:

- Optimizing kiln maintenance
- Minimizing unscheduled downtime and the associated expenditure
- Detecting hot spots due to deteriorating refractory lining and brick fall
- Assessing the coating thickness
- Observing the flame position and shape causing abnormal operating conditions

The system also has the option of monitoring tire slippage and migration.

It is important to note that the role of process camera and kiln shell scanner is increased with the recent developments in the field of image and video processing for prediction of process parameter with the help of AI and machine learning algorithms. Prediction of kiln parameters can be used in APC for the pyro system to improve the operational efficiency and combustion control while enabling safe compliance with pollution minimizing protocols. It helps in detecting snowmen, red rivers, and other anomalies and makes routine plant maintenance and shutdowns more predictable and less disruptive to the process.

c. Gas Analyzers

Analysis of gases generated in the cement process gives one of the most important information for optimizing the fuel consumption,

TABLE 5.7
Details of Gas Analysis

Measurement Location	Control Action	Gas Compositions Measured
Kiln inlet	• Optimizing the primary firing • Reducing the fuel consumption • Maintaining the clinker quality	CO, O_2, NO, CO_2, CH_4, SO_2
Calciner	• Optimizing the secondary firing • Reducing the fuel consumption	CO, O_2
Preheater outlet	• Implementing safety measures • Preventing explosion in ESP • Controlling false air in preheater	CO, O_2
Cooler stack	• Emission monitoring	CO, NO_x, SO_2, O_2, CO_2, HCI, VOC, HF
Kiln Stack	• Emission monitoring	CO, O_2
Coal bin	• Safety measurement • Prevention of smoldering (due to air ingress)	CO, O_2
Coal mill	• Safety measurement • Prevention of smoldering (due to air ingress)	CO, O_2

achieving the clinker quality, and safeguarding the working environment. Parameters measured, corresponding locations, and control actions are summarized in Table 5.7.

5.9 CONVENTIONAL APPROACHES FOR PROCESS OPTIMIZATION

The process optimizers are being used in the cement industries for more than the last two decades. These systems are based on either fuzzy logic or model predictive control (MPC), as discussed in Chapter 3 of this book. Some of the salient aspects of these systems are presented in the following sections in the context of this chapter.

5.9.1 FUZZY LOGIC CONTROL PHILOSOPHY

Usually the human thinking is vague. For example, one may find such qualitative descriptions as "The interest is rather great" or "The burning zone is not particularly hot". Yet the work goes on quite successfully in many cases. But in process control the conventional PLC technology is based upon exact values of a parameter, which follow the binary logic. The binary logic compares a measured value with a reference value and check whether the measured value is more or less than the reference. For example, if a person defined to be tall only when his height is 1.80 m, a person of 1.79 m height will be

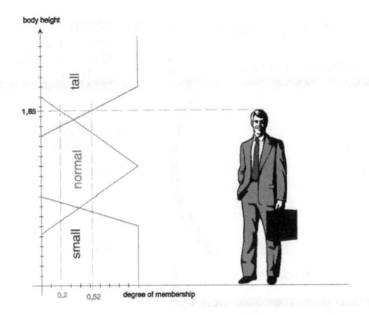

FIGURE 5.9   Fuzzification of body height.

considered "NOT tall". This constraint of PLC-based logic system has led to the application of Fuzzy logic, which eliminates this jumping effect.

A Fuzzy Logic Controller module allows the use of simple IF-THEN rules in order to map the processes that could not otherwise be expressed by mathematical equations. Complex systems which require multiple inputs and multiple outputs for decision making can be modeled using fuzzy logic. Each fuzzy block consists of a set of input and output variables. Each variable has a set of membership functions associated with it. Sets of rules (IF-THEN statements) are entered to simulate the decision-making process. This technology is used for various decision making and root-cause analysis of a process. In Fuzzy logic application, the measured value "body height 1.85 m" could have different degrees of membership, for instance, with a classification value of 0.52 to be classified as "tall" or with a value of 0.2 to be classified as "normal" as shown in Figure 5.9.

The detailed structure of a Fuzzy controller is shown in Figure 5.10. Raw signals from a process are first converted from crisp quantities to fuzzy quantities. Thereafter, the fuzzy quantities are processed by fuzzy rules (IF-THEN) to generate Fuzzy output. Subsequently the output will again be defuzzified in the final stage to convert to analog values. The application of Fuzzy rules may be exemplified as follows: "IF the temperature in the burning zone of a kiln is too high and IF the O_2 content at the kiln inlet is too low with respect to the respective pre-set values, THEN reduce the fuel feed".

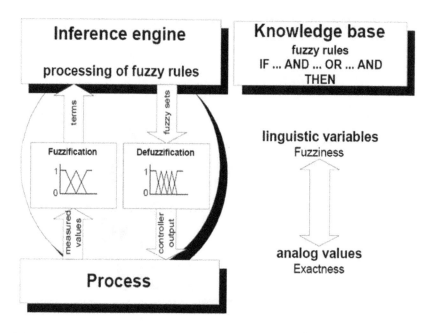

FIGURE 5.10   Functioning of a fuzzy controller.

5.9.2   MODEL PREDICTIVE CONTROL

MPC is often used to control and optimize industrial processes. Usually, the use of these techniques for real plants includes the development of mathematical models describing the process, and the selection/design of a suitable cost function, which takes into account the goals to achieve. The goal can be either economic or parameter based, i.e., it can be a profit-solving or error-reducing function.

MPC uses a model of the system to make predictions about the system's future behavior. MPC solves an online optimization algorithm to find the optimal control action that drives the predicted output to the reference. MPC can handle multi-input multi-output systems that may have interactions between their inputs and outputs. It can also handle input and output constraints. MPC has preview capability; it can incorporate future reference information into the control problem to improve controller performance. MPC can also handle process delay inherently.

The main idea behind MPC is better understood if it is compared with a chess game: at any time the player looks at the position in the board, calculates several sequences of moves ahead, and chooses the best one according to his evaluation of the resulting positions. The first move of the best sequence is applied. The cycle repeats once the opponent (resultant event) makes his move.

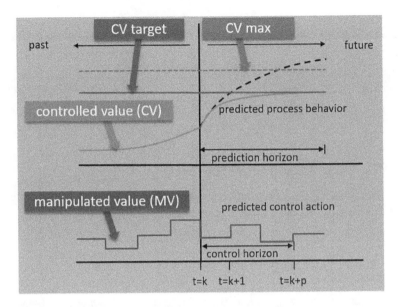

FIGURE 5.11   Principles of Model Predictive Control (MPC).

As in a chess game, when using MPC, a sequence of future optimal control actions is chosen according to a prediction of the evolution of the system. Each sequence is computed by means of an optimization procedure consisting of the performance improvement and protection from constraint violation. The principle steps in MPC are schematically shown in Figure 5.11, which can be further amplified as follows:

1: At time t = k, solve the open-loop process behavior within the prediction horizon.
2: Apply the optimal control actions within the control horizon to reach the target within the prediction horizon.
3: Obtain new measurements and update the process model and solve in the same procedure within the control horizon.

The MPC prediction trend has been illustrated in Figure 5.12 with the help of oxygen content in the preheater exit gases.

5.9.3 LIMITATIONS OF CONVENTIONAL AUTOMATION SYSTEMS

Most of the cement plants operate with a reasonable automation level with inter-connected production sections. The current technology solutions in automation establish a predefined hierarchy where the connectivity levels within the plant periphery exist. The sensors measure the process state and the software controls and optimizes the process. Although PLC, DCS, and

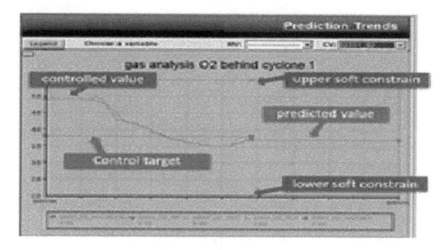

FIGURE 5.12   MPC prediction trend.

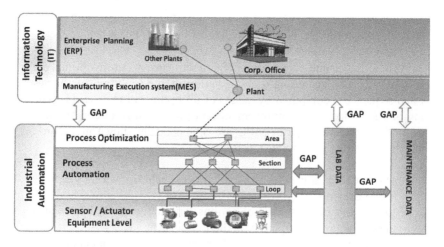

FIGURE 5.13   Limitation of Industry 3.0 automation systems.

IT-based solutions are used in conventional systems, still there is a gap between automation and IT systems as shown in Figure 5.13. As a result of inadequate integration of automation and IT, it is not possible to use the available data to optimize the process and to ensure quality and reliability of the cement plant in the best possible way.

As previously stated, most of the available pyro optimization systems are based on rules, fuzzy logic, MPCs, and knowledge-based models. The main idea of a rule-based system is to capture the knowledge of a human expert in a specialized domain and embody it within a computer system. Given the necessity of online prediction and online response capability of such systems, the total human knowledge could not be implemented into such rule-based systems yet. Lack of computational power and heterogeneity of physico-chemical phenomena do not allow such systems to operate fast enough to cover the whole width of complexity especially for coprocessing of AFs. The experience gained with these systems is that they were not sufficiently robust and precise. Often, when a situation other than the pre-set problems occurs, the systems get stuck and fail to control or assist further. The major limitation of the optimization software is that they work for few week or months, but after some time the system drifts apart or stops functioning with the changing conditions and requirements like the quality of the raw material, fuel, or technology.

5.10 IMPLEMENTATION PLAN FOR INDUSTRY 4.0 TOOLS IN CEMENT PLANTS

It is widely acknowledged that the recent industrial progress is primarily shaped and driven by the advancement in the information technology, which supports and enables better digitalization in plants. In Industry 4.0-based manufacturing, the fusion of real and virtual worlds will make a cyber-physical system, which may have a huge impact on every element of the manufacturing process due to the following advancements:

- Smart sensors and industrial IoT (Internet of Things) will allow real-time data collection during production processes.
- Fast broadband internet connection will facilitate the transmission of large amounts of data between people, machines, and plant sites.
- Cloud computing will allow the storage of big data and make it available at any location.
- Big data analytics will permit collaborative processing of huge volumes of data.

In comparison to Industry 3.0, the modern industry needs unprecedented degree of integration between information, communication and manufacturing systems at their disposal. Comparison to the development of the autonomous driving car reveals that the amount of data to be acquired, handled, and manipulated can easily turn out to be as much as 1000 times higher than in today's conventional automation approach. The additional data will have to be acquired by additional sensors and inter-communication among the several units. However, this will still not be sufficient for a complete autonomously operated cement plant, given the complexity of especially AF-coprocessing in a cement plant. The multitude of additionally required measurement sensors as well as their inherent inability to measure time-inconsistent, fluctuating (time and space), and sometimes even hidden process data will not be technical and economically feasible through hardware

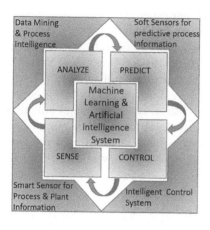

Data Mining & Process Intelligence

Soft Sensors for predictive process information

ANALYZE   PREDICT

Machine Learning & Artificial intelligence System

SENSE   CONTROL

Smart Sensor for Process & Plant Information

Intelligent Control System

- Installation of IOT based smart sensors for predictive and condition based monitoring of machine for reliable operation.
- Installation of special process instruments for reliable process information.
- Installation of Cross belt analyzer for stock pile and Raw material control.
- Installation of Automatic sampler and Robo Lab for Integrated Quality control
- Integration of all plant Data to cloudserver.
- Data management system based on AI and machine learning tools for prediction of condition of machines, development of soft sensor for quality and process values.
- Advance optimizers for Integrated Process and Quality control with help of Artificial intelligence and machine learning tools.
- Testing of process optimizer on digital twins (Cement plant simulation system)
- Collaborative operation by subject matter experts

FIGURE 5.14   Data-driven (industry 4.0) process optimization concept.

sensors. For that reason the concept of digital twins will allow to operate a digital twin of the plant in parallel to it in real time. This digital twin will generate and avail all those inherent transient, 3D, and internally hidden data for fast and effective modeling and control.

Cement companies are moving in carefully toward digital transformation by starting pilot projects to verify and implement the requirements to sense the benefits including economic viability. Data is the key for implementation of Industry 4.0. The data-driven plant control and optimization have four pillars, i.e. sense, analyze, predict, and control as shown in Figure 5.14.

Digital transformation begins with the evaluation of current sensor infra-structure and identification to unleash the latent needs of each functional department in the plant: operation, process, quality, maintenance, reliability, and safety. Based on the findings, the additional IIOT-based sensor, special process instrument and quality control system shall be installed to generate the needed data. The second step will be the integration of data on web cloud. Finally, the digital twin technology will enable access to the remaining data world which was not accessible through conventional hardware sensors and data communication. The system configuration of a cement plant under the Industry 4.0 standard is displayed in Figure 5.15.

A comprehensive cyber-security plan should be made to connect the process control network with the corporate network and the outside world via the web cloud. Clouds are nothing but a virtually unlimited number of servers with massive computing and storage capacity connected to the internet. The AI and machine learning software "apps" are used to analyze the raw data from the sensors in the plant. One of the big benefits of the cloud is its ability to combine data from the large numbers of different machines, even from the machines dispersed around the globe. The more data you can amass, the more

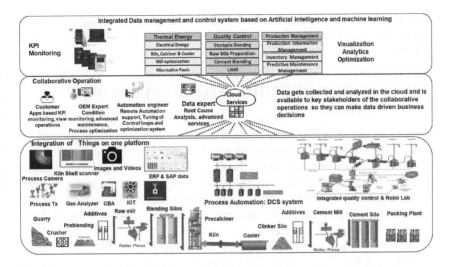

FIGURE 5.15 System configuration of cement plant based on Industry 4.0 [7].

precise machine learning algorithms can be in predicting process, quality, and maintenance needs and other actions. The software apps are specialized for the task at hand, such as monitoring the condition and performance of the equipment. The beauty of cloud computing is that the server machines need not be physically present in the plant, and they can be accessed from anywhere in the world. The basic building blocks of industry 4.0 are further explained in subsequent sections.

- Installation of IOT-based smart sensors for predictive and condition-based monitoring of machine for reliable operation
- Installation of special process instruments for reliable process information
- Installation of cross belt analyzer for stock pile and raw material control
- Installation of automatic sampler and Robo-Lab for Integrated Quality control
- Integration of all plant data to cloud server
- Data management system based on AI and machine learning tools for prediction of condition of machines, development of soft sensor for quality, and process values
- Advance optimizers for Integrated Process and quality control with help of AI and machine learning tools
- Testing of process optimizer on digital twins (Cement plant simulation system)
- Collaborative operation by subject matter experts

It may be relevant to mention here that the basics of and advancements in digital twins and smart sensors are discussed in Chapter 8 and Chapter 9 of this book, respectively.

5.11 INTEGRATED ROBOTIC LABORATORY FOR QUALITY CONTROL

Increasing use of AFR in the thermal process may affect the quality of the end product. Hence, it is essential to acquire promptly the quality data within the framework of process and production control. The integrated robotic lab for quality control is an important infrastructure for this purpose. Advances in representative sampling, transport, preparation, analysis, and optimization systems now make it possible to capture any deviations in standards and ensure reliable quality control.

A typical layout and system configuration of a robotic laboratory (Robo-Lab) is shown in Figure 5.16. The random or average samples are extracted from the material flow by automatic sampling devices and are transported to the plant laboratory via pneumatic tube mail. The sample is received at the receiving station. The sample preparation and analysis equipment are arranged in a circular configuration around a robot or rotating arm. The sample preparation for analysis is executed by the robot. A fully-automatic lab is very flexible and can use a variety of analyzing equipment such as X-ray fluorescence spectrometer (XRF), X-ray diffractometer (XRD), particle size analyzers, etc.

Figure 5.17 shows the raw material quality control loop and the key features of a typical controller. X-ray analyzers and cross belt analyzers transfer the analyzed data to a raw material controller, which directly adjusts the mixture proportion of the raw material components. The raw material controller

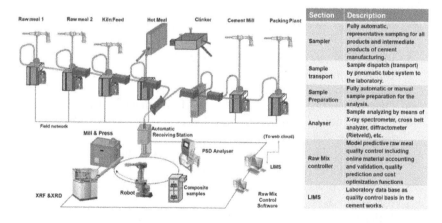

FIGURE 5.16 Typical layout and system configuration of a Robo-Lab.

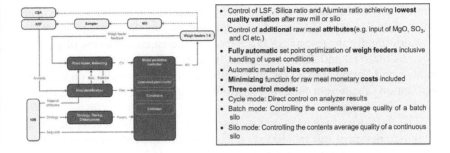

FIGURE 5.17 Raw material quality control loop and key features.

controls the raw material quality parameters (LSF, SM, and AM) by changing the proportionate ratio of weigh feeders as per the target and measure the value of quality parameter and integrated tonnage of material components.

The control mode can be selected on the basis of the required quality so as to minimize the variation of the kiln feed analysis.

- Cycle mode: Direct control based on the analyzer results
- Batch mode: Controlling the average quality of contents of a batch silo
- Silo mode: Controlling the average quality of contents in a continuous silo

Raw material controller is based on the MPC. It supports all the new generation analysis methods like online X-ray and online cross belt analyzers for raw material proportional control of raw mill and stock piles. It includes online material accounting and validation, quality prediction, and cost optimization functions.

Auto sampling can be configured in different process sections of a cement plant such as the limestone crusher, raw mill outlet, calciner for hot meal, kiln feed, clinker, and cement. Recent developments in the field of online analysis by various manufacturers have reduced the time lag in the measurement and analysis. These analyzers can link to Integrated Quality Management Server via field bus for further analysis and control.

Hot meal sampler and Loss-on-Ignition analyzer are used to collect sample from calciner to determine the degree of calcination of hot meal before it enters the kiln. The degree of calcination is an indication of how efficient the preheater and pre-calciner is being operated. This indication can be used to balance the fuel between the main burner and the pre-calciner and in the use of AFs [8].

Clinker sampler and Free Lime Analyzer: Free lime is a critical parameter for the optimization of a kiln and to ensure that good quality clinker is produced. Having frequent and accurate information on the free lime content of clinker will enable the operator to control the kiln in short intervals. This short-interval control will avoid over-cooking and under-cooking clinker [8].

Additionally, the role of XRD analysis has increased due to use of alternative raw material and AFs and their subsequent effects on the quality of the clinker and hot meal. XRD is effective for qualitative as well as quantitative estimation of different mineral constituents and phases, encountered during the thermal process.

The data-driven advance quality control system can be developed for the optimum design of raw mix based on the impact of used alternative raw material and AFs. In practice, a host of benefits can be derived from an automated laboratory as mentioned below:

- Frequent and quick data collection for decision making
- Automatic and representative sampling with reduced human error
- Automatic transport of samples to the laboratory
- Direct, frequent, and timely transfer of data to central control room
- Reduction in standard deviation in the process and in the end product via adaptive control of material composition
- Reduction in cost of raw material through optimum usage
- Optimization in thermal and electrical energy efficiency of the plant
- Increased laboratory efficiency and transparency in operation
- Better cost/quality relationship through integration of quality data with process data
- Reduced man power deployment for data collection, analysis and processing, and reporting
- Improvement in work place quality.

5.12   ADVANCED PROCESS CONTROL SYSTEMS BASED ON ARTIFICIAL INTELLIGENCE

AI is a collective term for a concept which strives to transfer patterns of human intelligence to machines. Although the concept itself is under intense discussion, basically two main branches of AI are being differentiated: strong AI and weak AI. While strong AI refers to behavioral patterns which would allow the machines to intelligently act completely on arm's length with humans in complex analytical and decision processes, weak AI refers to partial objectives of that, mostly focusing on the ability of analytical skills in complex and probabilistic situations. While strong AI still awaits breakthrough, the advances in weak AI are already very much operative in many applications. The backbone of these applications is machine learning and especially its most elaborate form called deep learning. Technically deep learning is based on artificial neural networks, which are a digitalized copy of the neural nets of the human brain. Made up of thousands of artificial neurons, the functional base unit comprising weighting input factoring, summation, and threshold functions to convert an input signal into an output signal, a neural network mathematically forms a system of mathematical equations comparable to a very complex multi-variable regression. Applying a know-how driven form of teaching, these mathematical systems can represent the behavior of the real system they digitally represent. In our application they can be taught to behave like a cement production plant or a part of such plant, and can then be used to analyze the behavior of that system. Being able to analyze that system and behaving like it, the next option to apply these neural networks is to predict the behavior of the real system (in our case the cement kiln) when certain changes are being introduced to its operation. Embedded into an appropriate automation system, these deep learning technologies can form base for a very powerful online optimization system.

Systems based on machine learning can very effectively identify and master automation tasks which they have been taught on. The weak spot of these systems is however that they cannot cope with unprecedented patterns in general. That is why the focus of present R&D work is on how these systems can also be equipped with a self-adaptation or self-learning ability.

So far these machine learning based APC will benefit from being complemented with rule-based APC systems. This combination represents to some extent the human behavior of information acquisition and decision making: There is one system which can handle very fast the situations which are known (the data-driven part), and another system which can think forward in order to solve those incidents which have not yet been experienced before (the rule-driven part).

The composite control structure of a system which applies different kinds of technologies for multiple purposes is shown in Figure 5.18. It comprises the following subdomains:

1. General AI to communicate with the process
2. Knowledge representation to act as its memory
3. Automated reasoning to use the stored information to answer questions and draw new conclusions
4. MPC for constrain satisfaction
5. Machine learning to detect patterns and adapt to new circumstances

It is important to understand the functioning of the various subdomains of AI and how these domains could be applied to enhance the result. Better results can be achieved by combining the complementary skills of these domains. For example, MPC is ideally constructed for a constrained optimization control which requires an accurate dynamic plant model whereas by using the ML an accurate plant model can be created so the ML and MPC architectures should be integrated to take advantage of both the algorithms to provide a robust control concept. In this way the best of these subdomains shall be used to achieve the optimum result.

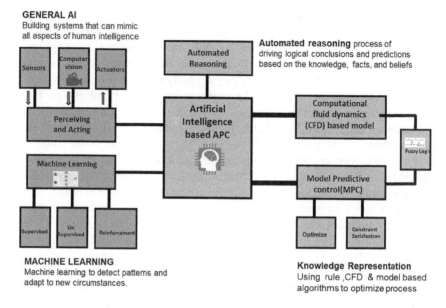

FIGURE 5.18   Building block of artificial intelligence. [8].

The AI subdomains are further elaborated in the following sections [9]:

a. **General AI (Perceiving and Acting)** systems mimic the human intelligence. It uses sensors, instruments, and computer vision to understand the status and actuator to take actions. Once a system understands what the sensors and images intend to communicate, it responds accordingly.

b. **Automatic Reasoning** is defined as the process of driving logical conclusions and predictions based on the knowledge, facts, and beliefs that are already available. It is a general process of thinking rationally to draw insights and conclusions from the data in hand. It is essential and crucial in AI so that machines can learn and think as rationally as a human brain. Developing reasoning within AI leads to the human-like performance of the machine.

c. **Machine Learning and Deep Learning** is the most important subset of AI which defines the ability to learn from experience, rather than just instructions. It identifies patterns and analyzes past data to infer the meaning of these data points to reach a possible conclusion. The ML also has various subsets like Supervised Machine Learning, Unsupervised Machine learning, and Reinforcement learning, etc. Supervised Machine Learning is a machine learning method in which models are trained using labeled (known) data. The algorithm can compare its predicted output with the known labeled output to modify its model to be more accurate. Unsupervised learning is another machine learning method in which patterns are inferred from the unlabeled input data. The goal of unsupervised learning is to find the structure and patterns from the input data. Unsupervised learning does not need any supervision. Instead, it finds patterns from the data on its own. A key benefit of this type of training is that we do not introduce our own human biases into the model. Reinforcement learning is dependent on the algorithms environment. The algorithm learns by interacting with the given data sets, and through a trial and error process tries to discover "rewards"' and "penalties" that are set by the programmer. The algorithm tends to move toward maximizing these rewards, which in turn provide the desired output.

Machine learning methods can be used to develop the data-driven models as shown in Figure 5.19. Data preparation is the process of reading data for training, testing, and implementation of an algorithm. It's a multi-step process that involves data capture, collection, cleaning and pre-processing, feature engineering, and labeling. These steps play an important role in improving the overall quality of the machine learning model. Machine learning requires a lot of training data (either labeled-supervised learning or not labeled-unsupervised learning).

FIGURE 5.19 AI- and ML-based model development steps.

Now, selection of the right model is the next step. There are many models that can be used for many different purposes so the programmer shall select a suitable model according to the need of the process.

The artificial neural networks based model works on similar principles as of human neural cells. They are a series of algorithms that capture the relationship between various underlying variables and process the data as a human brain does. The artificial neural networks provide a powerful and robust means to recognize a pattern and derive a relationship from the available data. Due to the availability of more data and more powerful CPU and GPU hardware, the modern ANN can use larger numbers of hidden layers with many more nodes in each layer. Therefore, the performance of modern ANN is substantially improved in comparison to the traditional ANNs, which were able to afford only one or two hidden layers owing to the limitation of computer hardware in the early days. ANN has self-learning capability and can be used as a soft-sensor to predict process parameters. ANN are useful for predicting any process variable's value, which nonlinearly depends on the other parameters.

Training process of the model is the main part of machine learning. The objective is to use training data and incrementally improve the predictions of the model. After training the model, the next step is to evaluate the model. The model is tested on independent data set for validating how well it will generalize with new unknown data. After evaluating the model, one should test the originally set parameters to improve the model. Increasing the number of training cycles can lead to more accurate results. Finally, the model should be checked for the prediction of process value against the real measured value.

5.13   AI-BASED APC FOR THERMAL PROCESS

The various suppliers have tried to use conventional APC like Fuzzy logic and MPC control concept to optimize the thermal process. Normally, conventional APC takes the decision based on the actual measurements and try to find the optimal set points by changing them in very small steps. This works well for certain, well-defined processes, but a thermal process is quite complex to control due to various unknown variables and changing process conditions, which results in a drift of APC after some time. The result is that most of these APC systems are out of service or require intensive fine-tuning and remodeling. The thermal process needs a more competent advanced process control system to handle this challenging situation more efficiently [10].

This is where the AI-based model makes a big difference because it uses both historical and real-time data and recognizes patterns in these trends. Based on historical data, the AI module compares the actual trends with the trends from the past and predicts what will happen in the future if the process continues to run with the same set points. Currently, the process may seem to run flawlessly but the AI module predicts the future situation, which might be any disturbance or inefficient operation. The AI module will proactively take timely corrective actions by adjusting the set point to the optimum level to achieve desired results [10].

As described in aforesaid selections, the AI-based APC will use a hybrid approach that draws on computer vision, data-driven models and knowledge-based models (Fuzzy/MPC/CFD) to generate predictions and prescribe actions that are more accurate.

The key step in developing an AI-based APC is to consider all influencing variables, constraints, and objectives in the design. The complete thermal process shall be divided into subparts in the form of control modules, e.g. kiln, calciner, and cooler. The next step is selecting the right combination of algorithm for each control module. Each algorithm (AI/ML, Fuzzy logic, MPC, CFD) has its own advantages and limitations. Therefore, different control module may need different combination of algorithms. This design and development process of AI-based APC shall be accompanied by both process experts and IT experts to develop the most suitable algorithm, which will be able to serve its best to optimize the thermal process and maximizing the utilization of the AFs in a cost-effective manner. It is also equally important to identify a bad process equipment or bad sensors, as in this type of situation even the most advanced process control system can not necessarily provide the expected results. This AI-based APC system shall be tested and fine-tuned before implementing in the real cement plant with the help of process experts and modern tools like digital twins. A structured and methodical approach to facilitate optimization for the development of the competent and robust APC solution for the thermal process is displayed in Figure 5.20.

The objectives of AI-based APC for the pyro system are as follows:

- Quality assurance: Enhance production quality and consistency
- Increase profitability
  - Maximization of plant output
  - Minimization of energy consumption and product waste
  - Maximization of use of AFR
- High reliability: Reduction of downtime
- Thermal Energy management:
  - Reduce preheater exit gas temperature
  - Operate at optimal low oxygen levels
  - Optimize burner flame shape and temperature
  - Minimize dust in exhaust gases by minimizing gas turbulence:
  - Lower clinker discharge temperature, retaining more heat within the pyroprocessing system
  - Lower clinker cooler stack temperature
  - Recycle excess cooler air
- Ensure of operational reliability

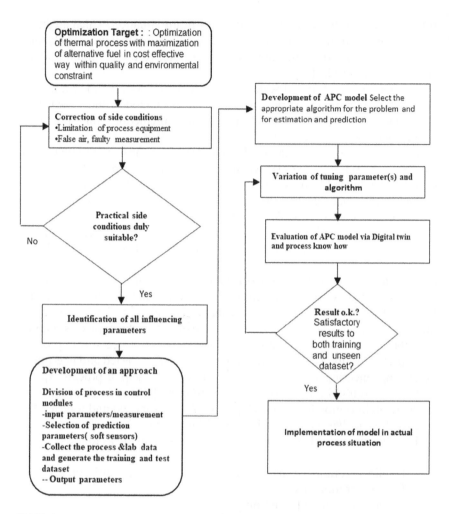

FIGURE 5.20 Structured and methodical approach to facilitate optimization.

- Environmental protection: Reduction of emission
- Improvement of operator's workload

The AI-based APC system will first stabilize the process by reducing variation and then it will move the set points close to constraint as explained in Figure 5.21.

As stated above, the process optimization is divided into control modules, i.e. kiln, calciner, and cooler. These control modules are controlled in the control strategy based on the priority management. The details of standard control modules for the pyroprocess along with the list of measured and manipulated variables are shown in Figure 5.22. The additional control

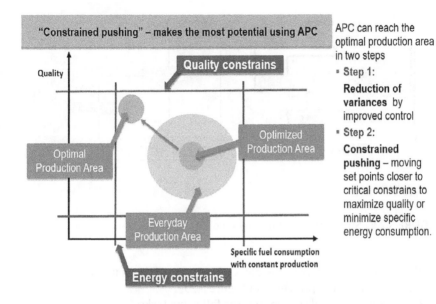

FIGURE 5.21 Implementation steps of modern AI-based APC.

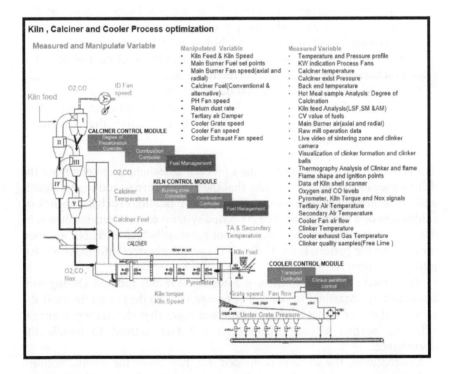

FIGURE 5.22 Control modules for the pyroprocess along with measured and manipulated variables.

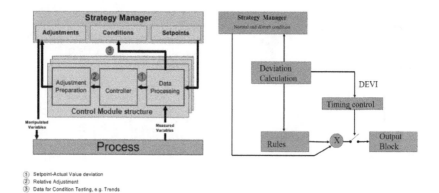

① Setpoint-Actual Value deviation
② Relative Adjustment
③ Data for Condition Testing, e.g. Trends

FIGURE 5.23 Structure of a control module.

modules can be prepared on the basis of specific requirements. The structure of a control module is presented in Figure 5.23.

Every control module carries out the following tasks:

- Data processing for the control of the concerned sub-processes and the necessary measured data, e.g. equalization of measured values, determination of real-time and historical data trends to calculate predictive process values (soft sensors), calculation of control deviation between set values and predictive values
- Determining the control action based on deviations
- Preparing first the relative adjustment requirements, e.g. through rounding off, then limiting, and finally calculating the absolute adjustment requirements

The AI-based APC uses various data processing techniques to eliminate the disturbance of signal and to ensure that the models correspond to real signals and not to noise. The data processing step has limitations such as the lack of calibration or a noisy signal. The AI-based systems can use various means of data processing so that such data are not necessarily "rejected" but are used for learning the pattern from it. This is the way the correct information is fed into the learning AI.

The strategy manager consists of a condition testing module. During condition testing, measured values will be compared with the pre-set limits. If the result of the limit values is within the pre-set limits then the strategy manager will take normal action, otherwise it will take actions to handle the disturbance.

Conceptually, the strategy is divided into two parts: the normal actions, which occur on a time cycle, and disturbance actions that are triggered by a specific event.

Disturbance actions are a prioritized set of controller modes that handle specific abnormal kiln conditions such as high CO, ring breakaways, hot kiln, raw mill start/stop, etc.

The core of the control strategy calculates the new set points necessary to control manipulated variables, e.g. kiln feed, kiln energy, fan speed, and kiln speed based on the measured inputs.

5.13.1 KILN CONTROL MODULE

The primary objective for the kiln control module is to produce a stable and desired quality of clinker along with minimum specific energy consumption and maximum production. The challenge of kiln optimization comes with the necessity to adjust heat input and retention time in the sintering zone constantly, happening due to the following reasons:

- Changing recuperation of heat from the cooler into the kiln
- Changing fuel properties
- Changing degree of calcination
- Changing chemical composition of the raw meal
- Changing product quality parameters
- Changing coating situations in the kiln
- "unpredictable" equipment failures

The main part of this module is to control the burning zone, combustion in kiln, fuel feed, and coordination with other modules as discussed below:

a. **Burning Zone Controller**

The objective of burning zone controller is to adhere to pre-set clinker quality (free lime content) by optimizing burning zone conditions, ensuring a sufficiently high burning of the kiln charge (clinker).

The variable "burning zone condition" is a target value which is determined from several measured variables (condition attribute), being used as measure for the heat available in the heating zone. The burning zone condition is estimated from the measurements like pyrometer, kiln torque, NOx, preheater outlet temperature, kiln camera image, and kiln shell scanner map.

b. **Burning Zone Combustion Controller**

The objective of the combustion controller in the burning zone is to adjust an optimum air volume in the kiln. The oxygen concentration of the kiln waste gas (i.e. oxygen concentration in the feed end chamber) is referred for the assessment of the combustion situation in the kiln. The 0_2 concentration shall be adjusted to a range of 1% to 2%, as the kiln operation with higher values will result in the following:

- Shifting the clinkering zone in the direction of the kiln inlet
- Increasing the gas temperature at the kiln inlet
- Cooling of the clinkering zone
- Higher dust load of the kiln waste gas (high gas velocity)
- Increased specific fuel demand

Kiln operation with reducing atmosphere, i.e. with formation of carbon monoxide (CO), must be avoided under all circumstances as the incomplete combustion of fuels may lead to the following dire consequences:

- Uncontrolled secondary combustion or even explosions (in dust collector)
- Detrimental effects in the burning process due to generation of less heat
- Increased formation of accretions near the feed-end chamber and the gas riser pipes
- Increase in the specific fuel demand

With the above tasks of the burning zone and combustion controller, the kiln control module automatically adjusts the fuel mix, the fresh feed rate, the kiln rotational speed, and the kiln fans to reach the required thermodynamic conditions, contributing to outstanding levels of process stability, clinker quality, and environmental compliance. The kiln control module includes functions that control the mixing of AFs to ensure consistent combustion conditions and heat distribution along the kiln to reduce the instabilities caused by the varying properties and difficult dosing of different fuels. Kiln control module also calculates targets for the calciner control module including an optimized calciner temperature target. Furthermore, the kiln control module will adjust the fuel mix to reach the required levels of oxygen in the calciner and the burning zone.

5.13.2 Calciner Module with Alternative Fuels Controller

The goal for the calciner control module is to achieve and maintain a stable and high degree of calcination. Generally, it is more important to keep it to the optimized level consistently, as any kind of fluctuation will cause a swing in the thermal process. The objective of the AFs control module is to maximize the utilization of AFs in a cost-effective manner without losing the quality of the final product.

The design of the AFs control system is quite challenging due to the following mixing and controlling problems:

1. Generally the real-time measurement of heat value, Na, Cl, S, and lump size and moisture of the AFs are unknown, though this information is necessary to control the process accurately.

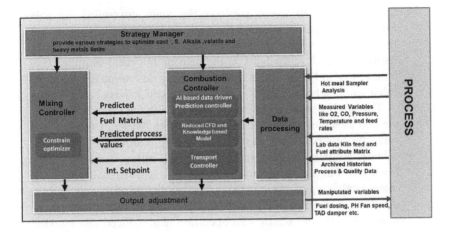

FIGURE 5.24 Structure of alternative fuel control module.

2. Measurement of the important quality parameters is discontinuous and is carried out at low frequency.
3. Transportation of AFs, their subsequent analysis, and the calcination process are not synchronous with substantial dead time, which result in difficulties of modeling the thermal process.
4. Process restrictions such as specific heat consumption, cement quality, environmental impact, volatile circuits (SO_3, Cl, alkalis in hot meal), ash absorption in clinker, and combustion properties of AFs.

Figure 5.24 shows the structure of AF control module. It will consist of the combustion controller and the fuel mixing controller. Calculation of the manipulated variables will be done on the basis of the process and quality data. Combustion controller is designed to maintain the calciner temperature, oxygen, and CO at suitable levels, while maximizing the usage of AFs. This block includes transport controller and predictive controller based on combination of AI- and knowledge-based algorithm to calculate optimum set point for calciner temperature and oxygen level. This mixing controller does the mixing of AFs to ensure optimum burning and stable operation of the thermal process.

The functions and constraints of various control blocks, described above, need further elaboration because of their multiplicity and complexity. The following are the important features.

a. **Calciner Combustion Controller**

As already stated, the combustion controller is designed to maintain the temperature of the calciner, oxygen, and CO at suitable levels. Chemical analysis of the hot meal is typically made once in 8 hours due to the sampling limitation of the hot meal. The calciner temperature is therefore used as the main controlled variable. Furthermore, combustion of the fuel needs to be guaranteed, and therefore oxygen level needs to be maintained above the predefined limits. Recent developments in the field of hot meal sampling and online "loss in Ignition" analyzer can help to determine the degree of calcination on hot meal before it enters the kiln at every half an hour.

The combustion controller coordinates with the kiln control module and uses the hot meal analysis along with oxygen, CO level, and other process parameters to find out the optimum set point for the calciner temperature to achieve the desired calcination level of the hot meal. Falling temperature level of the calciner will mean less intense calcining of the raw meal. The burning process may eventually be disturbed to such an extent that larger corrections will be required to restore steady kiln operation (e.g. reduced raw material feeding, increasing the fuel rate to the kiln burner). Exceeding the optimal temperature level may result in unwanted intense calcining. Too strong calcined raw meal assumes poor flow properties and tends to stick which may clog the meal ducts and cyclones.

The control strategy should first stabilize the process parameter to control the unknown disturbances and then gradually optimize the process parameters such as the calciner temperature as shown in Figure 5.25. It will be very helpful to optimize the AF in the calciner.

b. **Transport Controller to Handle Time Lags**
Normally primary fuels are transported via pneumatic means to the calciner and have fast and reliable dosing system. On the other hand,

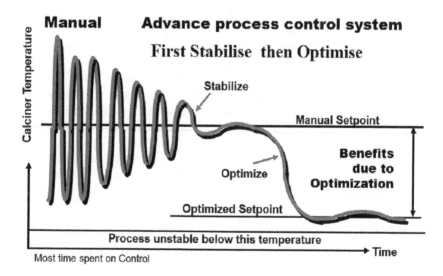

FIGURE 5.25 APC approach: first stabilize then optimize.

AFs are transported via conveyor belts. This dynamic behavior depends on the conveyors, feeding systems, the measuring devices, sampling, the sample preparation, and the thermal process of the calciner. These fuels have a transport delay up to many minutes before they reach the calciner. The control strategy of the fuel management shall include transport control in addition to the combustion control. The transport control shall model different transport delays of each of the feeders. Different feeder types are modeled according to the different characteristics [11]. The control features of the transport controller to eliminate the time lag or delay in conveying of fuels, sample preparation, and analysis are shown in Figure 5.26.

c. **AI-Based Prediction Controller**

It is difficult to measure the heat value and other attributes of the AFs at frequent intervals. Data-driven concepts shall be used to develop process indicators or so called "soft sensors". These process-dependent rules, which are designed or "trained" for each specific application, use all the available information to produce reliable signals of variables which are difficult or impossible to measure at high sampling rates. The historical archived operational and real-time data shall be used to establish and diagnose the relationship of variables.

Prediction of the attributes of the AFs shall be done on the basis of the initial analysis and process effects of the degree of calcination, calciner temperature, O_2, CO level, and pressure and temperature

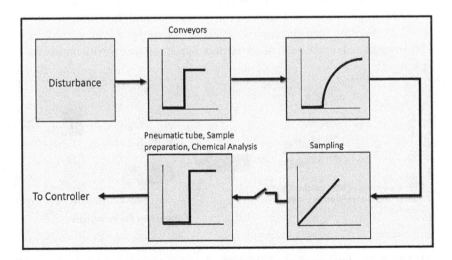

FIGURE 5.26 Control features of the transport controller [10].

profile across the various components of thermal process. With these results, it is possible to control the system even more precisely in order to get a more constant calcination of hot meal.

d. **CFD-Based Motivated Physical Model**

Although 10,000 to 30,000 signals are captured in the modern cement manufacturing process, in some locations the quantum of signals is grossly inadequate to provide a clear process picture. For example, the calciner is provided with only one sensor for temperature measurement, while the CFD (computational fluid dynamics) study of the calciner shows that temperatures vary sharply in the three-dimensional space. Hence, CFD-based motivated physical modeling of calciner and rotary kiln may be used to generate important inner process values (degree of calcination, gas flow rate, and free lime content) for a better controllability of the process. Integration of CFD-based modeling into the APC system is shown in Figure 5.27 [12].

It is important that an AF controller combines a data-driven AI-based model with a process model consisting of a real-time CFD model so that the advantages of the two modeling strategies are utilized and interlinked. The strength of the AI algorithm in the form of rapid and real-data-based predictive capability is contrasted with its weakness that lies in the fact that a machine learned model cannot recognize plant conditions for which it has not been trained. This is where the complementary strength of the deterministic CFD-based process model steps in to handle the unknown operating conditions reliably [11].

e. **Mixing Controller**: Implementation of Constrain Management and Mixing Strategy for Optimization of the AFs and Raw Mix.

Quantity of fuels required, and their mixing proportion, shall be calculated on the basis of the required product quality with improved

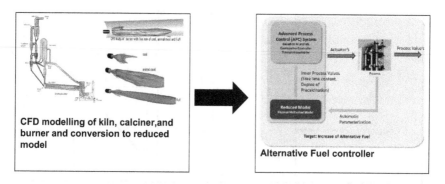

FIGURE 5.27 Integration of CFD-based physical motivated modeling into the APC system [12].

specific heat. To maximize the use of AF, careful consideration of material characteristics is needed as it impacts the quality of the product, environment, cost of raw material, and fuels etc. Alternative raw materials and fuels must be used in quantities and proportions with other raw materials in order to achieve the desired balance of material composition in the kiln product clinker.

The following parameters shall be considered to maximize the use of AFR:

1. Clinker Quality: Four primary components, namely alite (C_2S), belite (C_2S), aluminate (C_3A), and ferrite (C_4AF) are formed during the production of clinker. The Bogue calculation is used to calculate the approximate proportions of the four main components in Portland cement clinker. Each of these constituents has a specific effect on the properties of clinker. The quality control of the clinker needs to maintain the main oxide ingredients such as CaO, SiO_2, Fe_2O_3, and Al_2O_3 as well as other minor constituents such as sulfate (SO_3), potassium oxide (K_2O), sodium oxide (Na_2O), titanium dioxide (TiO_2), and phosphorous pentoxide (P_2O_5). Three basic oxide ratios called Lime saturation factor (LSF), silica ratio (SR), and alumina ratio (AR) are almost universally used to characterize the quality of the clinker in cement production.
2. The control of alkali, sulfur, chloride, volatiles content, and heavy metals, such as magnesium, cadmium, lead, mercury and thallium, is also of paramount importance: Excessive inputs of these compounds may lead to buildup and blockages in the kiln system.
3. Calorific value of fuels is the key parameter for mixing calculation of fuels.
4. Moisture content: Overall moisture content (of alternative and conventional fuels and/or raw feed materials) may affect the productivity, efficiency, and also increase energy consumption.
5. Ash control of the AF is another key requirement that shall be considered as the ash takes part in the clinker formation reactions. The chemical composition of the ash shall be checked to ensure that the final composition of the raw mix meets the necessary requirements for clinker production.

Mixing controller will calculate the optimal mixture of AFs based on prediction values, manipulated variables (output of the controller), information, and strategies given by directing station with the help of the simplex algorithm (an iterative method to approach the best solution).

To complete the modeling, it is necessary to know the input parameters data (raw materials, fuels, raw materials costs, etc.), and the same should be addressed as a function of the proposed objective results like cost optimization, environmental impact minimization. The flowchart given in Figure 5.28

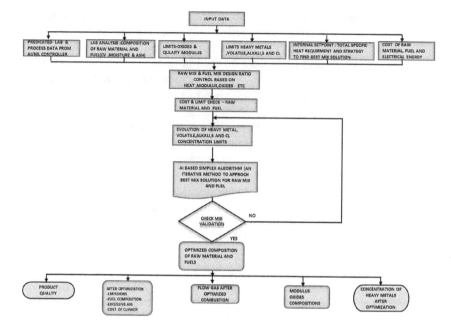

FIGURE 5.28   Stepwise procedure for raw mix and fuel mix optimization [13].

illustrates the procedure which is necessary to accomplish the model of the proposed system [13].

The mixture optimization should consider the stable operation of the rotary kiln, the quality of the clinker produced, the minimum cost of the used composition, and the electric power; all these variables are considered in the nonlinear model proposed through the following objective function, Eq. (5.1) [12].

$$C = \sum_{i=1}^{n} p_{i_c} \cdot X_i + p_E \cdot A \cdot \exp^{(B \cdot S)} \tag{5.1}$$

The objective function (C) of the model should try to obtain a minimum cost in the clinker production, considering the raw materials' cost as well as the consumption of the energy required for grinding. The first term (linear) represents the raw materials and fuel's (primary and alternative) costs used in the clinker production (pi, is the raw materials and fuels costs $i = 1,2\dots\dots n$ that participate in the burning process with their respective percentages X1, X2,……Xn). The second term (nonlinear) represents electricity cost (PE) and the energy consumption required in kWh/t for the grinding process of a certain specific surface (S is the area of the specific surface in cm^2/g, A and B are constants that depend on the clinker composition). Table 5.8 and Table 5.9 show typical information of raw materials chemical composition and fuels

TABLE 5.8
Chemical Composition (% Mass) of Raw Materials

Material	Notation	CaO	SiO$_2$	Al$_2$O$_3$	Fe$_2$O$_3$	MgO	SO$_3$	Na$_2$O	K$_2$O
Limestone	X1	50.66	5.04	1.19	0.67	0.78	0.1	0.1	0.3
Clay	X2	1.23	61.62	16.59	9.01	0.3	0.3	5	
Sand	X3	1.13	93.00	2.87	1.20	0.10	0.5	0.5	1
Iron ore	X4	0.71	7.60	1.13	82.97	–	–	–	–

TABLE 5.9
Composition of the Primary and Secondary Fuels

Component	Coal % Weight	Petcoke % Weight	Tires % Weight
Notation	X5	X6	X7
C	63.9	80–100	72.15
H	3.6	3.5	6.74
S	4.6	0.5–7.0	1.23
O	0.9	9.67	
N	1.8	1.5	0,36
Cl	–	0.149	
Zinc	0.04	1–85 ppm	
Cadmium	0.001	1 ppm	0.0006
Lead	0.027	1–10 ppm	0.0065
Thallium	0.0004	1–80 ppm	0.00001
Arsenic	0.00017	0.1–10 ppm	
LHV [kJ/kg]	25.392	32.447	32.100

composition, respectively, to understand the concept of optimization and modeling [13].

Based on the values in Tables 5.8 and 5.9 and Eq. (5.1), an objective function was set up, which represents costs minimization problem, relating to the operational and environmental costs [13].

$$\text{Min } Cos_1 X_1 + Cos_2 X_2 + Cos_3 X_3 + Cos_4 X_4 + Cos_5 X_5$$
$$Cos_6 X_6 - Cos_7 X_7 + Cost_{EE} \cdot \{(5,76(MS) - 5,82) \cdot e^{-(0,2(MS)+0,98)*S}\} \quad (5.2)$$

$$\text{where MS} = \frac{5.04X_1 + 61.62X_2 + 93X_3 + 7.6X_4 + 9.32X_5 + 1.93X_7}{1.86X_1 + 25.6X_2 + 4.07X_3 + 84.1X_4 + 12.29X_5 + 0.92X_7}$$

The Strategy Manager will provide the various strategies to take decision in different conditions so that the algorithm can find the best choice of fuel mixing with the minimum cost of fuel within the operational constraints. If the solution of one strategy is out of the admissible range then the controller will find another solution based on the next given strategy. The typical equations of the optimization model regarding operational constrains are given in Table 5.10 [13].

TABLE 5.10
Equations of the Optimization Model

Subject To:		Control Limit
$50.60X_1 + 1.23X_2 + 1.13X_3 + 0.71X_4 + 1.03X_5 + 0.93X_7 \geq 63.76$	Eq. (5.3)	CaO Limits
$50.60X_1 + 1.23X_2 + 1.13X_3 + 0.71X_4 + 1.03X_5 + 0.93X_7 \leq 70.14$	Eq. (5.4)	
$5.04X_1 + 61.62X_2 + 93X_3 + 7.6X_4 + 9.32X_5 + 1.93X_7 \geq 19.71$	Eq. (5.5)	SiO_2 Limits
$5.04X_1 + 61.62X_2 + 93X_3 + 7.6X_4 + 9.32X_5 + 1.93X_7 \leq 24.25$	Eq. (5.6)	
$1.19X_1 + 16.59X_2 + 2.87X_3 + 1.13X_4 + 5.08X_5 + 0.79X_7 \geq 3.76$	Eq. (5.7)	Al_2O_3 Limits
$1.19X_1 + 16.59X_2 + 2.87X_3 + 1.13X_4 + 5.08X_5 + 0.79X_7 \leq 6.78$	Eq. (5.8)	
$0.67X1 + 9.01X2 + 1.2X3 + 82.97X4 + 7.21X5 + 0.13X7 \geq 1.29$	Eq. (5.9)	Fe_2O_3 Limits
$0.67X_1 + 9.01X_2 + 1.2X_3 + 82.97X_4 + 7.21X_5 + 0.13X_7 \leq 4.64$	Eq. (5.10)	
$0.78X_1 + 0.10X_3 + 0,44X_5 + 0.12X_7 \leq 6.5$	Eq. (5.11)	MgO Limits
$0.762X_1 + 2.74X_2 + 83.64X_3 - 185.83X_4 - 18.96X_5 - 0.186X_7 \geq 0$	Eq. (5.12)	Equations (12) to (17) are the restrictions of the mixture control modules for the clinker quality
$-0.018X_1 + 7.5X_2 - 82.011X_3 + 219.47X_4 + 23.88X_5 + 0.554X_7 \geq 0$	Eq. (5.13)	
$0.319X_1 + 4.877X_2 + 1.31X_3 - 106.73X_4 - 4.29X_5 + 0.621X_7 \geq 0$	Eq. (5.14)	
$0.619X_1 + 7.737X_2 + 0.37X_3 + 222.88X_4 + 14.387X_5 - 0.439X_7 \geq 0$	Eq. (5.15)	
$38.24X_1 - 155.67X_2 - 173.6X_3 - 164.34X_4 - 37.86X_5 - 4.2X_7 \geq 0$	Eq. (5.16)	
$-35.48X_1 + 190.65X_2 + 212.43X_3 + 201.0X_4 + 46.51X_5 + 34.5X_7 \geq 0$	Eq. (5.17)	
$25392X_1 + 34436X_6 + 32100X_7 = 3600$	Eq. (5.18)	Total fuels feeding required for specific heat consumption
$0.046X_5 + 0.07X_6 + 0.0123X_7 \leq 0.05$	Eq. (5.19)	Sulfur control
$0.1X_1 + 3X_2 + 0.5X_3 \geq 0.20$	Eq. (5.20)	Sulfur trioxide control of raw material
$0.1X_1 + 3X_2 + 0.5X_3 \leq 2.07$	Eq. (5.21)	
$0.1X_1 + 3X_2 + 0.5X_3 \geq 0.03$	Eq. (5.22)	Alkalis in raw material control
$0.1X_1 + 3X_2 + 0.5X_3 \leq 0.33$	Eq. (5.23)	
$0.3X_1 + 5X_2 + X_3 \geq 0.31$	Eq. (5.24)	
$0.3X_1 + 5X_2 + X_3 \leq 1.76$	Eq. (5.25)	
$A_5X_5 + A_6 X_6 + A_7X_7 \leq 0.10$	Eq. (5.26)	High volatile and heavy metals, such as cadmium, lead, mercury and thallium are controlled.
$B_5X_5 + B_6 X_6 + B_7X_7 \leq 0.35$	Eq. (5.27)	
$C_6 X_6 \leq 0.05$	Eq. (5.28)	
$D_5X_5 + D_6 X_6 + D_7X_7 \leq 0.10$	Eq. (5.29)	

Table 5.11 shows the optimization model results. Looking at the above results, it is possible to foresee the raw mix composition and mix of multiple fuels, when it is decided to use the AFs and alternative raw material in cement rotary kiln. It is also possible to calculate the substitution levels of the primary fuel by AF derived from the input data of fuels to be used. This model

TABLE 5.11
Optimization Model Results

Objective Function C = Cost/ton	Oxides Compositions in Clinker (%)	Modulus	Specific Heat Consumption = 3600 (kJ/kg clinker)
Compositions (kg/kg Clinker)	Consumption of fuels (kg/ton clinker)		
X1 = 1.2175	CaO = 62.07		
X2 = 0.2007	SiO$_2$ = 20.13	MS =2.50	Petroleum Coke = 74.1
X3 = 0.0161	Al$_2$O$_3$= 5.22	MA =1.82	Used tires = 28.0
X4 = 0.0000	Fe$_2$O$_3$= 2.86	MH =2.20	
X5 = 0.0000	MgO = 0.95		
X6 = 0.0741			
X7 = 0.0280			

is an example of either to maintain the optimum raw mix and fuel for achieving the desired clinker quality or to optimize the cost of production [13].

This constrain management strategy shall be implemented in the mixing controller to control the dosing of AFs within the constrain limit.

5.13.3 COOLER CONTROL MODULE

The clinker cooler is a vital part of the kiln system and has a decisive influence on the performance of the plant. The clinker cooler represents the start of the pyroprocess for the cement clinker production if seen in direction of the gas transport. This means, that the performance of the clinker cooler influences largely the performance of the overall process.

Cooler Control Module is designed to maintain the material bed, under-grate pressure, and secondary air temperature at suitable levels, while maintaining an optimal air to clinker ratio.

The objectives of cooler control module are as follows:

1. To recover a large amount of heat with limited air quantity to cooler, resulting in. high secondary air temperature and low energy consumption
2. To cool down the clinker heavily to achieve low final temperature of the clinker
3. To cool the clinker rapidly.

The objectives will be achieved with the help of the following subsystems:

a. **Transport Control Inside the Cooler**
   Sequential ratio control of the number of strokes of Grates: The pressure in the first grate chamber is regarded as the indicator for the

layer height on the grate. Through altering the number of strokes, the transportation speed of the clinker and therewith the layer height will be controlled. The number of strokes of the grates of all subsequent grate chambers will be supplemented by means of a sequential ratio control.

Number of strokes of Grate 1 = f (pressure 1)
Number of strokes of Grate 2, 3 = f (number of strokes 1)

Separate control of the number of strokes: The elementary standard configuration leaves a different distribution of the clinker (layer height) in the longitudinal and cross direction above the grate plates. Through a separate consideration of the available pressure signals at the cooler and an appropriate control of the number of strokes, local differences in the layer height can be reacted upon.

The number of strokes of the grates is controlled separately, depending upon the pressures under the grate plates.

Number of strokes = f (pressure 11, pressure 12,)
Number of strokes = f (pressure 21, pressure 22,)
Number of strokes = f (pressure 31, pressure 32,)

b. **Clinker Aeration Control**

The objective of the aeration controller is to maintain an optimal air to clinker ratio. It adjusts the speed of the cooler fan of cooler compartment according to the clinker production rate and as per the optimal specific air volume per kilogram of the clinker.

The combination of the various functions available in the cooler control module will result in stable and improved heat recuperation, due to which the whole kiln system performance will be benefitted.

5.14   EVALUATION AND IMPLEMENTATION OF ADVANCED PROCESS CONTROL SYSTEM

The cement industry is in the middle of an ongoing change process which is driven by the advancement of information technology, machine learning, and AI concepts. The changes are aimed toward achieving further plant and production optimization. It is necessary to understand that adopting an only purely data-driven approach is probably not sufficient. It is equally necessary to rely on deep machine and process expertize to turn the available data to useful information and finally to obtain a robust performance optimization system.

It is known that the cement production process is quite complex and is influenced by thousands of variables from raw material attributes, physical

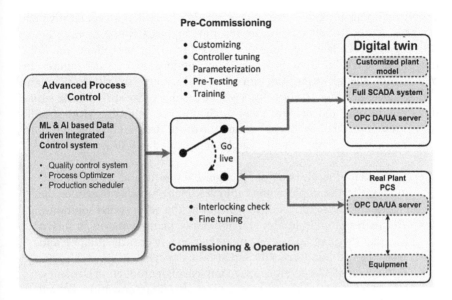

FIGURE 5.29 Evaluation of APC via digital twins of cement plant.

and chemical properties of fuel, different thermal and process conditions, and the desired quality and type of the end product. It is quite challenging for most of the cement plants to optimize these variables at each step in the process without extensive process knowledge and advanced tools. Therefore, it is of paramount importance to critically check the data-driven advanced process control system before implementing in a real cement plant with the help of process experts and modern tools like digital twins, as shown in Figure 5.29.

The virtual cement plant (digital twin) imitates the dynamic characteristics of entire cement production process through a digital model. Digital twins have become well known in many applications over recent years. Although it is considered by many as a new technological revolution, the digital twinning concept has existed for many years in modeling and simulation technology. It is merely the intensive incorporation of digital twins into active control systems that has led to this term being coined and widely used.

Model-driven digital twins make virtual commissioning more accessible. The following are the benefits of digital twins in evaluating the APC modules.

- Risk-free test environment for APC concepts
- Identification of most practical difficulties arising from APC concepts
- Deeper understanding of the possibilities of an APC solution
- Increased acceptance to plant operators
- Speeding up of the plant commissioning

The virtual cement plant (digital twin) can also be used as a simulator which can provide risk-free effective means for training the CCR operators and process engineers. Unlike the situations prevailing in the real plant, the limit ranges can be tried in digital twins without running any risk and failures. In other words, the operating staff can be trained without any risk to the plant and products. Further, the operators will be capable of locating critical situations well in time and manage these safely and economically. More specifically speaking, the benefits of simulator-based training, in addition to providing a risk-free opportunity, can be enumerated as follows:

- To evaluate the ideas of plant engineers and operators on simulator prior to the implementation in real plant, which will reduce the chance of adverse effect of implementing the new process solution.
- To enhance troubleshooting capabilities of plant operators in such a way that they can take decision on their own in critical situation and avoid the breakdown of kiln and grinding systems.
- To help qualified operators to accomplish high production efficiency, high availability of the plant due to short failure downtime as well as low emissions. It also helps in reaching production and quality targets so that the maximum production with target quality can be achieved.
- To meet the periodic need for plant optimization created by any new technology, upgrading of facilities, enforcing stricter pollution norms, diversifying the cement types, and handling multiple AFs, which are the constant challenges for the operators to cope up with.

5.15 COLLABORATIVE OPERATION IN DATA-DRIVEN ECOSYSTEM

The integration of data with web cloud will help to operate the plant in collaborative manner. Client as well as various domain experts can support the plant as shown in Figure 5.30 [13].

The following are the important points for sustainable operation of advanced process control system:

- Reliable sensors and data bases are critical for optimum and desired result of advanced process control systems.
- Domain experts' inputs are needed for optimized and trouble-free operation of plant.
- Monitoring of AI model performance is needed as the system can start drifting due to process changes necessitating fine-tuning for better results.
- Improving the model and replicating the best experience of other sites.

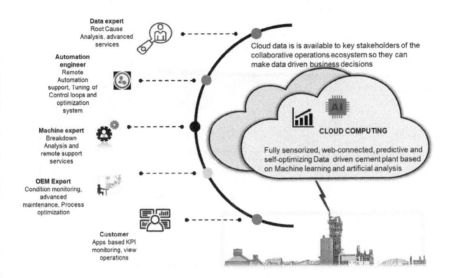

FIGURE 5.30 Collaborative operation in data-driven ecosystem [7].

5.16 CONCLUDING OBSERVATIONS

All over the world, cement producers are facing growing challenges to lower the production costs, conserve material and energy resources, as well as to reduce CO_2 and NOx emissions. The principal strategies to overcome these challenges are to increase the overall efficiency and use of alternative materials for sustainability as shown in Figure 5.31.

Sustainable development will result in delight of all stakeholders of the cement industry:

Industry Delight (Profit): A cost-effective substitution of natural resources thereby improving the competitiveness of an industry

Ecology Delight (Planet): Environmentally sustainable waste management and saving of natural resources

Society Delight (People): A long-term sound solution for treatment of different types of wastes produced by the society

The specific actions for the sustainability road map are the following:

1. Improvement in the overall efficiency of cement production and switching to AFR
   - Implementation of state-of-the-art process technology, suitable to handle AFR

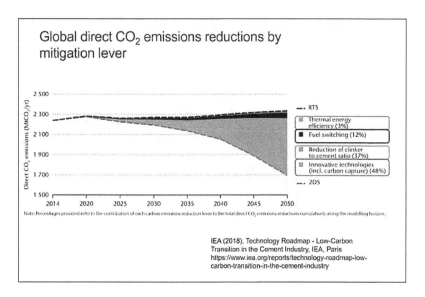

FIGURE 5.31 Technology roadmap for low CO$_2$ emission [14].

- Implementation of the waste heat recovery system
- Implementation of industry 4.0 digitalization concepts
- Guidance by process know how experts
- Collaborative operation
- Training of the plant staff

2. Reduction of clinker to cement ratio: The most suitable path to achieve sustainability in the cement production is the reduction of the clinker factor by adding Supplementary Cementitious Materials (SCM). The most common SCMs used today are fly ash, slag, and natural pozzolan. The reduction of steel production and phasing out of coal power plants has resulted in the significant decrease in the supply of these common SCMs, so it is very important to find a new resource for, e.g. calcined clay.

3. Implementation of innovative technology: The continuous efforts are made by technology suppliers to develop the new technology and process concept of carbon capture.

   - Oxyfuel (oxygen fuel) is an innovative technology. In this concept, the clinker production combustion process is operated with pure oxygen instead of the ambient air so that no nitrogen enters the combustion process, which results in production of highly concentrated CO$_2$ (80%–100%). The CO$_2$ gas is finally separated and not released into the atmosphere. The greenhouse gas can then be converted into a raw material, thus forming the basis for the production, for example, fuels, plastics, and fertilizers.

Sustainable development is a pathway to the future. Environment and the economy are really two sides of the same coin. In cement manufacturing also, we have to balance sustainable development, environment, and economy of the manufacturing technology of cement.

REFERENCES

1. Dave Evans, Whitepaper on "The Internet of Things", April 2011, 5–6, Available at www.cisco.com (docs › innov › IoT_IBSG_0411FINAL Dave Evans).

2. A Chatterjee and Tongbo Sui, *Alternative fuels – effects on clinker process and properties, Keynote paper at the 14th International Congress on Cement Chemistry, Prague, Czech Republic, Cement & Concrete Research*, Vol. 123, 2019, 105777.

3. Anjan Kumar Chatterjee, *Cement production technology: principles and practice*, CRC Press, Taylor & Francis Group, FL, 2018, 103–133.

4. Michael Brachthauser and Robert Mathai, *Secondary fuel solutions*, KHD Humboldt Wedag, Germany, ZKG, January 2005, 1–4.

5. Matthias Mersmann, *Burning alternative fuels in cement kilns, aixergee process optimization*, ZKG International, June 2014, Available at www.zkg.de/en/artikel/zkg_Burning_alternative_fuels_in_cement_kilns_2018848.

6. Report on "NOx Control Technologies for the Cement Industry", EPA (United States Environmental Protection Agency), Publication No. EPA-457/R-00-002, September 2000.

7. Collaboration in data-driven ecosystem: people make the difference, Available at: https://new.abb.com/pulp-paper/abb-in-pulp-and-paper

8. ITEC online analyzer for free lime and loss on ignition, Available at www.iteca.fr

9. Vaishli Advani, What is artificial intelligence, Available at www.mygreatlearning.com/blog, July 15, 2020.

10. Mark Yseboodt, *On the road to digitalization*, Siemens, World Cement, May 2020, 19–22.

11. Konrad S. Stadler, Burkhard Wolf, Eduardo Gallestey, Precalciner control in the cement production process using MPC, *IFAC Proceedings*, Vol. 40, No. 11, 2007, 201–206.

12. M. Mrsmann and M. Weng, *Optimizing cement production with the aid of digital twins, aixergee GmbH*, Germany, Cement International, May 2019, 20–27.

13. Ricaro C. Carpio, Rogério J Silva, Francisco de Sousa Júnior, Leandro dos Santos Coelho, Alternative fuels mixture in cement industry kilns employing particle swarm optimization algorithm, *Journal of the Brazilian Society of Mechanical Sciences and Engineering*, Vol. 30, No. 4, Rio de Janeiro, Oct./Dec. 2008, 335–340.

14. https://iea.org/reports/technology-roadmap-low-carbon-transition-in-the-cement-industry

REFERENCES

6 Control of Cement Composition and Quality
Potential Application of AI Techniques

Mohsen Ben Haha, Maciej Zajac,
Markus Arndt, and Jan Skocek
HeidelbergCement AG, Leiman, Germany

CONTENTS

6.1 Introduction: Quality Control in Cement Plants 200
6.2 Quality Control Practices in Cement Manufacturing 200
6.3 Data Collection Methods for Cement Production
    Quality Control 202
    6.3.1 Analytical Methods...202
    6.3.2 Physical Methods 205
6.4 Quality Control Alongside the Process 205
    6.4.1 Sampling Importance 205
    6.4.2 Quarry and Raw Milling 206
    6.4.3 Hot Meal and Clinker 207
    6.4.4 Cement..207
6.5 Essence of Artificial Intelligence or Machine Learning 208
6.6 Relevance and Limitations of AI...209
6.7 Quality Deviations: Causes and Potential Use of ML 210
    6.7.1 Materials From Own Quarries 211
    6.7.2 Purchased Materials and Combustibles...................................212
    6.7.3 Raw Mix...212
    6.7.4 Raw Meal..213
    6.7.5 Kiln Feed 214
    6.7.6 Hot Meal...214
    6.7.7 Fuel Preparation 215
    6.7.8 Clinker 215
    6.7.9 Dispatched Cement 215
6.8 Application of ANN to Final Product Quality.................................216

DOI: 10.1201/9781003106791-6

6.8.1   Strength Prediction: Traditional Statistics vs.
         Predictive ANN...216
6.8.2   Deviations in Ground Cement: Influence of SO_3 Level 218
6.9   Concluding Remarks...219
References...221

6.1   INTRODUCTION: QUALITY CONTROL IN CEMENT PLANTS

Quality control is a process aiming at reviewing the quality of all processes and materials involved in the manufacturing process. The industrial quality control systems are based often on the ISO 9000 [1]. The process is described through several quality management principles with a strong focus on customer, without losing the motivation and implication of the process approach on the continual improvement.

There are several objectives of quality control in the cement manufacturing process. They include the needs to ensure that the cement standard specifications are complied with, the market-relevant requirements are fulfilled, and the environmental protection requirements are met at each production step. It enables to deliver the customers with a product of high quality and uniformity. Furthermore, quality control allows to achieve an efficient operation of the processing equipment without major problems, e.g. a smooth operation of rotary kilns without stoppages. It enables using the input material, including the raw mix components (limestone, clay, alternative raw materials), fuels, and alternative fuels in an optimum way. It allows to reach the defined quality levels and high consistency throughout the entire manufacturing process. It is also observed that a combination of more strengthened environmental regulations, enhanced plant performance goals, and cement market quality demands has resulted in improved levels of quality control in many industries and more particularly for the cement producers [2].

Cement manufacturers are bound by the international and national cement standards, codes, and regulations, providing the performance requirements and compositional ranges of cements. Consequently, the quality control must strictly meet these requirements. The examples of these standards are ASTM standards in the USA, including ASTM C150; ASTM C595; ASTM C1157; EN 197 standard in European Union, and the national standards in different countries such as in the China, e.g. GB 175 for net Portland cements), India including IS 269, IS 8112, IS 12269, IS 445, IS 1489, and similar.

6.2   QUALITY CONTROL PRACTICES IN CEMENT MANUFACTURING

The core factor of the quality control is the laboratory where the analysis of the relevant materials and data is processed and stored [3]. There are two analysis practices: (1) performing the analysis in the field, i.e. the testing is undertaken in the vicinity of the sampling point (in situ) and (2) extracting the samples for transport to the central laboratory where they are tested.

The centralized quality control concept is the classical arrangement; all analyses are carried out at the central laboratory in the cement plant. This requires that all samples from the different sampling points are transported to the laboratory and that the laboratory has the appropriate capacity to analyze the samples at a reasonable time and cost [4]. In the modern cement plants, the sampling, the transport of the samples to the plant laboratory, the sample preparation, and testing are mostly automated [5], except for the physical testing discussed later. Several types of automated and programmable samplers exist, which are adapted to the process sampling point and material to be sampled. Manual sampling and manually operated sampling devices are presently used for materials tested at low frequency or for specific sampling campaigns. Pneumatic tube systems are the most common sampling transport where powdery or granular samples are filled into a capsule and then transported through a tube system to the laboratory [6]. However, for specific applications tailor-made solutions are applied. Nowadays, automated sampling preparation systems exist for size reduction and splitting of materials to be tested. These devices are typically combined with the corresponding sampling and transport equipment and can be directly integrated with the analytic equipment. The sample preparation solution contains mostly the sample splitting and grinding equipment. Furthermore, different solutions for automated preparation of pressed tablets or fused beads for chemical and phase analysis are available, respectively. Analytic equipment can be automated to high degree. Analytical instruments in most cases are connected to sample preparation units by small conveyor belts or with robots. Tasks scheduling and execution is fully automated, and the generated data are saved in the plant or the company's respective databases. In the older cement plants, the samples are taken and transported manually. These can be then prepared and measured manually or processed by the automated systems. Consequently, depending on the plant arrangement and adoption level of the state-of-the-art techniques, the quality practices can have a different degree of automation. Furthermore, the container laboratories can be installed in the cement plant for the specific purposes, e.g. during the plant modernization. The container laboratories present an intermediate solution between the automated central laboratory and decentralized online analyzers. Different analytical equipment can be installed in a container. In order to facilitate the operation, container lab is often located close to the sampling points. Analysis in the container laboratory can be fully automated. A photographic view of a modern laboratory is presented in Figure 6.1.

In a decentralized quality control setup, major quality control functions are carried out by control systems installed locally, i.e. close to the production equipment. In such arrangements, the distances between sampling places and the analytical equipment are short. Decentralized units are typically fully

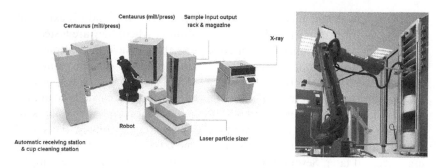

FIGURE 6.1 Fully automated cement laboratory preparation QCX ROBOLAB from FLSmidth [7].

automated. They include a sampling unit, a sample preparation unit and an analytical unit. However, there are existing systems that measure directly the material stream and thus do not require sampling and sample preparation arrangement. The results of the tests are automatically sent to the plant control unit and saved in the respective databases.

6.3 DATA COLLECTION METHODS FOR CEMENT PRODUCTION QUALITY CONTROL

Traditionally, the quality parameters of the clinkering process and of the cement clinker are assessed by determining the free lime content. The free lime content of clinker can be determined with the several methods including the titration methods [8], conductometry, and XRD technique.

Nowadays, cement plant laboratories are equipped with devices enabling treatment and basic characterization of the sample. The equipment for sample treatment includes facilities for weighing, drying, sieving, crushing, grinding, homogenization, and splitting. The equipment for characterization consists of the analytical instruments for determining the chemical composition of the samples and physical characterization of the particulate materials and also the facilities for the physical assessment of the cement performance.

6.3.1 ANALYTICAL METHODS

The modern cement plants employ two techniques based on the application of X-rays: X-ray fluorescence (XRF) and X-ray diffraction (XRD) for routine analysis of the materials during the cement production steps.

XRF is a fast method to obtain elemental oxides composition of material samples for controlling almost all stages of production [9]. XRF spectroscopy is capable to detect and quantify nearly all elements enabling a proper control of the cement production [10]. The sample is exposed to X-rays produced by an X-ray source (X-ray tube). On irradiation, the elements present in the sample emit characteristic "secondary" (or fluorescent) X-ray radiation. By measuring the intensity of the emitted radiation, it is possible to determine how much of each element is present (quantitative analysis). The method requires a special sample preparation which is applied to powdered solids: fusion (fused beads) or tablet pressing (pressed powder tablets or pellets). Nowadays, the sample preparation, and the measurement itself, are automated in most modern cement plants. It can be integrated into the central laboratory automation systems.

In cement industry, XRD technique is used to acquire qualitative and quantitative analysis of crystalline compounds in the tested materials [11]. In this method, a sample is exposed to X-ray waves that are scattered by the samples. These diffracted X-rays are then detected, processed, and counted. As a result, a characteristic diffraction pattern for a crystalline material is obtained. This is used to identify the phase by comparing it with patterns from a special database. Only crystalline phases produce distinctive patterns that can be used as fingerprints for identifying them. The amorphous phases, which for example occur in blast furnace slag, fly ash, or calcined clay, are more difficult to identify and quantify. For the quantitative information, the Rietveld refinement is frequently applied. It is a computational method to determine mineral concentrations in material samples from powder diffractograms. The Rietveld method fits a theoretically calculated pattern to a practically measured XRD pattern by a least-square refinement. It is a powerful method for phase quantification, allowing for example quantification of clinker phases on a routine basis.

The XRD technique is a non-destructive and rapid method. In order to conduct a measurement, the powdered solids are pressed in a special pellet. Like the XRF technique, the sample preparation and the measurement itself are automated in most modern cement plants. It can be integrated into the central laboratory automation systems.

Additionally, thermogravimetric analysis (TGA) can be used in the cement plant laboratory. This is mostly applied to the analysis of gypsum ($CaSO_4 \times 2H_2O$) and hemi-hydrate ($CaSO_4 \times \frac{1}{2} H_2O$) contents [12] in the cement and also for the determination of calcite [13] content in either cement or raw meal. Additionally, it can be used to test the pre-hydration of cement. TGA involves measurement of the change in weight of a sample as function of increasing temperature. A typical TGA curve shows the mass loss of a sample plotted against the temperature, with the characteristic effects of the phases mentioned above. The method requires powdered solids.

Chloride content in raw materials, raw meal, clinker, cement, and fuel samples can be evaluated by the potentiometric titration. In older arrangements, the titration is done manually; however, nowadays automated titration equipment is in use. Different types with different degrees of flexibility and sample throughput are available.

The use of particle size analyzer based on laser diffraction [14, 15] (laser granulometry) has become widespread in cement industry. This technique was initially introduced as supportive to Blaine surface area measurement method, but today it is replacing the traditional technique. Laser granulometry is used to analyze the fineness of the ground material like the raw meal and cement. The particle size distribution is calculated by comparing the scattering pattern obtained from the sample material falling through a laser beam with the help of a special optical model (Fraunhofer theory and the Mie theory). The measurements can be automated. This equipment can not only be used as a stand-alone equipment but can also be integrated into the central laboratory automation systems.

In addition to these most common techniques, some more instrumental analysis is possible in the cement plants. This includes elemental analysis for sulfur and carbon. The available devices also enable analysis of nitrogenl. Both sulfur and carbon can be easily and accurately determined with a special S/C analyzer, consisting of a furnace and infrared (IR) detection cells. The sample is burnt in pure oxygen atmosphere in which sulfur and carbon compounds are oxidized to gaseous SO_x and CO_2, respectively.

Furthermore, the spectroscopic techniques for (trace) elemental analysis can be applied depending on the needs. These instruments are useful when analyzing the trace elements in different samples. The methods include AAS (atomic absorption spectroscopy) or ICP-OES/MS (inductively coupled plasma optical emission spectrometry/mass spectrometry). These techniques offer detection limits up to the ultra-trace range (sub ppb level), relevant for some environmental analyses. However, they require the samples to be prepared in liquid form and there are very demanding requirements of operator skills and operating conditions. Consequently, these facilities are not considered as typical equipment of standard cement plant laboratories. They are either available at some central location of a company or the tests are subcontracted.

The common analytical methods (XRF, XRD, laser granulometry) can be installed either in the laboratories or as distributed facilities at the plant sites. The Prompt Gamma Neutron Activation (PGNA) technique is installed exclusively at the plant site. It enables an automatic bulk material analysis. In a PGNA online installation, the bulk material passes through a beam of neutron bombardment. Neutrons interact with elements present in the tested material, which emit secondary prompt gamma rays. These gamma rays are collected and analyzed with a gamma ray detector. This technology has become the state-of-the-art tool for real-time elemental analysis of bulk materials. PGNA technology allows measuring the elemental composition of bulk materials such as raw materials, minerals, and coal. In contrast to XRF analysis, it can analyze the full material stream in real time and does not need sampling nor sample preparation. PGNA analyzers are online systems installed on the conveyor belt, so materials are measured continuously and in real time.

6.3.2 Physical Methods

Modern cement plants are equipped with techniques allowing the physical and mechanical testing of the cement, cement paste, and mortar properties. This type of equipment is mostly specified in national cement standards and codes [16]. The following list presents a list of major equipment used for physical and mechanical testing of cement, cement paste, and mortar:

- Automated and manual Blaine apparatus for measuring the Blaine surface area
- Vicat apparatus of different configurations, manual or automatic, for determining the setting time
- Flow table for workability
- Soundness by Le Chatelier method consisting of molds and Le Chatelier water bath
- Mortar preparation and storage devices including mortar mixer with extended facilities (automated dosage of sand and water), dedicated mortar prisms, vibration tables, climatic storage chambers (humidity cabinet), wet storage tanks
- Hydraulic press for compressive strength and bending strength testing
- Differential calorimeters for heat of hydration

6.4 QUALITY CONTROL ALONGSIDE THE PROCESS

This section discusses the application of the different measurement tools and practices along the production process of cement.

6.4.1 Sampling Importance

Fluctuation is an important recurrent problem for quality parameters. Therefore, several quality control measures are implemented to assess these issues. The sampling places, the frequency, and the representativity are the most important parameters to specifically assign these fluctuations to the measurement and/or sampling errors or whether they are process related.

The analytical results of raw meal, clinker, cement and fuel are important for comparing the performance of different cement plants. These data are sometimes misplaced. New modeling and control quality techniques may play a big role to compare these data once available.

In cement plants, the quality data are and must be assessed over long periods, i.e. hourly, daily, weekly, monthly, and even yearly. Evaluation of quality data over longer periods of time traditionally required the use of statistical tools. Though individual fluctuations may potentially influence the quality of the cement, such simple variations are sometimes irrelevant for a sufficiently stable production at reasonably low variable cost. If problems occur with these simple variations, the traditional parameters, such as standard deviations from the targeted main control parameters in the cement production line, may not often provide a satisfactory answer to the quality variation. Profound data analysis offers, on the other hand, a more global understanding of the data along with its interactions that are not captured through traditional statistical tools and trend values. Thus, the traditional approach of targeting or avoiding an absolute range of values (e.g. for free lime) may sometimes be used as the main quality indicators, but they may not always be satisfactory to resolve a problem of deviation in quality [17].

6.4.2 Quarry and Raw Milling

The first step in the cement process is to extract limestone and clay from the earth. The key elemental components for cement are Ca, Al, Fe, and Si. Additional elements such as Mg, Na, K, and S are introduced with limestone and clay. In the limestone quarries, the quality control is mostly performed by means of analysis of drill cuttings. However, depending on the quarry geology, additional samples may be tested. Standard methods used are XRF, XRD, and TGA methods. If needed a detailed sample analysis can be conducted to test content of heavy metals, for example. In general, the quarried limestone rock is crushed in the quarry. For the complicated geological situation there is a need for quality control of the main raw components after crusher. Nowadays, as already discussed above, the most popular method for analyzing raw components after crushing is the PGNA technique, which provides quasi-continuous chemical data of the material on the conveyor belt.

Corrective materials are added to the raw mix after crusher or at the raw mill in order to achieve the desired chemical composition of the raw mill feed material. These materials may be tested to get the chemical composition and physical characteristics. The frequency depends on the uniformity and may be limited to verification of the respective specifications on delivery.

To reach the favorable raw meal composition, the raw materials are blended in front of the raw mill according to the targeted proportions. The blend is controlled in terms of the chemical parameters—Lime Saturation Factor (LSF), silica modules, iron and alumina modules, and the fineness parameters, e.g. residue on 90 μm sieve. XRF is the typical method for chemical analysis of raw meal. Sieving is the standard method for fineness determination, while in automated laboratory systems laser granulometry can be applied.

The raw meal is transported to a storage hall or silo in order to store the ground material, both for building up the reserve and for homogenizing the raw meal. From the storage it is transported to the kiln. Monitoring of the kiln feed composition is very important and has a direct impact on the clinkering process. Typically, the chemical composition is controlled by XRF and the fineness by either sieve analysis or laser granulometry.

6.4.3 Hot Meal and Clinker

The kiln feed is fed into a rotary kiln, a large chemical reaction chamber composed of the two main parts: pre-heater tower that can be equipped with the pre-calciner and the rotary kiln itself. The purpose of the kiln operation is to form cement clinker containing the main clinker components C_3A, C_4AF, C_3S, and C_2S. Analysis of the hot meal from the pre-heater tower provides information about the tendency to form build-ups and consequently about the stability of the cement clinker production. The hot meal can be analyzed by XRF, XRD, and chemically for the chloride content.

Clinker is the most important intermediate product in the cement production process. While the quality control upstream of the is focused on the chemical composition and proportions of the raw materials used to produce the cement clinker, the materials evaluation downstream of the kiln focuses more on the clinker phase assemblage. The main information concerns the degree of burning of the clinker (completion of the chemical reactions to form cement clinker phases), the quality of the clinker, and the burning conditions. This information is usually provided by the free lime content of clinker and the XRF and XRD (wih Rietveld calculations) methods are employed. Additional analysis may be considered depending on the specific plant needs.

6.4.4 Cement

From the cement kiln, the clinker is transported to the clinker silo. Thereafter, the cement clinker is transported to the cement mill where it is ground with calcium sulfate (setting time regulator). In this way the neat Portland cement is produced. Furthermore, the clinker can be ground with the supplementary cementitious materials (SCMs) or these can be blended with the ground cement clinker to produce composite cements. Then cements are transported to the respective cement silos.

The characteristics of the SCMs that can be used for the cement production are defined in the national standards and codes. Quality control of SCMs and the setting time regulator starts with an initial qualification of the material sources in order to verify conformity with standard codes and plant-specific regulations. This may include the complete chemical, phase analysis and performance in the cement. The extent of the control is decided case by case, considering the type, source, and volume of the material. In other words, this part of quality control is plant- and material-specific.

The quality control at the stage of cement grinding aims to test if the cement leaving the plant and shipped to the customer fulfills adequately the customer demands of the product, in addition to conforming to the required specifications. The methods applied here include analytical and physical methods. The arrangement of the quality control is plant-specific and depends on the quality control concept, arrangement of the plant, product portfolio, and local specifications and requirements. The methods include XRF and

XRD for the chemical and phase composition. Furthermore, Blaine surface area and particle size distribution by laser granulometry are typically tested. Finally, the mechanical performance (in terms of water demand, setting time, compressive strength gain over time) is investigated according to the prevailing cement testing standards.

The testing scope, procedures, and frequency depend not only on the local cement standards but also on the local habits.

6.5   ESSENCE OF ARTIFICIAL INTELLIGENCE OR MACHINE LEARNING

It is difficult to decrypt all the talk about artificial intelligence (AI) because so many different terms are used—some of them interchangeably—and AI's capabilities seem to span so many possible scenarios.

The best way to think about AI is as a large umbrella of technologies, methodologies, and algorithms that help humans perform tasks easier, faster, and more efficiently [18]. Under this umbrella resides a large—and growing—collection of techniques such as machine learning (ML), image recognition, neural networks, speech recognition, deep learning, natural-language processing, handwriting recognition, and cognitive computing, among others, many of which overlap or complement one another to help enterprises resolve challenges.

For example, ML focuses on real-world problems by processing—and learning from—large amounts of data. Deep learning, which many consider a subset of ML, uses neural networks to be able to sort through nearly unimaginable volumes of structured and unstructured data to come to conclusions [19]. Cognitive computing is a subset of AI that attempts to mimic the way humans think in a way that addresses more complex scenarios for decision-making.

With the large amount of data present and over years of experience in cement plants as described in the previous paragraphs, ML appears to be the most suitable way to process them. Processing the data in classical statistical way helps users to define boundaries for production and to control the day-to-day production [20]. However, producing at a specific requirement may need a deeper look into the data. Here comes the advantage of applying AI or ML. The unimaginable number of combinations of structured and unstructured data to conclude on the production data is only possible via ML.

The quality control of clinker production is well designed for ML approaches, as the numerous data needed for decision-making are processed in a cement plant both in batch and real time [21, 22]. The difficulties lie most of the time in coordinating between the process data collected with the batch of clinker tested. This information will be clearer in the later sections of this chapter.

The possibility of using ML with the data obtained from the process and quality control along the whole production chain, from raw materials input through the production of intermediate products to final product output is tentatively described below. Possible reasons for quality deviations and how ML may intervene with corrective measures are also presented.

As the production process of clinker and cement is complex, it is not possible to cover all potential scenarios. The principal rules and guidance on corrective actions cannot only be seen by ML, as will be described later, and they need the input of practitioners as well in some typical situations.

6.6 RELEVANCE AND LIMITATIONS OF AI

It may be pertinent to point out that despite the use of the terminology "intelligence" in AI, certain aspects are not considered in its application in the present context, e.g. individual plant situation and configuration. The technical parameter such as the raw mix homogeneity, for example, is not considered in the actual tools. The specific aspects of the plant setup look irrelevant in actual models, despite the application objective being the minimization of deviations from target.

It might be necessary to define this information, when treating the data, as additional parameters and to introduce them as fictive values.

All information, regarding correctives, LSF, SR, AR of clinker, etc. are treated as number in total chemistry calculation or raw meal formulation comparable to the fineness. Impacts observed in kilns such as dust and fuel variations cannot be captured fully except if they are traced online which is commonly not done in today's production process. It makes thus more sense to focus on fluctuations of clinker and cement properties rather than on those of raw meal and kiln feed when using ML techniques.

ML data analysis must be based on individual results from single plants with continued sampling procedures and frequencies. Unlike the statistical analysis, they do not require averages of averages as these types of calculation hide the singularities and particularities of each measurement.

Using ML for the prediction of the cement properties (mainly compressive strength) does not require an in-depth understanding of the process behind in comparison to traditional modeling concept based on thermodynamic and micromechanical approaches. All what is required is a large volume of already available quality data. The complexity of the processes in cement plant is hidden behind the correlations determined under controlled conditions. These hidden correlations can influence the compressive strength. However, these influences are not measured or not yet pointed out. Traditional statistical methods often lead to poor correlation coefficients between observations. Hence, the large volume of data acquired and gained for quality control purposes can still be analyzed by neural network and predictive model for cement strength can be developed.

Neural network does not require any prior knowledge about cement. By analyzing a large number of inputs and outputs, the underlying patterns are identified and can be applied for the new inputs [17]. Figure 6.2 explains the

Database   Neural network   New QC record

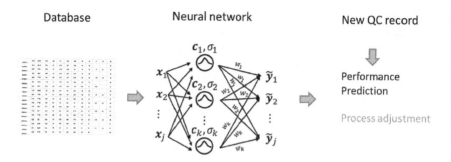

FIGURE 6.2   Basic simplified approach involving artificial network principles.

basic strategy. The neural networks present a brain-like learning process with all advantages and limitations: similar input leads to similar output: the networks show a high degree of adaptability: learning requires time and enough data, respectively: the more data are available, the better the model will be. Any change in the process composition not known before requires a learning before it could be predictable.

The simplified mathematical approach behind the neural network can be explained with a simple one-dimensional example. Neural networks are robust, self-adaptable, user friendly, and of course not limited to strength prediction models. It is indeed a valuable tool to predict process data, given that the process did not change over the years.ML can help to react as quickly as possible in cement production process [23]. Depending on the control parameter, the situation might also be critical to process or even to environment. Here, the time factor plays a dominant role. In addition in a continuous cement manufacturing process, the amount of cement produced with deviating quality is also critical [24]. Depending on the case-by-case situation, the urgency and the extent of the corrective action have to be evaluated, in the worst case even resulting in a production stop [25].

6.7   QUALITY DEVIATIONS: CAUSES AND POTENTIAL USE OF ML

Different possible situations within the tolerance range around a target value are usually present in production process [26]. The quality deviation detected by any analysis method is assumed to be real, i.e. the result of the measurement is assumed to be correct. If a particular parameter deviates from the target value and is outside of the operating range, there may be several potential reasons for such deviations, which can be listed for identifying the root causes.

The different actions to be implemented, when such deviations occur, can be related to the deviation zones as described below. Adjustments can be done according to the local situations. Defined process control parameters and ranges of operation in a cement plant are given as a practical example.

The zones of deviations are described as follows:

- Zone 1: No action is needed, if consecutive test results just fluctuate around the target value within the defined operating range.
- Zone 2: The operator must take corrective actions in order to come back to the target value. A corrective action might be initiated in advance, if consecutive test results show a clear trend that predicts the likelihood of future results falling outside the tolerance range. If the next regular results after the initiation of corrective actions come back to zone 1 or show a possible trend of coming back into zone 1, no further action is required. But if the test results continue to be still in zone 2, additional samples have to be taken to confirm the effect of the corrective actions. If the problem persists, the shift fore-man must be informed in order to investigate the problem. If nec-essary, he will inform the Quality Manager and/or the Production Manager.
- Zone 3: If one result occurs in this zone, immediate action is required, and the result needs to be cross-checked by additional sampling and testing. If the problem persists, the shift foreman must be informed. The production unit might need to be shut down, if the problem can-not be fixed in a timely fashion. In that case, Quality and Production Managers must be informed.

6.7.1 MATERIALS FROM OWN QUARRIES

The quality of cement clinker is directly linked to the chemistry of the raw materials used [27]. Limestone represents 80%–90% of the raw materials fed to the kiln. The availability of good raw materials is the main factor for decid-ing the location of cement plants. Generally large quantities of a uniform source of calcium oxide, silica, alumina, and iron oxide are used for cement production. Quality of raw materials from the quarry can vary due to tecton-ics like faults, folds, or karstifications if these geological aspects remain unde-tected at the time of initial explorations or are not detected in the analysis of the drilling samples. Further, deviations may take place if mining does not follow the agreed quarrying plan (selected area to be mined, fixed ratios from different selected areas). This needs to be checked on site with the responsible quarry manager. If a cross-belt analyzer is used, care must be taken to evalu-ate accuracy of data via dynamic correction. However, the accuracy control and adjustment for a pre-stacker analyzer is not as easy as for pre-mill analyz-ers; therefore, dynamic correction can be a critical operation. If it is not prop-erly carried out, there can be undesirable quality deviations. ML can help identifying these types of deviations and thus correcting the situation in case it is needed [28]. Direct correction on the raw meal can be predicted and thus changed before measuring the 28-day strength of cement.

6.7.2 PURCHASED MATERIALS AND COMBUSTIBLES

In order to produce high kiln temperatures, fossil fuels such as coal, petroleum coke, and natural gas are regularly purchased and used. Large amounts of combustibles are used on a daily basis. With the rising cost of energy, fuel costs comprise high percentage of a manufacturer's budget and the incentive to seek less costly fuels is a tradition in different cement plant.

Alternative fuels are usually l used and their quality differs from one source to the other [29]. Modern facilities typically use alternative fuels to partially replace fossil fuels at high rates. Alternative fuels are typically waste products from other industries that are destined to be land-filled or incinerated [30]. The incineration of these wastes at cement plants is cheaper than installing new incinerators. Temperatures used to produce the clinker meet the requirements for incinerating hazardous wastes. Different deviations and error can be observed on this supply chain. Their origins can be summarized as follows:

- Error on supplier side (wrong material, wrong source, or provenance change without advance information of the client and his agreement, e.g. fluctuation of the moisture level of fuels, limestone, or slag on delivery).
- Deviation due to a material or production change on supplier side without information and agreement of the customer.
- Deviation due to a non-detected failure in the production process chain on supplier side, including storage and transport to the cement plant (e.g. possibility of contamination of the material during the transport).

All these issues need generally to be clarified with the supplier and carrier. Failures or weaknesses in the inspection methods or systems for the incoming raw materials can happen. Adequate delivery control needs to be in place. ML can generally not detect these errors as these inputs generally are not reported, and a full analysis of materials supply is not generally undertaken systematically. But once this processes is systemized, it can help even to further reduce the environmental impact of the cement plant [31, 32]. Data collection and systematic reporting of fuels analysis is therefore important as it allows to satisfactorily predict the kiln response to the firing of alternative fuels before there usage. This may ultimately help in the prevention of environmental problems due to the polluting emissions.

6.7.3 RAW MIX

Raw mix preparation aims to minimize the feed chemistry fluctuations [26]. The raw mix should satisfy in practice the stoichiometric requirements of the LSF, Silica Modulus (SM), and Alumina Modulus (AM) [33]. The previously mentioned quarry fluctuations are smoothened early in the raw meal preparation process, namely at the pre-blending beds. Optimum proportioning of the different raw materials is performed there. The Raw Mix Optimization program reduces and controls short-term fluctuations to the target values. This permits implementation of predictive actions rather than reactive ones. Models can detect if:

- The blending of raw components was not carried out according to the fixed ratios of each material.
- The dosage of one component was defective. This can be due to the following reasons:
  - A failure in the dosing system itself, e.g. of the weigh feeder, or, due to a non-detected interruption of the material flow dosed volumetrically, because of blockages or material bridging in the silo/bin
  - A failure in automatic raw mix control
  - An error in manual raw mix control
  - A deviation in the composition of one component due to uncontrolled or undetected contamination or intermixing with another material occurring on the way to the storage silo or dosage bin

The consequences of poorly prepared raw meal are well known. Predicting it allows to have all quality targets reached at all stages of the raw mix blending, from the quarry to its grinding, at the lowest possible cost [34]. Controlling raw meal proportioning creates the basis for the modular growth adapted to the plant's and customer's needs [35]. However, ML as described previously can only be applied if real figures and numbers are attributed to these processes [36]. The online estimation of process parameters is important to define the quality of the clinker and appears as the most valuable factor in preparing a good cement. The raw data need to be however well analyzed and pre-processed to remove the outliers. The estimation capabilities of ML are proven by simulation of different developed models. Depending on the data in hand, some network models can produce better estimation capabilities than others.

6.7.4 RAW MEAL

Similar to the raw mix, a quality deviation in the raw meal composition can occur, when the dosage of one component is defective or if some pre-blending of constituents was not carried out as instructed. In case of deviation of fineness, it is more likely related to the following:

- The separator of the closed-circuit mill and/or to the performance of the filter including dust return
- A possible failure in automatic control or an error in manual control

Here again ML can help only if this type of operation is reported continuously and documented [37, 38]. More efforts are needed at the industry level to introduce in future models that would help intervene at this stage.

6.7.5   Kiln Feed

Information like the deviations in quality or fineness of kiln feed are indicators that something is wrong in the homogenizing process, which needs to be controlled by production staff (e.g. silo aeration, silo extraction points, material flow out of different silos, kiln dust return in direct or compound mode, etc.).

As long as this information is available, ML can process these data without any problem [39]. Application of ML to a lime kiln [40] has proven that the kiln feed quality control could contribute to an enhanced performance of the kiln and of the product quality.

However, other information such as dust management, the enrichment of fine particles in dust, resulting in a LSF of dust deviating substantially from raw meal LSF are not possible to treat by ML despite their huge impact on the chemical variability and burnability of kiln feed. Depending on the dust composition and volume, this operation always requires inputs from the plant operators.

Another important factor is the way how dust is returned: In the worst case, the dust is returned to the raw meal silo, creating the risk of layer formation, in the best case, a separate dust silo is available, from which dust can be added to kiln feed in a controlled way. If no separate dust silo is available, dust addition to kiln feed will be the most reasonable alternative. Anyway, this effect can cause substantial differences in raw meal and kiln feed composition. These operations are missed most of the time and, thus, not captured in the ML database, as these are dependent on the operator's strategy at the cement plant.

6.7.6   Hot Meal

Variations in the quality of hot meal are primarily connected to the burning process itself. The reasons are multiple and can be adjustment of kiln operation parameters, feed rates of kiln feed and combustibles, changes in fuel mix, circulations inside the kiln, build-ups in the preheater and at the kiln inlet, and bypass rate. ML can only help if these operations are detailed and reported in the plant operational database.

Temperature in the kiln is often reported. However, most of the process data are not correlated with the clinker quality or hot meal quality. Tentative attempts on applying ML have been made successfully [41]. However, most of the tests are done on the final products and only in few cases cement plants could report exactly the conditions of production of different clinkers.

6.7.7 FUEL PREPARATION

As the main control parameter during fuel preparation is its fineness, a deviation of it is most likely related to the process rather than to fuel quality, except in case of non-detected intermixing of materials. Such parameters if available could be easily incorporated in ML quality control tool [42].

Given the differences in ignition temperatures of different fuel types, it is important that fuels are prepared and introduced through an appropriate installation at the correct points in the process to ensure complete combustion or incorporation and to avoid unwanted emissions. Therefore, in most cases, a specific additional installation needs to be adjusted to allow wide use of these materials and ensure product quality.

6.7.8 CLINKER

In the cement manufacturing process, the quality of the clinker exiting the rotary kiln determines the eventual quality of the cement produced. The process (input) parameters define the clinker quality parameters.

Real-life data of clinker are often not available as it remains confidential with the cement manufacturers for their clear competitive advantage. The only available data that is shared outside relates to the normal operating conditions of the plants. More extensive operating conditions are available for some plants where changes in process characteristics are reported. ML helps to detect whether changes in the operating conditions really happened in the plants [43]. Indeed, similar to hot meal, deviations in clinker quality are related to the burning process and changes in the burning conditions (adjustment of kiln operation, feed rates, fuel mix, reducing atmosphere, circulations, build-ups, and ring formation, etc.). If reported, all these parameters together can be treated through ML [44].

6.7.9 DISPATCHED CEMENT

It goes without saying that detection of a quality deviation at the end of the complete cement production chain is the most critical one and needs to be followed up without delay.

If quality control during the cement grinding process showed no deficiency and deviation, we can distinguish between two situations:

- Cement quality deviates in its composition and possibly in its fineness, indicating that most likely something happened on the way between cement mill and storage silo. ML may offer possibilities to define the strength that may be achieved by the cement at the specified ages.
- Cement quality deviates in the physical or mechanical properties, probably due to insufficient or deficient evaluation of the cement

quality within the auto control, no or late recognition of the deviation trends, inadequate or late adjustment of quality targets – possibly in the early stage of the production process.

In the first case, the following root causes may be considered:

- The cement has been ground and transported to the wrong storage silo (e.g. due to human error and missing the interlocking of transport routes from the mill to the silo for different types of cements).
- Missing the transport route control, including the position of valves (open/closed), operated manually or monitored from the control room.
- Material flowing through a leaking valve to the wrong silo.
- Inadequate or too long purge procedure during type change in the cement mill.

6.8 APPLICATION OF ANN TO FINAL PRODUCT QUALITY

There are immense possibilities of applying models based on artificial neural networks to predict the cement properties. Two examples are presented below.

6.8.1 STRENGTH PREDICTION: TRADITIONAL STATISTICS VS. PREDICTIVE ANN

On the left side of Figure 6.3, the relation between the Blaine surface area of cement of a given plant with its 2-day, 7-day, and 28-day compressive strength values is shown, in which no correlation is detectable between cement fineness and compressive strength. However, by analyzing the complete quality control data with ML, a predictive model could be trained and used for

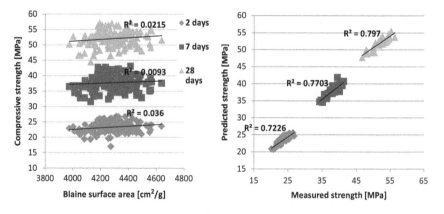

FIGURE 6.3 Comparison between the traditional approach (left) and predictive modeling with ANN (right).

process control (Figure 6.3 right), in which there is a visible correlation between the measured and predicted strength values.

The approach to the development of the predictive modeling consists of the following steps:

- In the first step, the input data, e.g. 2-day compressive strengths, Blaine surface area values, CaO, SO_3, Cl⁻ contents, etc. and the output data, e.g. 7- and 28-day compressive strength, need to be defined. The input and output data can be any type of data, which is available.
- Within the scope of ANN approximation and data analysis, outliers and wrong data are eliminated.

Then the following modeling strategy is adopted:

First, the capability of the model is checked; all compressive strengths and setting times were predicted "at once". This means that the model could predict the 28-day compressive strength without taking into account the data on early-age strengths, i.e. the strength data at 1, 2, and 7 days. The predictions were based exclusively on data other than compressive strength and setting time.

The intention of the quality control model is to predict the performance of cements, whose composition etc. is already available, but their strength needs to be predicted to optimize and steer production. It has to be noted that only outliers with absurd values are removed and that models are not optimized during evaluation.

The model is set up with large data points, using the oldest data points obtained from the same process and production technology.

Building the model requires the following procedures to be repeated until all data points in the data set are used in the model:

- The model is trained using the data points (N) introduced initially to the model.
- Strengths of the N points are predicted.
- Strengths of the N+1st data point, i.e. of the next data point that will be included, are predicted.
- Finally, this N+1st data point is added to the model, and this procedure is continuously repeated.
- Results of such an approach are shown in Figure 6.4 below.

Initially, the model was trained with only 20 data points, and future points were added as progressively generated. The strengths were predicted, once the input parameters were available, and compared to the measured values (a step known as "forward prediction"). The measured values were then included into the training set. As can be seen in Figure 6.4, the average absolute error is below 1 MPa. Some forward predictions had resulted in significant error. This corresponded to situations when the input data was significantly out of

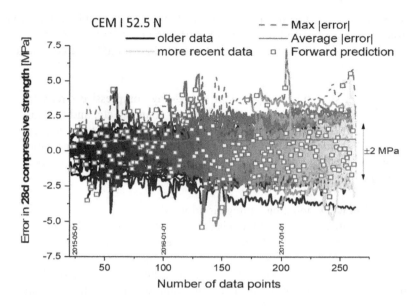

FIGURE 6.4 Accuracy of the prediction model with time. Forward prediction denotes situations where input variables of a given point are already available, but its strength not yet determined.

the training range and yet fed to the model. The black curve at the very bottom of the graph in Figure 6.4 is a typical outlier.

6.8.2 Deviations in Ground Cement: Influence of SO₃ Level

In this example, the reasons for the early strength fluctuations in the ground cement could not be assessed from the parameters measured by QC either using numerical statistics or expert judgment. The parametric correlations were hidden in the common fluctuations in the production process (Figure 6.5 left). The use of ML helped to create virtual scenarios with all other parameters at constant SO_3 levels in a way that the impact of SO_3 content on the 1-day compressive strength of the cement mortar could be revealed (Figure 6.5 right). It helped to understand the production process better, to find the root cause, and to finally adjust the process to avoid the fluctuations in the early-age strength of the cement.

Deviations in the quality of cement, its composition as well as its fineness (not only Blaine but also particle size distribution) can be due to similar situations as listed in the sections addressing raw mix and raw meal, namely defective dosage of constituents, failure in the separator, filter or dust return, failure or error in the control loop. Furthermore, proper dosing of grinding aids/quality improvers and their composition need to be checked regularly. Water spraying and cement mill temperature can also play a crucial role.

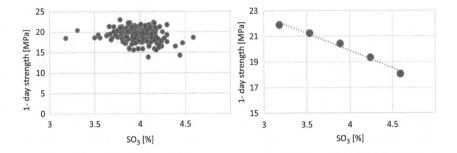

FIGURE 6.5 Impact of the sulfate content in cement on its 1-day strength

If no silo for the control of intermediate products is available, frequent cement type changes at a specific mill can also contribute to cement quality fluctuations. All the parameters are hard to assess. However, treating the chemistry, mineralogy, and sulfate contents enables most of the time ML to find out measures to have a consistent quality.

6.9 CONCLUDING REMARKS

The way how to search for the root cause of any deviation and the initiation of corrective actions are to be found individually case-by-case in the plants, depending on the local situation and the concerned equipment, material or production step. A quality control tool based on ML must include the corrective actions, which should be taken step by step by the operators, if results are deviating from the normal operating range.

Depending on the magnitude and duration of the quality deviation, the risk can grow regarding:

- Health and safety as well as environmental issues
- Process operation and control
- Equipment and material
- Product quality and the impact of its use after production or application
- Costs

Urgency of the reaction as well as level of measures strongly depend on the risk level mentioned above, the stage in the production chain, where the deviation happens (the more critical, the later the production step and the higher the impact on the final product), the volume of material concerned and the rising costs related to any corrective action to be taken. Applying the new technologies can help to reduce costs and overcome these problems. The volume of deviating material produced will largely depend on the output capacities of the production units and storage installations and especially of the duration of the deviation. Tonnages vary considerably in different cement plants.

Cement plants are in increasing need of quality control that require more and more analysis of samples and involve complexity of applied methods. This is a result of environmental regulations placing stricter controls on cement production and market competition. This evolution, however, can be taken care of by the developments in the measurement techniques, their miniaturization, convenience to use, and finally with the advanced automation. These developments allow to gather significantly more data along the cement production and application of the modern numerical tools to analyze them.

A typical fluctuation in cement plant can occur while preparing raw meal especially on the particle size distribution, which is an essential parameter for raw meal burnability. Analyzing raw meal data via ML can help to see the impact of the variation of measurement on deviations from the target properties of the clinker whether it is the content of C_3S or free lime, and so on. It can demonstrate the need to clearly define under which circumstances a corrective action has to be taken. The extent of the correction will depend on the trend of the last consecutive results. The operator can define from which value he considers that the deviation and the situation become critical to clinker quality and the model will help to define the limits of PSD that should be respected.

In order to ensure manufacturing of products adjusted to customer needs, smooth operation at all production steps, economic utilization of raw materials, and compliance with regulations and standards, a quality assurance system must be established at every cement plant. It covers the entire production process from raw materials via intermediate goods to the final product cement. The basis of the quality control concept is the sampling and testing plan, which must be drawn up according to the applicable guides, taking the individual plant situation into account. All testing results should be documented and checked manually or automatically for plausibility. Depending on the magnitude of deviation from the fixed quality target, on detected trends or fluctuations of quality, corrective actions are initiated or decided.

ML as described can detect in the data analyzed deviations due to incorrect (non-representative) sampling, erroneous measurements, deviating measurement conditions, and incorrect calibrations or settings of automatic control. If sampling or sample preparation is not done properly, it will be worthless to look at statistics using traditional methods. With the use of neural networks it is possible to detect the anomalies in comparison with the correct data [45]. However, if the sampling was always done wrongly, ML cannot help to detect and rather will consider potentially correct data as wrong.

The decision whether corrective actions are initiated or not depends on the output from neural networks, and thus it is important to have a large set of correct data. There, the acceptance limits around the fixed target, where corrective actions are to be taken, need to be clearly defined.

REFERENCES

1. P. L. Johnson, *ISO 9000: Meeting the new international standards*. McGraw-Hill New York, 1993.
2. Q. A. Nguyen and L. Hens, "Environmental performance of the cement industry in Vietnam: the influence of ISO 14001 certification," *Integrating Cleaner Production into Sustainability Strategies*, vol. 96, pp. 362–378, June 2015, doi: 10.1016/j.jclepro.2013.09.032.
3. C. W. Moore, *Control of Portland cement quality*, 2007.
4. R. Jacobs and S. Regis, "Automated cement plant quality control: in-situ versus extractive sampling and instrumentation," 2005, pp. 11–22.
5. R. Jacobs and S. Regis, "Automated cement plant quality control," *IEEE Ind. Appl. Mag.*, vol. 12, no. 2, pp. 21–28, Apr. 2006, doi: 10.1109/MIA.2006.1598023.
6. W. Gecks and S. T. Pedersen, "Robotics-an efficient tool for laboratory automation," *IEEE Trans. Ind. Appl.*, vol. 28, no. 4, pp. 938–944, Aug. 1992, doi: 10.1109/28.148461.
7. "QCX Robolab selection guide: Options from standalone to fully automated cement laboratory preparation," *FLSmidth Brochure*.
8. O.-W. Lau, S.-F. Luk, N. L. N. Cheng, and H.-Y. Woo, "Determination of free lime in clinker and cement by iodometry. An undergraduate experiment in redox titrimetry," *J. Chem. Educ.*, vol. 78, no. 12, p. 1671, Dec. 2001, doi: 10.1021/ed078p1671.
9. M. Bouchard et al., "Global cement and raw materials fusion/XRF analytical solution. II," *Powder Diffr.*, vol. 26, no. 2, pp. 176–185, 2011, doi: 10.1154/1.3591181.
10. J. Anzelmo, "The role of XRF, inter-element corrections, and sample preparation effects in the 100-year evolution of ASTM standard test method C114," *J. ASTM Int.*, vol. 6, no. 2, pp. 1–10, Jan. 2009, doi: 10.1520/JAI101730.
11. G. Walenta and T. Füllmann, "Advances in quantitative XRD analysis for clinker, cements, and cementitious additions," *Powder Diffr.*, vol. 19, no. 1, pp. 40–44, 2004, doi: 10.1154/1.1649328.
12. D. Jo, R. S. Leonardo, F. K. Cartledge, O. A. M. Reales, and R. D. Toledo Filho, "Gypsum content determination in Portland cements by thermogravimetry," *J. Therm. Anal. Calorim.*, vol. 123, no. 2, pp. 1053–1062, Feb. 2016, doi: 10.1007/s10973-015-5078-y.
13. S. A. Bernal et al., "Characterization of supplementary cementitious materials by thermal analysis," *Mater. Struct.*, vol. 50, no. 1, p. 26, Aug. 2016, doi: 10.1617/s11527-016-0909-2.
14. B. Osbaeck and V. Johansen, "Particle size distribution and rate of strength development of Portland cement," *J. Am. Ceram. Soc.*, vol. 72, no. 2, pp. 197–201, Feb. 1989, doi: 10.1111/j.1151-2916.1989.tb06101.x.
15. Zhongjing Ren, "On-line monitor technique of cement particle size distribution," Oct. 1993, vol. 2066, doi: 10.1117/12.162103.
16. V. Viswanathan, "Quality control in cement plant," *Cem. Concr. Sci. Technol.*, vol. 1, pp. 140–173, 1991.
17. Y. Zhao, B. Ding, Y. Zhang, L. Yang, and X. Hao, "Online cement clinker quality monitoring: A soft sensor model based on multivariate time series analysis and CNN," *ISA Trans.*, Feb. 2021, doi: 10.1016/j.isatra.2021.01.058.
18. N. J. Nilsson, *Principles of artificial intelligence*. Morgan Kaufmann, 2014.

19. I. Arel, D. C. Rose, and T. P. Karnowski, "Deep machine learning - a new frontier in artificial intelligence research [research frontier]," *IEEE Comput. Intell. Mag.*, vol. 5, no. 4, pp. 13–18, 2010.

20. P. E. García-Casillas, C. A. Martinez, H. C. Montes, and A. García-Luna, "Prediction of Portland cement strength using statistical methods," *Mater. Manuf. Process.*, vol. 22, no. 3, pp. 333–336, Mar. 2007, doi: 10.1080/10426910701190352.

21. S. Akkurt, G. Tayfur, and S. Can, "Fuzzy logic model for the prediction of cement compressive strength," *Cem. Concr. Res.*, vol. 34, no. 8, pp. 1429–1433, Aug. 2004, doi: 10.1016/j.cemconres.2004.01.020.

22. S. Akkurt, S. Ozdemir, G. Tayfur, and B. Akyol, "The use of GA–ANNs in the modelling of compressive strength of cement mortar," *Cem. Concr. Res.*, vol. 33, no. 7, pp. 973–979, Jul. 2003, doi: 10.1016/S0008-8846(03)00006-1.

23. D. Damljanovic, "Success factors for the implementation of artificial intelligence in the cement industry: Mannersdorf plant case study," 2019.

24. R. King, "Intelligent control in the cement industry," *IFAC Proc. Vol.*, vol. 21, no. 19, pp. 303–307, 1988.

25. Z. Yuan, Z. Liu, and R. Pei, "Fuzzy control of cement raw meal production," 2008, pp. 1619–1624.

26. A. K. Chatterjee, "Chemistry and engineering of the clinkerization process—incremental advances and lack of breakthroughs," *Cem. Concr. Res.*, vol. 41, no. 7, pp. 624–641, 2011.

27. P. Hewlett and M. Liska, *Lea's chemistry of cement and concrete*. Butterworth-Heinemann, 2019.

28. M. Fangqing and L. Yunxia, "Mind-evolution-based machine learning for dynamic quality control of raw materials in a cement plant," 2004, vol. 3, pp. 2409–2414.

29. U. Kääntee, R. Zevenhoven, R. Backman, and M. Hupa, "Cement manufacturing using alternative fuels and the advantages of process modelling," *Fuel Process. Technol.*, vol. 85, no. 4, pp. 293–301, 2004.

30. C. Horsley, M. H. Emmert, and A. Sakulich, "Influence of alternative fuels on trace element content of ordinary portland cement," *Fuel*, vol. 184, pp. 481–489, 2016.

31. M. Mirmozaffari, M. Yazdani, A. Boskabadi, H. A. Dolatsara, K. Kabirifar, and N. A. Golilarz, "A novel machine learning approach combined with optimization models for eco-efficiency evaluation," *Appl. Sci.*, vol. 10, no. 15, p. 5210, 2020.

32. E. Marengo, M. Bobba, E. Robotti, and M. C. Liparota, "Modeling of the polluting emissions from a cement production plant by partial least-squares, principal component regression, and artificial neural networks," *Environ. Sci. Technol.*, vol. 40, no. 1, pp. 272–280, 2006.

33. H. F. Taylor, *Cement chemistry*, vol. 2. Thomas Telford London, 1997.

34. J. Gómez Sarduy, J. Monteagudo Yanes, M. Granado Rodríguez, J. Quiñones Ferreira, and Y. Miranda Torres, "Determining cement ball mill dosage by artificial intelligence tools aimed at reducing energy consumption and environmental impact," *Ing. E Investig.*, vol. 33, no. 3, pp. 49–54, 2013.

35. X. Lin and Z. Qian, "Modeling of vertical mill raw meal grinding process and optimal setting of operating parameters based on wavelet neural network," 2014, pp. 3015–3020.

36. Y. Boukhari, "Application and comparison of machine learning algorithms for predicting mass loss of cement raw materials due to decarbonation process," *Rev. Intell. Artif.*, vol. 34, no. 4, pp. 403–411, 2020.

37. J. Qiao, X. Zhao, and T. Chai, "Soft measurement model of raw meal decomposition ratio based on data driven for raw meal calcination process," 2019, pp. 2857–2862.
39. X. Zhang and J. Zhao, "Prediction model for rotary kiln coal feed based on hybrid SVM," *Procedia Eng.*, vol. 15, pp. 681–687, 2011.
40. M. Järvensivu, K. Saari, and S.-L. Jämsä-Jounela, "Intelligent control system of an industrial lime kiln process," *Control Eng. Pract.*, vol. 9, no. 6, pp. 589–606, 2001.
41. Z. T. Xue and Z. Li, "Application of fuzzy neural network controller for cement rotary kiln control system," 2012, vol. 457, pp. 531–535.
42. Y. D. Aleksandrovich, M. V. Zalmanovich, and D. E. Pavlovich, "Machine vision system for assessment of firing process parameters in rotary kiln," *World Appl. Sci. J.*, vol. 24, no. 11, pp. 1460–1466, 2013.
43. W.-N. Wu, X.-Y. Liu, Z. Hu, R. Zhang, and X.-Y. Lu, "Improving the sustainability of cement clinker calcination process by assessing the heat loss through kiln shell and its influencing factors: a case study in China," *J. Clean. Prod.*, vol. 224, pp. 132–141, 2019.
44. Q. Liu, H. Yu, X. Wang, and S. Lu, "Research on online monitoring of energy consumption of Rotary Kiln for Cement Clinker formation heat," 2020, vol. 1, pp. 469–474.
45. I. Mahdavi, B. Shirazi, N. Ghorbani, and N. Sahebjamnia, "IMAQCS: Design and implementation of an intelligent multi-agent system for monitoring and controlling quality of cement production processes," *Comput. Ind.*, vol. 64, no. 3, pp. 290–298, 2013.
38. J. Qiao and T. Chai, "Soft measurement model and its application in raw meal calcination process," *J. Process Control*, vol. 22, no. 1, pp. 344–351, 2012.

7 Asset Performance Monitoring and Maintenance Management in Cement Manufacturing

P. V. Kiran Ananth and K. Muralikrishnan
Confederation of Indian Industries, CII – Sohrabji Godrej
Green Business, Hyderabad, India

Anjan Kumar Chatterjee
Conmat Technologies Private Limited,
Kolkata, India

CONTENTS

7.1 Introduction 226
7.2 Basics of Asset Performance Approach in Industrial
Environment...227
7.3 Practices of Technical Performance Monitoring in
Cement Plants 230
    7.3.1 Large Cement Manufacturing Groups 230
    7.3.2 Small- and Medium-Sized Cement Groups 232
7.4 Key Assets in Cement Manufacturing and Their Performance
Monitoring Aspects..232
    7.4.1 Preheater with Precalciner...233
    7.4.2 Rotary Kiln..233
    7.4.3 Clinker Cooler 234
    7.4.4 Refractory Lining...234
    7.4.5 Bag Filters..235
    7.4.6 Mill Separators 236
    7.4.7 Process Fans...236
    7.4.8 Power Transformers 237
    7.4.9 Motors and Drives...238
    7.4.10 Power Distribution System...238
    7.4.11 Variable Frequency Drives 238

	7.4.12	Gearbox	239
	7.4.13	Belt Conveyor	240
	7.4.14	Compressors	241
	7.4.15	Process Pumps	241
	7.4.16	Bucket Elevators	242
	7.4.17	Couplings	243
	7.4.18	Bearings	243

7.5 Current Monitoring Practices for Energy Efficiency Assessment 243

7.6 Recent Application of AI-Based Components for Asset Performance Monitoring 244

7.7 Advances in Maintenance Strategies and Practices...........................248

7.8 Near-Term Prospects..251

7.9 Concluding Observations 253

References..254

Annexures...254

    Annexure 1. Limestone Crusher Section..254

    Annexure 2. Raw Mill Section (VRM) 255

    Annexure 3. Raw Mill (Ball Mill) 256

    Annexure 4. Pyro Section 256

    Annexure 5. Cement milling – Ball Mill 258

    Annexure 6. Utilities 260

    Annexure 7. Captive Power Plant – Process..................................260

    Annexure 8. Captive Power Plant – Electrical...............................261

7.1 INTRODUCTION

The cement manufacturing is known to be highly capital-intensive with diverse range of structure and equipment. In addition to the main machinery, such as crushers, raw and finish grinding mills, kiln systems, packing and bulk shipping facilities, a cement plant is strewn with various structures, which include silos, preheater towers, kiln supports, mill foundations, transfer towers and conveyors – some constructed with steel but mostly with concrete. Furthermore, the plants are provided with the grid power, internal electrical networks and captive power generating facilities. A single million-ton production line has more than 700 motors to drive a large number of equipment. Sensors, instrumentation, process automation and control devices, all of electronic nature, are extensively dispersed throughout the production facility. Added to all the above features are the fleets of mobile vehicles for mining and material transport.

Because of the diverse and complex infrastructure, maintenance has been a part and parcel of cement manufacturing. The maintenance activity has been multi-disciplinary, ranging primarily across chemical, mechanical, electrical and electronics, automotive, civil and structural engineering. In economic terms, the maintenance cost has generally been about 15% of the production cost or 3%–5% of the plant and machinery cost.

The asset performance and maintenance have, therefore, been in focus, all along the history of cement production, for technological development so as to improve the competitiveness of the cement plants. Over time, the asset performance has turned into a multi-disciplinary subject that refers to a business's ability to manage its assets and to secure profitable returns from them. It is typically used to compare a company's performance over time or against its competition. Simultaneously, the asset maintenance function progressed from physical inspection and attending to defects, repair and replacement to computerized maintenance management systems (CMMS). In recent years, with the advent of artificial intelligence (AI)-based tools and techniques, the asset performance management (APM), and maintenance strategy have become more and more reliability-centric and predictive.

In the above background, this chapter reviews the status and trends of asset performance and maintenance in cement manufacture and the application potential of emerging concepts of AI to improve asset utilization.

7.2 BASICS OF ASSET PERFORMANCE APPROACH IN INDUSTRIAL ENVIRONMENT

The capital-intensive industries have been progressively realizing that even a small improvement in asset performance results in perceptibly higher returns on investments or ensuring the same levels of return from fewer assets. Hence, both the business managers and industry analysts monitor the performance of assets created as a block with the help of financial matrices and ratios. The most widely used parameter is the 'fixed assets turnover ratio'. While this can be regarded as a sound management approach, in practice the asset performance needs to be understood as an onion structure with enveloping shells of maintenance philosophy, operations technology and business strategy as schematically depicted in Figure 7.1.

Nowadays the APM has turned out to be a buzzword and there are various tools and solutions available with differing functionality. However, they all aim to improve equipment availability and reliability while reducing risk and cost. They typically capture data related to asset condition, which include work order history, condition data via plant historians, laboratory results and inspection results to provide a holistic technical view of the asset performance. This data is used for reliability analytics and asset health visualization and also life-cycle costing. The multifarious APM capabilities are summarized in Figure 7.2 [1].

The first and foremost requirement is to undertake data analytics geared towards reliability analysis through querying, extracting, visualizing and analysing both the structured and unstructured data from production systems. The modelling step follows thereafter to develop component level models of

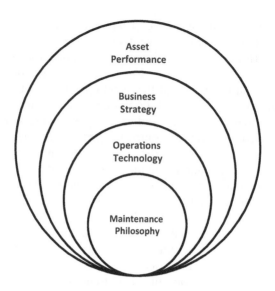

FIGURE 7.1   Onion structure of asset performance.

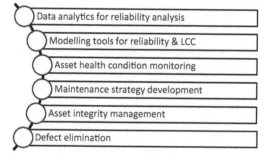

FIGURE 7.2   APM functions supported by various tools (adapted from [1]).

mobile and fixed assets so as to undertake benchmarking, repair/replacement decisions and to consider execution alternatives. The health condition monitoring tools support to identify abnormal physical conditions of equipment so as to trigger operator intervention and also to institute changes in maintenance strategies in order to avoid recurrence of such conditions. This leads to the development or modification of maintenance strategy for enhanced availability of equipment and reducing the operating expenses. The asset integrity management refers to the calculation of risk and residual life of assets on one side and adoption of optimal inspection strategies on the other. Finally, the APM solution must support identification and elimination of recurring and chronic defects that might be impairing the equipment performance, development of execution plan and tracking of its completion. It may be relevant to

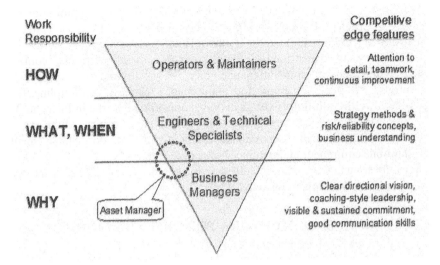

FIGURE 7.3 Pattern of techno-managerial roles and responsibilities for implementation of APM [2].

mention that safety management also forms an important part of the APM solutions.

The interfacing of APM solutions with a production enterprise equipped with enterprise resource planning (ERP)/enterprise asset management (EAM) and condition monitoring systems is shown in Figure 7.3 [1]. Where the ERP/EAM systems provide the transactional platform to support the organizational management process, the APM system may have an interface with it in order to enable optimization of the maintenance strategies in the ERP. When the asset health data is used to trigger work orders, these are also pushed to the ERP/EAM for execution. In addition, the APM solution also ingests data from the production systems, which can range from output details, laboratory test results, inspection reports, sensor data from the machines and even alarm and event data. Some APM solutions may even offer the mobility tools for recording inspection reports and site photographs. Several APM software packages such as IBM Maximo, Upkeep, eMaintCMMS, Fractal Asset Cloud, Infor CloudSuite EAM, etc. are available, but the correct choice for particular applications remains an important issue that demands domain knowledge and IT expertise.

In this emerging working environment, it is important to understand the roles and responsibilities of different techno-managerial groups operating in an organization, an illustration of which is given in Figure 7.3 [2]. As shown in the inverted pyramid, the leadership vision gets transmitted through different levels of engineering managers and domain specialists to the field operators and asset maintainers, who basically execute the tasks as necessary.

Monitoring by asset managers paves the way towards better exploitation of assets. The preparedness of workforce and business managers for their changing roles has to be ensured in the industrial enterprises.

It is pertinent to mention here that though the cement industry has continued to be focused on maintenance since the 1970s and has relentlessly improved the plant reliability and maintenance-related cost discipline, the AI-based APM solution is yet to make its appearance in the industry in its modern form and concept. Extensive process monitoring, adoption of CMMS, reliability-centric maintenance, maintenance of cost management and mobile equipment maintenance have been addressed all over the industry sector for long and are being effectively pursued [3]. Some of these aspects are discussed in the following sections.

7.3 PRACTICES OF TECHNICAL PERFORMANCE MONITORING IN CEMENT PLANTS

In early 1990s, performance monitoring of equipment in cement plants was done largely by manual methods, with little intervention from computers, which were used mainly for data storing. Lack of automation and data management tools made data capturing and management quite cumbersome. However, the level of automation in cement plants kept growing and in time became extensive, and by now, entire plants are controlled through a centralized control system. Presently, all the data that flows through the centralized control system is managed and stored using various associated software responsible for different performance verticals. Operators of the cement plant analyse the captured data and formulate necessary actions to improve performance. The type of action varies from plant to plant, depending on standard operating procedures (SOP), guidelines and safety measures followed by the plant. On broad classification, cement plants APM strategies and methodologies can be grouped into two categories: large groups and small groups.

7.3.1 Large Cement Manufacturing Groups

The large cement manufacturing groups have extensive automation and high-end ERP systems. The ERPs in large cement plants manage all the data regarding the plant, including plant parameters, daily report generation, maintenance schedules, maintenance job scheduling and assignment and resource (man power, material, etc.) management. However, certain critical parameters are still measured manually, such as flow measurements, transformer measurements and motor health assessment (except performance parameters (V/I/kW), temperature, vibration, gearbox and bearing wear and thickness). Continuous measurement of these parameters is not possible. Some manual measurements cannot be done with equipment in running condition. Because of these and other challenges, some large plants have formulated a strategized classification of these parameters, based on the possibility of measurement, as follows:

R11: Running in inspection (No tool required)
S11: Physical inspection in shutdown (Without tool)
R22: Running in inspection (Tool required)
S22: Physical inspection in shutdown (With tool)

The parameters that can be inspected in running condition through observations (lubrication oil level or sound) come under the R11 category. A typical example is the false air inspection across preheater through noise. The parameters that can be inspected during the shutdown without the help of any tools, such as refractory conditions inside the preheater are designated as S11. The parameters that can be inspected in running condition with the help of tools, such as measuring the vibration/current/static pressure at preheater fan by manometer, etc. are categorized as R22, while the parameters that can be inspected during shutdown with the help of tools, such as the bearing clearance check through filler gauge/residual thickness available in the liner/classifier blades, etc. are identified as S22. Depending on the above activities, the task list is made for each section (Process, Mechanical, and Electrical and Instrumentation), and the list is uploaded in ERPs in a format illustrated in Table 7.1.

After uploading the task list, if any deviation is observed from the permissible limit, it is first classified as high, medium or low, depending on the severity and criticality, and an automatic notification of abnormality is generated in the system. Finally, the responsibility is allocated and scheduled to a particular engineer or section in-charge to rectify the problem at the site. After completion of work, the notification can be closed in the system.

ERPs have enabled interconnectivity using internet between multiple plants belonging to the same group. With this interconnectivity, large groups manage the inventory not at plant level but at group level. When a plant in a group needs a critical spare/component/equipment that is available in another plant of the same group, the spare/component/equipment is quickly moved, hence saving time rather than waiting for repair or delivery of new equipment

TABLE 7.1
Task List Format

Activity	Range or Permissible Limit	Category	Severity	Observations
False air inspection	8%	R11	Medium	12%
Fan vibration	7 mm/s	R22	High	Less than 5 mm/s
Refractory lining	110 mm (50% of brick thickness)	S22	High	40 mm

TABLE 7.2

Maintenance Schedule in Practice for the Critical Electrical Equipment

MCC	Monthly
Transformer	Yearly
VFD	Monthly
VFD (Offline)	Quarterly
DC Drives	Monthly
HT Motors	Quarterly

by the technology provider. This way, large groups reduce overall downtime of their plants.

7.3.2 SMALL- AND MEDIUM-SIZED CEMENT GROUPS

Small- and medium-sized cement groups in some countries, including India, are following condition-based APM today. Major barriers in predictive and prescriptive approaches are the lack of extensive online data measurement instruments and reliable online analytics tools. Still, the majority of the plants have well-defined SOPs or maintenance manuals for asset management and maintenance scheduling.

Asset performance monitoring activities are carried out with the help of process and mechanical parameters such as temperature, pressure, colour (visual inspection), viscosity, hardness, V-belt tension and thickness. Monitoring of these parameters is not on a continuous real-time basis, but on fixed schedules, such as weekly, monthly, or semi-annually, etc. A typical maintenance schedule in most cement plants for the critical electrical equipment is shown in Table 7.2.

For non-critical equipment such as auxiliary bag filters, smaller blowers and fans, plants still continue to follow the reactive approach.

7.4 KEY ASSETS IN CEMENT MANUFACTURING AND THEIR PERFORMANCE MONITORING ASPECTS

The key assets of a cement plant can be categorized into three main functional groups:

- Process
- Electrical
- Mechanical

The key assets of the above three groups are listed in Table 7.3.

TABLE 7.3

List of Key Assets in Three Functional Groups

Process	Mechanical	Electrical
Burner	Gearbox	Transformer
Refractory	Belt conveyor	HT motors
HAG	Bucket elevator	LT motors
AFR handling system	Compressor	HT/MV VFD
Bag filter	Pump	LT VFD
Separator	Bearings	Cables/cable tray
Rotary Kiln		MCC/substation
Cooler		PLC
VRM		EMS
Ball mill		IT system
Roller press		
Fan		
De NOx/Cl bypass/FGD		
CEMS		

The functional features, parameters monitored and the current monitoring practices for the critical equipment listed in Table 7.3 are discussed below in detail.

7.4.1 PREHEATER WITH PRECALCINER

The feed to the preheater in a cement plant, consisting of multistage (4–6 stages) cyclones, is heated by the exhaust gases from the kiln and cooler, and by firing additional fuel in the calciner to help the material to reach the temperature of 850°C, which results in calcination of around 80% of the feed. The parameters of the preheater like cyclone draft, material and air temperature, excess O_2, PH fan inlet draft and PH fan RPM are measured via various sensors and are fed into the centralized automation system (DCS), where the parameters are monitored by the plant operators. Some cement plants have also installed PID loops in the calciner fuel firing. The fuel firing is controlled on the basis of the calciner temperature alone, while ignoring all the other parameters of the pyro section.

7.4.2 ROTARY KILN

The heated material from the preheater is fed to the kiln, where the rest 20% of the material is calcined and the whole material is transformed into clinker in the burning zone at a temperature of around 1,400°C. The key parameters in the rotary kiln are volumetric loading, secondary air temperature and hood draft. The secondary air temperature depends upon the heat recuperation in the cooler and the hood draft is controlled by preheater fan suction. The volumetric loading of the kiln is controlled by the kiln rpm.

In most of the cement plants, all the kiln parameters are measured via sensors and fed into the DCS. Through DCS the parameters are monitored by the plant operators. Some cement plants have PID loop for maintaining the volumetric loading of the kiln. The PID loop manipulates the rpm of the kiln with respect to the kiln loading.

7.4.3 CLINKER COOLER

The hot clinker is fed to the cooler, where the clinker is cooled with the help of air. The major part of the heated exhaust air from the cooler is fed to the kiln and calciner for heat recovery. The key parameters in the clinker cooler are pressure drop across clinker bed, cooler cross bar strokes, vent gas temperature, clinker temperature at the outlet, cooler fan rpm and water spray. All the above parameters are measured through sensors and are fed to the DCS. Through DCS, the operators monitor the parameters and manipulate the control variables accordingly. However, in some plants, the cooler fan speed is controlled via PID loops. The PID loops control the rpm of the cooler fans on the basis of the pressure drop across the clinker bed, ignoring the other process parameters.

While all the aforesaid parameters of the pyro system are monitored through the plant DCS, there are still important parameters like the preheater gas flow, cooler loading, residence time in preheater and kiln and the specific energy consumption of the pyro system (Thermal and Electrical) that are still measured and monitored manually by the plants. Due to the manual monitoring, these critical parameters are measured and monitored on weekly, monthly or even half yearly basis.

7.4.4 REFRACTORY LINING

Refractories are a key component of cement industry. Because of the very low thermal conductivity, refractories inhibit heat flow from inside of the kiln and other components to the outside, due to which refractories are able to contain high temperature within the kiln and other components on a continuous basis. This enables, the temperature of the metal structure to be maintained within the tolerable limits, thus making manufacturing possible on an industrial scale. The inhibition of heat flow also helps in conserving energy, thus making the process more economical.

Certain parameters decide the refractory selection, such as raw mix impact on refractory, kiln diameter, impact of fuel, coating formation, etc. (Table 7.4).

Previously the refractory lining building and dismantling was done manually. But presently the majority of plants follow screw jack method aided by a

TABLE 7.4
Important Parameters of Kiln Refractory

Process Parameter	Parameters for Procurement	Parameters Considered During Operation	Current Practices of Monitoring
Refractory application, which depends on zones inside the kiln	Burning zone: 70% Al_2O_3 or Basic brick depending on the raw mix; transition zone: 40% Al_2O_3; cooling zone: Almag bricks; diameter of the kiln	Shell temperature Minimum life expectancy for alumina bricks should be six months for continuous operation, and basic bricks around 12 months; optimum coating formation inside the burning zone	Kiln temperature profiling through shell scanner or manually done via pyrometer

brick machine for refractory building and for dismantling broke machine is used, which takes just 1–2 days to complete the entire process.

7.4.5 BAG FILTERS

Bag filters are extensively used in the cement industry to maintain the work environment and avoid material loss from the process. It recovers the product after grinding and controls dust emission. There are certain parameters to help select the right bag filter, such as ventilation volume requirement for de-dusting the equipment, right selection of the bag material with respect to dust, gas composition, nature and process condition, air-to-cloth ratio, emission level control to be achieved (inlet and outlet dust concentration limits), life expectancy of bags, process gas temperature, site conditions and type of equipment – whether it is opened or closed circuit (Table 7.5).

TABLE 7.5
Important Parameters of Bag Filters

Equipment	Process Parameter	Parameters for Procurement	Parameters Considered During Operation	Current Practices of Monitoring
Bag Filter	Ventilation volume, pressure drop, types of dust, emission level control	Inlet dust load, PSD of product and composition, ventilation volume and pressure drop	Outlet emission, pressure drop, new bag replacement cost, other spares cost	Check ventilation volume and pressure drop across filter through flow measurements, Air permeability

TABLE 7.6
Important Parameters of Separators.

Equipment	Process Parameter	Parameters for Procurement	Parameters Considered During Operation	Current Practices of Monitoring
Separator	PSD,% residue on product	Product residue, material characteristic, MOC, operating conditions such as concentration, gas and particle velocity	Seal gap, static vanes gap, dynamic vanes condition, blade thickness, bearing temperature	Check static vanes gap, measure seal air gap, inspect dynamic vanes

7.4.6 MILL SEPARATORS

Separator plays a vital role in achieving productivity in the overall grinding process in the cement industry. Certain parameters help in finding the right separator, be it for VRM or ball mill. Factors to consider during procurement are as follows (Table 7.6):

7.4.7 PROCESS FANS

The cement industry is one of the major users of industrial fans. Fans in the cement industry are heavy duty and perform two basic functions: supply of air or removal of exhaust gases, and material handling. There are many process fans in the cement industry, such as raw mill fans, induced draft fans, cooling fans, raw mill exhaust fans, coal mill fans and cooler exhaust fans. In addition to these key process fans, the cement industry also has many non-process fans. These fans work with bag filters for de-dusting of numerous conveying systems. The fan parameters are given in Table 7.7.

TABLE 7.7
Important Parameters of Fan

Process Parameters	Parameters for Procurement	Parameters Considered During Operation	Current Practices of Monitoring
Pressure, Temperature, Flow, control mechanism	Design parameters based on consultant/ OEM PG/maximum requirement	Vibration, Sound, loading, Damper position, Speed control	Efficiency, spare cost (impeller, bearing, and others), clearance, margin available in flow, head and efficiency

7.4.8 Power Transformers

Transformer data (such as voltage, current and temperature readings) should be recorded on a regular basis in order to determine the operating conditions of the transformer.

The typical maintenance activities for the distribution transformer are as follows (Table 7.8):

The typical maintenance activity of Dry type transformer is as follows (Table 7.9):

TABLE 7.8
Maintenance Activity for Oil Filled Power Transformers

Maintenance or Test	Recommended Interval
Review equipment ratings	5 years
Preventive maintenance	Per manufacturer's recommendations
Transformer physical inspection	Annually
Bushings – visual inspection	Quarterly and 3–5 years
Bushings – check oil level	Weekly
Bushings – cleaning	3–5 years
Transformer and bushings – Doble test	3–5 years (6 months to 1 year for suspect bushings)
Transformer and bushings – infrared scan	Annually
Insulating oil – DGA, physical and chemical tests	Annually after first year of operation
Core – Megger® test	If DGA indicates
Leakage reactance, Turns Ratio tests, SFRA test	If problems are indicated by other tests
Cooling fans – inspect and test	Annually
Oil pumps and motors – inspect and test	Annually

TABLE 7.9
Maintenance Activity of Dry Type Transformer

Maintenance or Test	Recommended Interval
Review equipment ratings	5 years
Infrared scan	Annually
Temperature alarm check	Annually
Visual inspection/cleaning	Annually
Check fan operation	Annually
Clean fans and filters	Annually
Turns ratio test	3–6 years or if problems are suspected
Megger® windings or Hipot	3–6 years or when problem is suspected

TABLE 7.10

Maintenance Schedule for Motors

Maintenance or Test	Recommended Interval
Insulation resistance (Megger®)	Annually
Infrared scan	Annually

7.4.9 MOTORS AND DRIVES

Motors drive pumps, fans, gates and valves in the cement industry. Critical motors should be tested on a routine basis.

The maintenance schedule for motors is mentioned below (Table 7.10).

7.4.10 POWER DISTRIBUTION SYSTEM

The relays and protective circuit are part of the electrical distribution system, and they sense the abnormalities in the system and equipment. The health of the relays and the protective equipment plays a major role for the critical assets in the system.

The typical maintenance schedule for the relays and protective circuit is mentioned below (Table 7.11):

7.4.11 VARIABLE FREQUENCY DRIVES

The VFDs are an integral part of the cement industry, where each critical equipment is controlled precisely and efficiently with the help of variable frequency drives.

TABLE 7.11

Maintenance Schedule for the Relays and Protective Circuit

Maintenance or Test	Recommended Interval
Fault/load study and recalculate settings	5 years
Electro-mechanical relays calibration and functional testing	Upon commissioning and every 2 years
Solid-state relays calibration and functional testing	Upon commissioning 1 year after commissioning and every 3 years
Microprocessor relays calibration and functional testing	Upon commissioning 1 year after commissioning and every 8–10 years
Protection circuit functional test, including lockout relays	Immediately upon installation and/or upon any changes in wiring and every 3–6 years
Check red light lit for lockout relay and circuit breaker coil continuity	Daily
Lockout relays cleaning and lubrication	5 Years

TABLE 7.12

Maintenance Schedule for the VFDs

Maintenance or Test	Recommended Interval
Filter mats (cleaning or replacement)	Every 2 years
Cleaning of inlet and outlet meshes	Every 2 years
Cleaning of sensitive parts inside the converter	Every 2 years
Environmental and operational data (since last inspection)	Every 2 years
DC link capacitors (capacity)	Every 2 years
Insulation test	Every 2 years
Safety circuits	Every 2 years

The maintenance schedule for the VFDs is mentioned below (Table 7.12):

7.4.12 GEARBOX

Since gears are mechanical machines to transmit motion or change speed, they should be inspected regularly and lubricated with gear oil in order to ensure their proper working. Internal parts of the gearbox can be checked visually during the shutdown for any damage. During regular maintenance, the following can be checked:

- Clamping is proper and no abnormal sound is heard.
- Oil temperature is within the permissible limit.
- Vibration is within limit (Use vibration analyser).
- Proper lubrication as per schedule, and oil is maintained above minimum level.
- No internal damage (During visual inspection).

The important gear box parameters are listed in Table 7.13 and the maintenance and remedial actions are presented in Table 7.14. The frequency of inspection and repair differs according to the operating condition and type of

TABLE 7.13

Important Parameters of a Gearbox

Process Parameter	Parameters for Procurement	Parameters Considered During Operation	Current Practices of Monitoring
Operating capacity (kW)	Loading capacity (kW); Type of gearbox based on application – driving end (input shaft – high speed) and non-driving or driven end (output shaft – low speed)	Vibration, lube oil temperature, lube oil level, sound check, bearing conditions	Lube oil quality by sample test, wear debris analysis

TABLE 7.14
Failure Conditions, Causes and Remedial Measures for Gearbox

Failure conditions	Possible Causes	Remedial Measures
Gear tooth is broken	Bounce in process operation, further damage	Replace gear
Gear is worn	Abnormal sound and bouncing exist in gear	Replace gear if gear thickness exceeds a specified value
Crack exists in gear rim and hub	Damaged gear	Replace gear
Key is damaged, gear bounces on an axle	Broken key	Replace key, proper alignment of a gear assembly

TABLE 7.15
Typical Frequency for Inspection and Repair for an Inline Helical Gearbox

Inspection/Repair	Frequency	Necessity of Parts Replacement
Re-stretch of chain	6 months	Tension of chain is loose, re-stretch it
Improper tightening	6 months	Tightening of bolts is loose, tighten additionally
Replacement of oil seal	1–2 years	In every overhaul/when oil leak, a seal is damaged, replace the oil seal
Replacement of bearing	5 years	Abnormal noise occurs, replace the bearing

gearbox used for the application. The typical frequency for an inline helical gearbox is illustrated in Table 7.15.

7.4.13 Belt Conveyor

Conveyors can serve well only when operated correctly, maintained properly and lubricated regularly. Proper maintenance of conveyors is important since it involves the care of many complex pieces of equipment. Failure of just one conveyor can shut down the entire operation and cause heavy loss and downtime. Generally, the conveying equipment consists of belting, a structure to support, a drive mechanism and pulleys to provide tractions, idlers to support and guide the belt, tensioning unit and a number of accessories for safe and good running of the system. The belt conveyors are designed and formed considering many factors, e.g., to suit the site conditions, nature of load and method of loading, etc. For keeping the conveyors working efficiently, a strict maintenance schedule should be drawn out and enforced effectively. Generally, the belt conveyors are provided with the following protection switches:

- Pull-Cord Switches – Pull Cord (manual reset type) switches are along both sides of the entire length of all conveyors. They are used as an emergency stop.

TABLE 7.16

Failure Conditions, Causes and Remedial Measures for Belt Conveyor

Failure Conditions	Causes & Possible Consequences	Measures
Side slip of belt	Support idler of the belt is not properly installed	Adjust with self-aligning idler, observe along the reverse rotation direction
	Non-parallel between transmission pulley and rear pulley	Adjust brackets at ends of the pulley to make tension equal
	Coal/Dust on pulley surface	Remove coal/dust, improve cleaner
	Belt is not properly jointed	Re-joint the belt
	Feeding of material is not in place	Adjust material dropping device, make feed (coal) to the centre of belt conveyor
Belt slips	Insufficient friction between belt and pulley, problem of water existing on a pulley	Increase tension, dry the pulley
	Tensile travel of belt is not sufficient	Re-adhere rubber belt

- Belt-Sway Switches – Belt Sway limit switches are placed at the head and tail part of each conveyor to limit belt sway to a permissible extent. This is to avoid spilling of material in the belt.
- Zero speed switches – The belt conveyor is provided with a zero speed switch at the tail end for sequential operation of the system. In case of belt speed falling below a tolerable limit, the switch stops the conveyor.

The maintenance of belt conveyor and remedial actions are indicated in Table 7.16.

7.4.14 COMPRESSORS

Compressed air is an essential utility in numerous process industries including the cement manufacturing. But in reality, compressed air is a very expensive utility because only 10–20% of energy invested reaches the point of end-use and the balance (80–90%) is wasted (generally) in the form of heat, friction, noise and misuse. The important parameters of a compressor are listed in Table 7.17.

7.4.15 PROCESS PUMPS

The cement industry uses water pumps (mainly in captive power plants), and centrifugal pumps are commonly in use. The maintenance of a pump is important to extend its service life and to ensure that it runs efficiently. The important parameters of a pump are given in Table 7.18.

TABLE 7.17
Important Parameters of a Compressor

Process Parameters	Parameters for Procurement	Parameters Considered During Operation	Current Practices of Monitoring
Pressure, Flow, Differential pressure of bag filter, Valve or Damper open/ close, control mechanism	Size based on design CFM, Pressure to be generated and kW motor, type of compressor based on application	Oil temperature, Element temperature, Pressure drop across the filter, Vibration, Sound, loading, Speed control	FAD, kW/CFM, spare cost (screw element or cylinder or impeller, bearing, and others), margin available in loading, Pressure required at user end, Drain Valve and trap operation

TABLE 7.18
Important Parameters of a Pump

Process Parameters	Parameters for Procurement	Parameters Considered During Operation	Current Practices of Monitoring
Tank or Sump level, Temperature, Flow, control mechanism	Size based on design flow and head required, kW rating motor	Vibration, Sound, loading, Valve position, Speed control	Efficiency, spare cost (impeller, bearing, and others), margin available in flow, head, and efficiency

7.4.16 Bucket Elevators

Different types of bucket elevators such as the bucket chain type elevators and bucket belt type elevators are used in the cement industry for the vertical conveying of bulk materials. Bucket elevators with centrifugal discharge are used normally, and most are of belt type. The bucket elevator has advantages such as large conveying capacity, high hoisting height, stable and reliable operation and long service life.

To improve maintenance, reduce defects, minimize the replacement of parts and increase the lifespan of bucket elevators, the following preventive maintenance measures are adopted:

- Eliminate air and material leaks.
- Conduct internal inspection of elevator (condition of links, buckets). Change worn out organs if necessary.
- Change buckets and deformed links and replace rings, bushings, and alignment as per the scheduled frequency.
- Replace zipper tracking for tension adjustment and alignment.
- Check if lubrication is proper and belt is properly adjusted.

- Clean the workplace and elevator components.
- Operate the hoists on elevators before the long stop (or replace them).

7.4.17 COUPLINGS

A coupling is a device used to connect two shafts at their ends to transmit power. While rigid coupling is a mechanical fastening of shafts connected with the axes directly in line, the flexible coupling is used to connect two shafts and to accommodate misalignment. The variable speed fluid coupling provides step-less speed variation in a wide range when connected to a fixed speed electric motor. The speed variation is obtained by varying the oil filling in coupling through a sliding scoop tube when operational.

The selection of the coupling is based on coupling type, either rigid or flexible, and coupling size, known as 'Sizing Coupling'. The selection criteria for either sort of coupling are based on the torque carrying capacity of the coupling.

The type of high-speed coupling between motor and gearbox in general is as follows:

- For motors of rating up to 30 kW – Resilient type flexible coupling.
- For LT motors above 30 kW – Traction type fluid coupling.
- For HT motors – Actuator operated scoop type fluid coupling. These couplings are air/water-cooled type depending on application/ requirement.

7.4.18 BEARINGS

Rolling bearings in cement production applications are subjected to extremely harsh conditions like dust, high temperatures and shocks. To satisfy these requirements, such bearings have really high load carrying capacities because of the use of the maximum possible number of rolling elements and coating systems matched to the application. If the condition monitoring is used in bearings for critical applications, such as mill drives or fans/pumps in cement plants, reduction in unplanned downtime can be achieved.

7.5 CURRENT MONITORING PRACTICES FOR ENERGY EFFICIENCY ASSESSMENT

It is known that cement manufacturing is a highly energy-intensive process and, therefore, salient energy monitoring parameters of the assets are integral to assessing the overall asset performance. Broadly speaking, energy monitoring is the process of determining the existing pattern of energy consumption and finding out the deviations from the norms and targets of the operating system, if any. The parameters that are generally used by the concerned energy managers and process engineers for optimizing the output and power consumption in different sections are summarized in Annexures 1–8.

7.6 RECENT APPLICATION OF AI-BASED COMPONENTS FOR ASSET PERFORMANCE MONITORING

The AI-based components have already appeared in the cement industry in a limited way, essentially in certain sections of the manufacturing process. Several applications are seen for the electrical installations and some in the specific unit processes. A few examples are presented below.

A. **Smart Motor Sensors**

A compact sensor unit is easily mounted on motors without the need for wiring. These sensors continuously monitor signals from the motor, precisely measuring key parameters at systematic intervals. A wide range of parameters are measured, which include vibration, skin temperature, magnetic field, acoustic signals and run hours, and insights are given in terms of overall motor condition, bearing condition, winding health, alignment and more. The system architecture of the smart motor sensors of a major technology provider is shown in Figure 7.4 [4]. The compact sensor unit transfers data using built-in Bluetooth Low Energy technology to a smartphone or gateway and to a secure cloud-based server. Data communications use industry standard encryption protocols, and all data is stored in the cloud in an encrypted form. The dataset is analysed using advanced

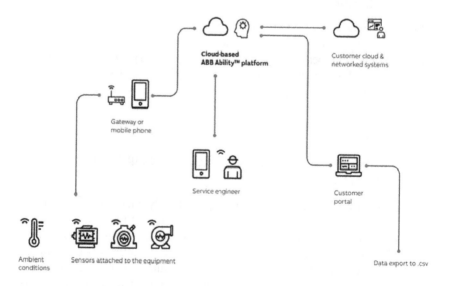

FIGURE 7.4   A typical smart motor system architecture *Source:* ABB) [4].

algorithms to produce meaningful information. The server sends this information directly to the user's smartphone and to a dedicated web portal. Plants can now plan maintenance according to actual needs rather than on the basis of predefined time intervals or operating hours alone. This drastically reduces the maintenance costs and even eliminates unplanned stops. There are also opportunities available to optimize the energy consumption of motors using these smart motor sensors. By combining data on the energy consumption levels of individual motors with plant operating information, it is possible to select the most appropriate motors to cut down energy costs.

B. Transformer condition monitors

A wide range of transformer monitoring and diagnostics equipment using AI and IoT is available in the market to empower asset managers with real-time information. With this broad range of asset information – connected online and available continuously, better decisions can be made to improve reliability, reduce maintenance expenses, proactively manage performance and delay or decrease capital expenditure.

These condition monitors always keep a check on a few basic parameters of a transformer, such as liquid temperature, winding temperature, load current, oil level, etc. This information provides real-time comparison of direct and calculated winding temperatures via integration with fibre optic temperature measurement. Maximized cooling efficiency with more accurate hot spot temperature measurement is possible with the help of these condition monitors.

These systems can perform in-depth monitoring of the Tap Changer, including differential temperature, tap position, tap changing motor performance, contact wear and logging of tap position history. They provide alarms for contact loading, excessive tap counts over time, false tap movements, false or no motor movement, motor and mechanism problems, overloading, breakage, binding and worn tap contacts or coking.

Apart from these, large amounts of both operational and breakdown data is collected centrally, and analysed using intelligent analytical algorithms to automatically determine the risk level and prioritize the need for a human intervention further into the data.

C. Intelligent Partial Discharge Monitoring for Cables

Partial discharge (PD) is a perfect indicator of insulation breakdown and occurs in electrical equipment under high voltage stress, usually greater than 2,000 V. The higher the voltage, the higher is the potential for damage and eventually even the downtime. Specifically, PD is a localized electrical discharge in an insulation system that does not completely bridge the electrodes. As insulation systems age, they become more susceptible to breakdown. Continuous monitoring provides early warning alarms to help prevent equipment damage or downtime. Technology is available to detect and notify facility engineers of severe insulation breakdown.

The PD data is synchronously gathered by high-frequency current transformers installed at the grounding or cross-bonding links of each monitored cable accessories and is transmitted to a data acquisition unit for pre-processing. Multiple data acquisition units are connected to a central computer with fibre optic cable. This ensures safety by providing galvanic isolation between the high-voltage area and the control room, where the central computer is located.

HV power cables, terminations and joints are generally factory-tested before installation to ensure their quality and reliability. However, mechanical forces during cable laying, hidden imperfections and flaws caused during the onsite installation of cable accessories can create PD when left undetected. If allowed to continue, PD will erode the insulation and eventually result in the complete breakdown and in-service failure of the entire cable system. Such failures cause unplanned power outages, loss of plant production, adjacent equipment damage and in the worst case, personal injury. By continuously detecting and trending PD activity with a PD monitoring system, it is possible to observe its development over time even for underground cables, where manual access is limited.

D. Radiometric Thermal Imaging for Substations

Electrical substation is a key installation for the entire manufacturing industry. Any outage or equipment failure at substation will lead to major production downtimes. Previously, predictive maintenance teams used to carry thermal imaging cameras to survey equipment, based on requests or scheduled maintenance visits. Accessing every part of a station is relatively complex and time consuming. Now, with the advent of radiometric thermal cameras, maintenance teams can remotely and proactively inspect equipment and get real-time temperature data for assets [6].

Radiometric thermal cameras can be installed at the best possible spots to cover a substation. The cameras continuously survey specific machinery, focus on transformer connections, incoming power transmission lines and other equipment. The Smart Sensor Gateways installed along with these cameras control the scanning functions to collect temperature data on assets, activating an alarm if the temperature is above the pre-determined threshold (Figure 7.5).

Equipping facility operators with this information allows senior management to assess how to respond to a problem component before it breaks out and decide about proactive steps if needed. If maintenance is conducted on a specific component, then live data

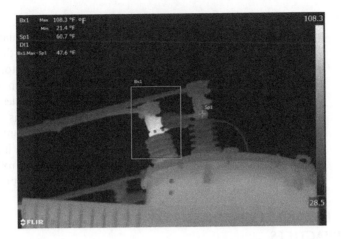

FIGURE 7.5   Radiometric thermal imaging (*Source:* FLIR).

and smart analytics on assets, through this intelligent radiometric cameras and analytics solution, help maintenance crews verify if the issue was fixed through repairs completely or whether the issue remains partially somewhere. Apart from this, thermal security cameras also protect assets from damage caused by intruders.

E. **Mill Optimization System**
   Mill optimization system, based on AI, provides the much-needed solution to the problem of poor operating methods and unplanned shutdowns. The mill optimization system measures and collects all the necessary data of the mill. It has two main parts:
   - Process modelling
     The process modelling system consists of smart algorithms, through which it is able to predict the fineness of a product, based on the following data collected by the system:
     1. Separator rpm
     2. Feed
     3. Swirl Valve position
     4. Mill filling level
        a. Circulating bucket elevator power
        b. Mill drive output
        c. Electric ear/Vibration sensor for chamber 1
        d. Electric ear/Vibration sensor chamber 2
     5. Material temperature
     After predicting the fineness of the product, the process modelling system compares the predicted value with actual value calculated in the laboratory and feeds to the control system the deviation between predicted and actual fineness [7].

- Model-based predictive system
  Control system mainly comprises PID loops that manipulate the mill feed and separator rpm, based on the deviation from the predicted and actual fineness, and the return from the separator. The transfer functions of the PID loops are optimized in such a way that they are able to best manipulate the control variables. The optimization is done through a series of step tests and is determined on the basis of the identification tool. Thus, the controller has complete internal dynamic process model of the mill, including all interactions. Based on this state model, the controller is able make predictions about the future movement of the process.

7.7 ADVANCES IN MAINTENANCE STRATEGIES AND PRACTICES

During the last five decades or so, the maintenance practices in the cement industry have hovered around corrective or breakdown maintenance, planned maintenance and preventive maintenance. The **corrective maintenance** was quite prevalent in the fifties and sixties in the last century and included all actions to return from a failed state to an operating or available state. The corrective maintenance actions were obviously unplanned and reactive in nature, when failures occurred. Even now the practice of running to failure continues for the kiln systems in the cement plants, which happen to be at the heart of the manufacturing process, despite the fact that some online monitoring facilities are available these days to indicate the refractory lining conditions inside a kiln. In the cement industry, **planned maintenance** has also been followed for all major equipment with the scheduling of the shutdown being fixed either by the management plan or determined by equipment failure. Such shutdown tasks generally include lubrication, filter cleaning, inspection and measurement of wear parts, replacement of minor components, inspection of facilities that cannot be approached while the system is in operation, such as the clinker cooler. The concept of **preventive maintenance** came into practice more and more after 1980s. The attempt was to undertake equipment inspection and testing to avoid premature equipment failure and to ensure all planned maintenance to extend equipment life.

Over the last three decades, the cement industry has mostly depended on **CMMS**. There are hundreds of proprietary systems in use, many of which are developed in-house. Broadly speaking, these systems possess the following features or functional capabilities [3]:

- an equipment database which stores descriptive and specification information on all machines and components
- a database for preventive maintenance tasks together with a scheduling function relating to operating time or to throughput

- a system for generating and logging work orders for repair or maintenance
- a database recording the history of maintenance of all items of equipment
- store room inventory management and procurement initiation
- repair and maintenance cost tracking
- safety record-keeping

The underlying efforts of all CMMS are to ensure cost effectiveness of actions taken, prevention of recurrence of defects, and availability of a knowledge base of all failure modes and effect analysis. The main focus has all along been to make maintenance reliability-centric, which is often expressed in terms of equipment uptime. Maintenance of mobile mining equipment and transport vehicles are included in CMMS. On the whole, these systems have yielded the following benefits:

- reduction in emergency repairs
- reduction of unscheduled repairs
- more planned and scheduled work
- reduced downtime and maintenance expenditure
- better manpower utilization
- preservation of assets to an extent

However, it is important to note that when the preventive maintenance policies are in place, the maintenance is generally performed well in advance of any potential failure. In addition, unnecessary corrective actions are performed, leading to inefficient use of resources and increased operating costs. Hence, the search for a more effective maintenance policy has been continuing.

With the advent of IIoT technologies, data analytics, cloud services and machine learning tools and techniques, the maintenance approaches have veered towards '**predictive maintenance**', capable of detecting the equipment abnormalities and failure patterns ahead of the actual event, giving out early warnings to plan the necessary maintenance action. The machine learning based predictive maintenance can be divided into two main classes – supervised and unsupervised, depending on the nature of data and information. The availability of maintenance information depends on the existing management practice of an organization. In the supervised dataset, the information on the occurrence of failures is present, while in the unsupervised data, the logistic and/or process information is available but no direct maintenance related data exists. Both kinds of datasets are usable in doing the modelling, though the first type containing data on fault history, maintenance and repair history and machine conditions will have an advantage in analytics relating to predictive maintenance. It must also be borne in mind that the data should be time-series. It includes timestamps, sensor reading sets collected at the same time as timestamps, and device identifiers.

A machine learning architecture for predictive maintenance, based on Random Forest Approach, has been suggested in [8]. The system was tested on a real industry example. The data was collected by various sensors, machine PLCs and communication protocols. The data was made available on the Azure Cloud architecture. The results showed a proper behaviour of the approach on predicting different machine states with high accuracy. The scheme of activities to be undertaken in predictive maintenance approach is shown in Figure 7.6.

The predictive management can be formulated in one of the two ways: classification approach and regression approach. The classification approach is a Boolean computing, using a less volume of data, and predicts whether there is a possibility of failure in next n-steps. The regression approach requires more data and predicts how much time is left before the next failure, the prevalent acronym of which is RUL (Remaining Useful Life). While the classification tools are a natural choice for distinguishing between faulty and non-faulty process iterations as observed from the process data, they do not naturally show the way to health factors that can help in maintenance-related decision-making and determining RUL. To overcome this limitation, studies are being carried out to apply 'multiple classifications (MC)' as a solution [9]. A schematic diagram of an MC PdM module is shown in Figure 7.7.

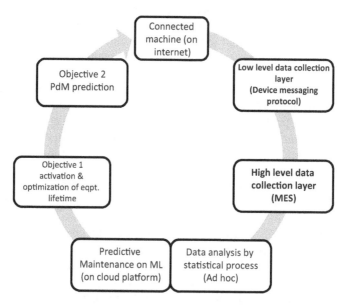

FIGURE 7.6 Scheme of activities to be undertaken for the predictive maintenance approach (adapted from [8]).

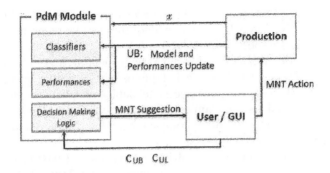

FIGURE 7.7 Configuration of a PdM module with classifiers [9].

The frequency of unexpected breaks (UB) refers to the percentage of failures not prevented and provided new data to update the model. The amount of unexpected life-time refers to the average number of process iterations that could have been run before failure. The costs of unexpected breaks (C_{UB}) and unexpected life-time (C_{UL}) were provided by the user. Performance of the MC PdM module increased with the number of classifiers, since each classifier provided more information on the health status of the process. The decision-making logic provided the maintenance suggestions to the graphical user interface (GUI), which in turn directed the maintenance action to production. According to the authors, the MC PdM methodology presented did not impose any restrictions on the classification algorithm that could be adopted for the individual classifiers.

The above studies are only to illustrate the expanse of development of predictive maintenance for industrial applications. It has already been mentioned earlier that equipment and plant maintenance forms the core of a mature APM programme. The predictive maintenance methods will provide the proper data-driven strategy for improving the asset performance and asset life. Since the cement industry has already moved into IIoT regime with smart sensors and since the opportunities for data analytics and cloud services are widely available, it is appropriate for the cement industry to delve into predictive maintenance more and more.

7.8 NEAR-TERM PROSPECTS

Predictive maintenance and condition monitoring will be the future of preventive maintenance and asset performance monitoring strategies. As already stated, predictive maintenance strategies make effective use of the latest technologies such as IoT and AI, optimize both cost and time and enhance reliability.

The AI and Big Data algorithms in predictive maintenance strategies provide the methodology to find patterns within large amounts of data. It is possible to outfit a factory with data-tracking IoT devices that can update factory datasets frequently. With the best dataset available, it is easy to build the best possible predictive algorithm. Such an algorithm will be accurate in helping factory managers respond even faster to equipment related issues.

Unlike traditional methods that might fail to identify the root cause of a problem or equipment failure, the IoT philosophy of management reaches conclusions fast, with real-time access to data and quick communication between team members and plant equipment. As a result, a plant can operate efficiently, potentially preventing complete equipment failure and unnecessary equipment replacement.

The visible opportunities for applying the IoT- and AI-based technologies to the key assets of the cement plants are mentioned in Tables 7.19 and 7.20.

TABLE 7.19
Near-Term Opportunities for Monitoring the Asset Performance

Key Asset	Technological Opportunities
VRM	Selection of optimum model for the given material based on cluster data/simulation using AI Feed size control, air flow control based on season/moisture, dam ring replacement, hydraulic pressure based on AI, product residue control/variation depending on requirement using AI
Ball mill	Circuit operation including mill feed, separator performance based on AI for each variety with data of clinker quality, feed size and additive quality and PSD, digital twin technology for material behaviour inside separator, mill. Smart sensors for cement temperature, product PSD, Grinding media selection and make up using AI, separator flow profiling and product PSD control based on variety and product performance
Pyro section: preheater, kiln, cooler, burner	Kiln operation based on past data analysis and AI, Digital Twin application for fuel/AF burning, emission control, clinkerization quality, cyclone efficiency estimation and control, clinker quality based on final product requirement, clinker production (output) estimation, PID control and optimization using AI techniques
Process fans	Use of AI to vary the best operating point depending on the process requirement/product requirement, new design specification based on the past data analysis using AI during retrofit/spare
Process Bag filters	Use of AI and IoT to optimize the use of compressed air with respect to the dust load
Gear box	Use of AI and IoT to predict breakdown at primary stage
Separator	Use of AI and IoT to predict breakdown at primary stage
Belt conveyor	Smart sensor for loading, optimizing power consumption, understanding overall connected circuit performance using AI, avoiding idle running and optimizing aux consumption
Motors and Drives	Smart sensor for loading, optimizing power consumption, understanding process operating conditions, use of AI to analyse process performance equipment
Transformers	Smart sensor for calculating transformer losses and optimal loading of the transformers
Compressors	Use of AI and IoT to calculate leakages and unnecessary pressure drops in compressed air network

TABLE 7.20
Near-Term Opportunities for Preventive Maintenance

Key Asset	Improvement Opportunities for the Maintenance Functions
Pyro section: preheater, kiln, cooler, Burner	Refractory maintenance including selection and installation – Use of AI to avoid problems of installation and commissioning, kiln shell maintenance/repair, cooler maintenance, dip tube replacement smart sensors for kiln shell assessment, cooler plate assessment, Load, girth gear, supporting roller performance
Crusher	MOC selection based on past data analysis using AI/cloud, replacement/ordering of liners and other spares, maintenance based on AI, smart sensors for lube oil temperatures, vibration, international data analysis for breakdown, maintenance
Process fans	Smart sensors for monitoring vibration, performance – bearing, power, thickness using IoT, customized model selection and manufacturing. Use of AI to avoid problems of installation and commissioning
Gear box	Smart sensors for monitoring vibration, performance – efficiency, power, using IoT, customized model selection and manufacturing. Use of AI to avoid problems of installation and commissioning
Separator	Smart sensors for monitoring vibration, performance – bearing, power, thickness using IoT, customized model selection and manufacturing. Use of AI to avoid problems of installation and commissioning
Belt conveyor	Smart sensors for current condition, residual life estimation
Motors and Drives	Smart sensors to enhance the performance
Transformers	Smart sensors to enhance the performance

7.9 CONCLUDING OBSERVATIONS

Over time, the cement plants consisting essentially of moving and rotating large-capacity mechanical, electrical, and electronic machinery and equipment have evolved into sturdy and stable operating systems. The plant availability has significantly increased. The benchmarks of productivity, energy conservation, resource utilization, and emissions of GHG have been continuously raised in the industry. Now the industry is at a threshold, where the results of traditional tools and techniques of continual improvement are flattening out. Hence, there is a need to explore the potentials of IoT, data analytics, and machine learning with the help of cloud platforms. The industry has the basic infrastructure of convergence of IT and OT but its domain knowledge in AI and data science need strengthening even for effective outsourcing of the cloud services and selection of software packages necessary to move forward.

The cement industry has always been conscious about effective maintenance for steady and efficient operation. There is still a high potential to adopt the tenets of reliability engineering and to make the maintenance methodology to be data-driven and predictive. The maintenance strategy must include analysis of failure modes and their consequent effects so that the recurrence of failures are avoided either by improved design of the failing equipment or changes in operational strategies.

Despite significant technological advances, the work environment within the cement plants still continues to be in departmental silos. Now it is time to integrate the APM including predictive maintenance not only with the plant operation but also with the business strategy of the enterprise. There is an imminent need for collaborative working between maintenance, operations, supply chain management, inventory management, safety–health–environment functions, finance and accounting systems. It is also important to bear in mind that the plant personnel will have to undergo significant reskilling and retraining in emerging fields of data science and machine learning. With AI-based APM at the doorstep, a paradigm shift is in the offing in cement plant management.

REFERENCES

1. https://assetivity.com.au/article/reliabilityimprovement/asset-performance-management-what-is-an-asset-management-performance-system.html
2. https://reliabilityweb.com/articles/entry/asset_management_concepts_practices
3. Philip A Alsop, Hung Chan, Herman Tseng, *The cement plant operations handbook, International cement review*, Tradeship Publications Ltd. UK, 2007.
4. ABB Ability™ Smart Sensor, Condition monitoring for motors, Product Note. https://search.abb.com/library/
5. https://flir.in/discover/security/radiometric/the-benefits-and-challenges-of-radiometric-thermal-technology
6. https://fliemedia.com/MMC/THG/Brochures/T820264/T820624_EN.pdf
7. Siemens AG, Cementability: SICEMENT – solutions for the next level of productivity, Brochure, 2018.
8. M. Pedanti, L. Romeo, A. Felicetti, A. Mancini, E. Frontoni, J. Loncarski, Machine learning approach for predictive maintenance in Industry 4.0, *14th IEEE/ASME International Conference on Mechanotronic and Embedded systems and Applications (MESA)*, 2018.
9. G.A. Susto, A. Scirru, S. Pampuri, S. McLoone, A. Beghi, Machine learning for predictive maintenance: a multiple classifiers approach, *IEEE Transactions on Industrial Informatics*, 11(3), 2015, 812–820.

ANNEXURES

ANNEXURE 1. LIMESTONE CRUSHER SECTION

Serial No.	Paraeter	Purpose	Preferred Monitoring Frequency
1	Crusher output size	To ensure crusher and raw mill output	Weekly
2	Crusher feeder speed (rpm) and running hours	To ensure optimum crusher output and loading	Online Daily

Serial No.	Paraeter	Purpose	Preferred Monitoring Frequency
3	Crusher output, TPH, BDP and actual	BDP and actual	Daily
4	SEC, BDP and actual	Deviation and improvement	Daily
5	Main bag filter DP	Optimum venting and power	Online continuous
6	BF venting specific air flow, m³/TPH	Identify excess air flow	Monthly
7	Moisture content of material	Too high wet material adds up to energy consumption. Monitoring and controlling moisture at Crusher product shall be more effective to control energy conservation in mining, transportation and raw grinding sections. To control by mine dewatering programme/plan the mine block operation/surface drying	Daily average sample or online continuous

ANNEXURE 2. RAW MILL SECTION (VRM)

S. No.	Parameter	Purpose	Preferred Monitoring Frequency
1	False air from mill inlet to mill fan outlet	Optimizing fresh air in RABH/ Kiln bag house fan and its power	Monthly
2	Mill fan Inlet pressure	Pressure drop across circuit	Online continuous
3	Mill outlet dust loading gm/m³	Optimize flow accordance with output	Monthly
4	Cyclone pressure drop	Achieve lowest SEC	Online continuous
5	Pressure drop across Mill fan inlet damper	Damper condition	Monthly
6	Louvre velocity	Optimize mill DP	Monthly
7	Mill reject%	To optimize mill fan SEC	Online continuous
8	Mill load (avg kW) to allowable kW SEC	Optimize output	Monthly
9	Mill drive	Monitor and maintain SEC	Online continuous, daily
10	Mill fan	Monitor and maintain SEC	Online continuous, daily
11	Mill fan efficiency	To achieve best tech possible, monitor and maintain	Monthly
12	Mill feed size	Optimize output	Weekly
13	Mill product residue target and actual	Optimize mill and kiln operation	Hourly
14	Feed moisture	For mill efficiency monitoring	Daily average
15	Mill internal water spray rate	For mill efficiency monitoring	Daily average

ANNEXURE 3. RAW MILL (BALL MILL)

S. No.	Parameter	Purpose	Preferred Monitoring Frequency
1	False air from mill inlet to mill fan outlet	Optimizing fresh air RABH/Kiln bag house fan	Monthly
2	Mill fan Inlet pressure	Pressure drop across circuit	Online continuous
3	Separator dust loading g/m³	Optimize flow accordance with output	Monthly/Online
4	Cyclone pressure drop	Achieve lowest SEC	Online continuous
5	Pressure drop across mill fan inlet damper	Damper condition	Monthly
6	Mill grinding media filling level	To achieve optimum grindability in mill	Online continuous
7	Circulation load	Ensure better separator efficiency	Online Continuous
8	Mill reject < 90 micron sieve SEC	Monitor separator performance	Shift-wise
9	Mill drive	Monitor and maintain SEC	Online continuous, Daily
10	Mill fan	Monitor and maintain SEC	Online continuous, Daily
11	Mill fan efficiency	To achieve best tech possible, monitor and maintain	Monthly
12	Mill feed size	Optimize output	Weekly
13	Mill product residue Target and actual	Optimize mill and kiln operation	Hourly
14	Mill load (avg. kW) to allowable kW	Optimize output and decide on grinding media make up charge	Daily
15	Piece weight in first chamber	To achieve optimum grindability in mill	Monthly
16	Grinding media surface area in second chamber	To achieve optimum grindability in mill	Monthly
17	Size of slot opening in the partition wall grates/ cleanliness	To achieve optimum material and gas/air flow through mill	Fortnightly
18	Pressure drop across mill	To monitor the material and air/gas flow and identify the blockages if any in the grates (partition and discharge diaphragm)	Online continuous

ANNEXURE 4. PYRO SECTION

S. No.	Parameter	Purpose	Preferred Monitoring Frequency
1	Kiln feed LSF SD	Kiln stability, optimum heat of reaction, clinker grindability	Daily
2	Preheater outlet oxygen	To maintain optimum excess air	Online continuous
3	Preheater outlet CO	To maintain optimum excess air	Online continuous

S. No.	Parameter	Purpose	Preferred Monitoring Frequency
4	Preheater outlet pressure and temperature	Maintain and monitor preheater thermal loss	Online continuous
5	Preheater fan inlet damper pressure drop	Damper condition	Monthly
6	False air across preheater (from kiln inlet to preheater fan outlet)	Optimize electrical and thermal sec	Monthly
7	Kiln inlet No$_x$ level	Burning Zone excess air level	Online continuous
8	Each cyclone ΔP and ΔT (BDP and actual)	Optimize electrical and thermal sec	Monthly
9	Dust concentration in down comer duct (BDP and actual)	Optimize electrical and thermal sec	Yearly
10	RABH DP Fan efficiency	Optimize bag life and fan power	Online continuous
11	Preheater fan	To achieve best tech possible, monitor and maintain	Monthly
12	RABH fan	To achieve best tech possible, monitor and maintain	Monthly
13	Cooler vent fan	To achieve best tech possible, monitor and maintain	Monthly
14	Cooler fans	To achieve best tech possible, monitor and maintain	Monthly
15	Temp drop across TAD	Reduce radiation loss and false air entry	Monthly
16	Cooler fans suction pressure	Optimize fan power	Monthly
17	Pressure drop across silencer in cooler fans	Ensure optimum power	Monthly
18	Damper pressure drop (if any)		Monthly
19	Preheater fan	Damper condition	Monthly
20	Cooler vent fan	Damper condition	Monthly
22	SEC		
	Preheater fan	Monitor and maintain SEC	Online continuous and daily
	Cooler fans	Monitor and maintain SEC	Online continuous and daily
	Cooler vent fan	Monitor and maintain SEC	Online continuous and daily
	RABH fan	Monitor and maintain SEC	Online continuous and daily
	Coal conveying blower	Monitor and maintain SEC	Online continuous and daily
23	Specific air flow		
	Cooling air	Monitor and maintain thermal & Electrical SEC	Monthly
	Cooler vent air	Monitor and maintain thermal & Electrical SEC	Monthly
	Preheater fan flow	Monitor and maintain thermal & Electrical SEC	Monthly
	RABH fan flow	Monitor and maintain thermal & Electrical SEC	Monthly
	Tertiary air flow	Monitor and maintain thermal & Electrical SEC	Monthly

S. No.	Parameter	Purpose	Preferred Monitoring Frequency
24	Coal phase density		
	Kiln	Optimize blower power and sp heat consumption	Monthly
	PC	Optimize blower power and sp heat consumption	Monthly
25	Primary air%		Monthly
26	Cooler bed height	To achieve cooler recuperation efficiency	Online continuous
27	Temperatures		BDP and actual
	Cooler vent	Monitor and maintain specific heat consumption	Online continuous
	Clinker	Monitor and maintain specific heat consumption	Online continuous
	Preheater outlet	Monitor and maintain specific heat consumption	Online continuous
	Tertiary air	Monitor and maintain specific heat consumption	Online continuous
	Secondary air	Monitor and maintain specific heat consumption	Online continuous
	Kiln exit gas	Monitor and maintain specific heat consumption/Volatile circulation phenomena	Online continuous
28	Water spray quantity		
	Cooler	Water, energy conservation, specific heat consumption	Online continuous
	Down comer/top cyclone	Water, energy conservation, specific heat consumption	Online continuous
29	Free silica (quartz) in kiln feed%	Kiln stability, optimum heat of reaction, clinker grindability	Hourly
30	Free lime in clinker%	Kiln stability, optimum heat of reaction, clinker grindability	Hourly
31	Kiln Feed Fineness – Residue on 212 micron sieve	Control of Free Lime and optimize energy consumption	Hourly

ANNEXURE 5. CEMENT MILLING – BALL MILL

S. No.	Parameter	Purpose	Preferred Monitoring Frequency
1	Circulation Load	Optimize separator performance	Online continuous
2	Separator loading (g/m³)	Optimize fan power	Online continuous/ monthly/variety-wise
3	Velocity inside mill	Avoid over grinding	Mill vent volume can be alternative
4	Specific grinding media weight for first chamber	Optimize grindability	Monthly/regarding half yearly

S. No.	Parameter	Purpose	Preferred Monitoring Frequency
5	Specific GM surface area for second chamber	Optimize grindability	Monthly/regarding half yearly
6	% filling level	Optimum output	Online continuous
7	Residue on 45 micron in the reject	Monitor separator performance	Shift-wise
8	Roller press BDP KW and actual loading	Optimum grinding	Online continuous
9	Product residue or Blaine target and actual	Optimum output and power	Hourly
10	Separator vent flow as% of circulating air flow	Control false air in the circuit, cooling of cement and optimize power	Monthly
11	Pressure drop across cyclone SEC	Optimize fan power	Online continuous
12	Mill, HPRG drives	Monitor and maintain SEC	Online continuous, daily
13	CA fan	Monitor and maintain SEC	Online continuous, daily
14	Mill vent	Monitor and maintain SEC	Online continuous, daily
15	Sept vent	Monitor and maintain SEC	Online continuous, daily
16	Bag filter DP		
	Sept vent	Optimize bag life and fan power	Online continuous
	Sept fan inlet	Optimize bag life and fan power	Online continuous
	Mill vent	Optimize bag life and fan power	Online continuous
17	Fan efficiency		
	CA fan	To achieve best tech possible, monitor and maintain	Monthly
	Mill vent	To achieve best tech possible, monitor and maintain	Monthly
	Sept vent	To achieve best tech possible, monitor and maintain	Monthly
18	Feed composition/recipe	To monitor consumption of additives and extenders	Online/continuous
19	Feed moisture	To monitor SEC	Daily
20	Pressure drop across mill	To monitor SEC	Online/continuous
21	Size of slot opening in the partition/end wall grates/ cleanliness	To achieve optimum material and gas/air flow through mill	Fortnightly

ANNEXURE 6. UTILITIES

S. No.	Parameter		Preferred Monitoring Frequency
1	Compressor (HP) SEC	Monitor and maintain power	Daily
2	Up to clinkerization		
3	Cement grinding		
4	Compressed air generation pressure	Optimize power and indication of leakage and pressure drop	Online continuous
5	Compressor loading%	Ensure optimum utilization	Monthly
6	Compressed air leakages%	Unproductive power	During every shutdown
7	Compressor SEC	Condition of compressor	Monthly where standby is available, otherwise during stoppages
	Compressor discharge air temperature	Monitor and maintain efficiency of compressor/cooling system/FAD capacity	Daily
	Screw compressor – oil pressure	Monitor and optimize no load power	Periodical
8	Cooling water circulating flow		
	Pyro section	Water consumption and power saving	Monthly
	Cement mill	Water consumption and power saving	Monthly
9	Cooling water inlet and return temp	Effectiveness of heat exchangers, process heat load and cooling tower effectiveness	Online continuous
10	COC	Water consumption	Monthly
11	Pump efficiency	Optimum power	Monthly
12	Pump discharge pressure	Line condition, requirement and valve throttling	Online continuous in case of common header or monthly
13	Fly ash unloading pressure	Optimize compressor power	Daily
14	Air conditioning SEC (kW/TR)	Optimize air cooler performance	Daily

ANNEXURE 7. CAPTIVE POWER PLANT – PROCESS

S. No.	Parameter	Purpose	Preferred Monitoring Frequency
1	Boiler exit oxygen	Monitor and maintain excess air	Online continuous
2	Id fan inlet oxygen	Monitor and maintain false air	Monthly
3	DP across BFP	BFP power	Online continuous
	Flow control valve		
4	Efficiency		

S. No.	Parameter	Purpose	Preferred Monitoring Frequency
	BFP	To achieve best tech possible, monitor and maintain	Monthly
	CEP	To achieve best tech possible, monitor and maintain	Monthly
	CWP	To achieve best tech possible, monitor and maintain	Monthly
	ACW	To achieve best tech possible, monitor and maintain	Monthly
5	Compressor SEC	Monitor and maintain power	Daily
6	Inst compressor pressure	Optimize power and indication of leakage and pressure drop	Online continuous
7	Ash conveying pressure	Optimize power and indication of leakage and pressure drop	Online continuous
8	Compressor loading	Ensure optimum utilization	Monthly
9	Cooling tower inlet and outlet temp	Effectiveness of heat exchangers, process heat load and cooling tower effectiveness	Online continuous
	Approach to wet bulb temperature	Monitor the efficiency of cooling tower	Monthly
10	Temp in ARC line (after valve)	Optimize BFP power, identify ARC valve life	Online continuous
11	Id fan inlet pressure	Optimize fan power	Online continuous
12	FD fan suction pressure	Optimize fan power	Online continuous
13	Fan efficiency		
	FD fan	To achieve best tech possible, monitor and maintain	Monthly
	Id fan	To achieve best tech possible, monitor and maintain	Monthly
14	SEC kW/MW (BDP and actual)		
	Pumps	Monitor and maintain SEC	Online continuous, daily
	Fans	Monitor and maintain SEC	Online continuous, daily
	Compressor	Monitor and maintain SEC	Online continuous, daily
15	Coal – moisture	Monitor and control parasite consumption	Daily
16	Heat rate	Monitor the boiler efficiency	Daily
17	Coal – proximate analysis	Monitor the boiler efficiency	Periodical/ shipment-wise
18	Gas turbine inlet air temperature	Monitor the turbine efficiency	Hourly

Annexure 8. Captive Power Plant – Electrical

S. No.	Parameter	Purpose	Preferred Monitoring Frequency
1	Transformer losses	To calculate efficiency	Monthly
2	Transformer winding temperatures	To eliminate or interlock with winding temperature	Online continuous, daily

S. No.	Parameter	Purpose	Preferred Monitoring Frequency
3	Transformer incoming voltage	TO minimize the operation of OLTC by manual/auto mode	Online continuous, daily
4	Transformer tap position	To optimize distribution voltage	Monthly
5	Motor loading	To Improve the efficiency	Monthly
6	Motor voltage	To reduce the voltage loss and for maintain optimum voltage	Online continuous, daily
7	Power factor	To reduce the distribution losses and increase the capacity (KVA)	Online continuous, daily
8	Capacitor power	To reduce the loss	Monthly
9	Captive power plant –frequency in Island mode	To minimize the frequency and saving power in centrifugal loads	Online continuous, daily
10	Captive power plant –power factor in island mode	To improve turbo generator efficiency	Online continuous, daily
11	Lighting voltage (210 V)	To save power and increase lamp life	Online continuous, daily
12	Distribution losses	To reduce cable losses	Online continuous, daily
13	Maximum demand	To avoid any penalties	Online continuous, daily
14	Temperature of major feeders	To avoid any shut downs (using thermograph)	Monthly
15	Voltage drop	To minimize distribution losses	Monthly

8 Digital Twin and Its Variants for Advancing Digitalization in Cement Manufacturing

Anjan Kumar Chatterjee
Conmat Technologies Private Limited, Kolkata, India

CONTENTS

8.1   Introduction 263
8.2   History of the Digital Twin Concept..264
   8.2.1   Manifestations of the Digital Twin Concept 265
   8.2.2   Indicative Developmental Trends of Digital Twins as
       Reflected in a Set of Publications 267
8.3   Interlinking Digital Twins and Product Lifecycle 268
8.4   Adopting Digital Twin Technology in Manufacturing.....................269
   8.4.1   Strategic Approach for Adoption 270
8.5   Functionality and Structural Configuration of
   Digital Twins in Manufacturing 271
   8.5.1   Process and System Optimization..274
   8.5.2   Tentative Structural Configuration of Digital Twins 275
8.6   Digital Twins Driving the Pilot Production Environment.................275
   8.6.1   Modelling Approach in the Pilot Facility 277
8.7   Relevance of Digital Twins for Cement Manufacturing...................278
   8.7.1   Information Density Consideration..279
   8.7.2   Digital Twin Options in Cement Manufacturing 280
8.8   Technological Preparedness with Enabling Tools 285
   8.8.1   Salient Technology Requirements...285
   8.8.2   IoT Platforms and Enabling Tools for Digital Twins 288
8.9   Digital Twins for Learning and Training 289
8.10 Concluding Observations 290
References..291

8.1   INTRODUCTION

The digital twin technology is an offspring of the digital world and has emerged in recent years as a strong hand-holder of further accentuation of

DOI: 10.1201/9781003106791-8

digitalization in different spheres of social, urban, and industrial growth. It is one among the top ten strategic technology trends named by Gartner Inc. in 2017. In common parlance it represents the convergence of the physical and virtual world where every industrial product or a physical entity will have a dynamic digital representation in its specific field of application. The fundamental concept is that throughout the lifecycle of the product or entity, right from design to disposal, organizations can have a complete digital footprint of their products in order to manipulate them all though the life in the most useful and productive mode. Initial applications of digital twin are already noticed in product manufacturing, industrial systems design and use, automobile industry, retail business infrastructure, healthcare industry, smart city construction, etc.

It is now widely perceived by the experts that the fourth industrial revolution, Industry 4.0 as it is called, has embraced digital twin technology to change the traditional approach of 'first build and then improve' to the virtual system based design to roll out any equipment or system by understanding its unique features, performance, or potential issues. With digital twin an operator can get trained on a look-alike virtual machine. Further, during the foreseeable advent of autonomous world of industrial machines, the digital twin technology will turn out to be the prime drivers of process optimization, coordination with the connected smart devices, self-diagnosis of the systems, and self-repairing of faults detected, with minimal manual intervention by the operating personnel.

It is obvious that in the backdrop of the above potentials, the digital twin contends to be a strong technology candidate for cement manufacturing, which, in reality, is a set of interconnected unit processes and a huge number of connected equipment and machines.

This chapter is intended to explain the broad features of the technology, important steps of implementation, and the visible potential for the cement manufacturing process.

8.2   HISTORY OF THE DIGITAL TWIN CONCEPT

Although the advent of digital twins was anticipated in the 1991 book entitled *Mirror World* by David Gelernter as mentioned in the Wikipedia on digital twin, the practical concept, captioned as 'conceptual ideal for PLM' was presented by Grieves to an industrial audience in 2002 in the context of setting up a Product Lifecycle Management Centre. For almost a decade thereafter the concept was variously renamed as 'Mirrored Spaces Model' and 'Information Mirroring Model'. The core concept was that a digital informational construct about a physical system could be created as an entity on its own and this digital information would be a 'twin' of the information that was embedded within the physical system itself and be linked with it through the entire lifecycle of the system. In 2011, the terminology of 'digital twin' became prevalent for the concept [1, 2].

8.2.1 Manifestations of the Digital Twin Concept

In the original concept of digital twins, a step-up process of starting from a prototype to an aggregate in a digital twin environment was conceived as depicted in Figure 8.1 [2].

A DTP is the digital description of the prototype of a physical entity. The digital construct contains informational sets necessary to describe and produce the corresponding physical version. The informational sets may include such inputs as a 3D model, a bill of materials along with their specifications, a list of processes and services to be deployed, the mode of disposal, etc. A DTI differs from the DTP in the sense that it is more specific in nature. In other words, a DTI describes a specific physical entity, which the digital twin will remain linked to throughout the life of that physical entity and may contain data in the form of a fully annotated 3D model with dimensions and tolerances of the physical instance, the bill of material that lists the present and past components, the process information listing the operations performed in creating the physical instance, results of tests carried out on the physical instance, a service record of past services performed and components replaced, operational features as obtained from the sensors, and so on.

While both the DTP and DTI appear as independent data structure, a DTA is the aggregate of all the DTIs and may be a computing construct that

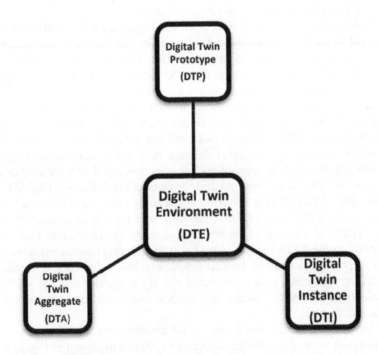

FIGURE 8.1 Digital twin forms as conceived originally.

has access to all DTIs with the capability of putting queries to them on an ad-hoc or proactive basis. The DTA with large inputs from multiple DTIs may correlate the sensor readings with failures, which would facilitate more reliable prognostics. All digital twins will operate in a space, called DTE, for a variety of purposes including predictive and interrogative functions. While the predictive function refers to knowing the future behaviour and performance of the physical versions, the interrogative function primarily relate to questioning the DTIs for all characteristics that are within the regime of instrumentation of the physical versions.

However, in the course of practice, the actual manifestation of the digital twin concept has often got tilted to 'digital models' and 'digital shadows', which differ from a digital twin in the mode of data exchange between them. When there is a digital representation of an existing or planned object without any automatic data exchange between them, it is termed as the digital model. The data exchange, however, can be manually executed. Digital models can be developed for a wide range of physical systems like factories, machinery, devices, components, products, etc. at the planning or designing stage by using the digital data of the existing systems. It is understood that any change of state in the physical object has no direct reflection in the in the digital version and vice versa. When there is one way data flow between the state of an existing physical object and its digital version, it is called a digital shadow. In this form, a change in the state of the physical object leads to a change in the digital version but not vice versa.

In the context of diversified technological approaches for digital twins, it is relevant to mention about the development of 'digital ghost' technology by the GE research team [3]. Most of the critical infrastructural assets such as the power plants, power transmission and distribution networks, transportation systems, water processing plants, etc. have 'control systems' at the core for their efficient and safe operation. These control systems read the sensors and send command signals to the actuators. These control systems are extremely vulnerable to cyber-attacks and unpredictable occurrence of faults. To provide a new layer of cyber-defence, in addition to other existing safeguards, the digital ghost technology has been developed with the help of digital twins, domain knowledge, AI and the latest control theories. Digital Ghost determines if the plant is behaving abnormally due to a cyber-attack (even when perhaps the operator's user interface says everything is okay) by using digital twins to understand the physics and controller-dependent asset behaviour and seeing if there are any significant departures from the expected nominal response. As claimed by the developers, the technology is capable of detecting, localizing, and neutralizing the abnormalities and the validation studies have shown 98% success in detection and localization, and about 50% in neutralization.

Considering all the above trends and developments, the diversity of the digital twin technology can be collectively depicted as shown in Figure 8.2.

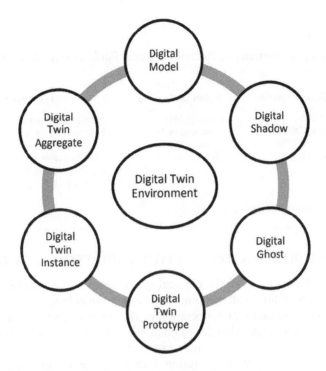

FIGURE 8.2 Different manifestations of digital twin concepts.

8.2.2 INDICATIVE DEVELOPMENTAL TRENDS OF DIGITAL TWINS AS REFLECTED IN A SET OF PUBLICATIONS

From a content-wise review and classification of a set of publications on topics related to manufacturing an attempt has been made to evaluate the current trend of focus in applying the digital twin technology in this field [4]. A large number of publications on digital twins in manufacturing published from 2014 onwards were classified, using the criteria of data integration level, focus areas and contents. A summary of results obtained is presented in Table 8.1, in which the rows and columns are independent of each other.

In a nutshell the current literature mainly consists of concept papers within the class of digital shadows and focuses on the plant process control. Sharing of case studies has made a beginning but the experiential data and information specifically in the construction and application of digital twins with two-way data exchange are rather limited. Further, the authors of several papers have highlighted the need for deeper research in many areas of digital twin technology. Bearing this in mind, the technical features and dimensions of digital twins are discussed in detail in the following sections.

TABLE 8.1

Classification Summary of Recent Technical Publications on Digital Twin Theme

Level of Data Integration	Nature of Contents	Focus Areas
Undefined 19%	Concept 55%	Production planning & control 49%
Digital model 28%	Case studies 26%	Maintenance 14%
Digital shadow 35%	Reviews 14%	Product lifecycle 12%
Digital twin 18%	Definitions 5%	Manufacturing in general 9%
		Layout planning 9%
		Process design 7%

Note: Information extracted from [4]

8.3 INTERLINKING DIGITAL TWINS AND PRODUCT LIFECYCLE

The virtual construct and the physical system are conceived to be connected with each other during all the four phases of a product lifecycle, e.g., creation, production or manufacture, operation or the stage of sustainment and support, and finally, disposal or decommissioning. In this interconnection lies the difference between the digital twins and computer-aided design (CAD) and also with sensor-enabled IoT solution. Both the latter systems are not capable of interactions between the devices and their full lifecycle processes. A digital twin can be fundamentally differentiated from them in being an evolving digital profile of the historical, current and future behaviour of the corresponding physical object or process that ultimately helps optimize the manufacturing performance. As conceived earlier [1], the primary interconnections between the real and virtual versions are depicted in Figure 8.3.

As depicted in Figure 8.3, the physical system may not exist at the 'create' phase. The system starts to take shape in virtual space as a DTP. This is the phase when the following four emergent aspects should be introduced into the virtual construct: predicted desirables (PD), predicted undesirables (PU), unpredicted desirables (UD), and unpredicted undesirables (UU), and this can be done by varying the simulation parameters across the possible range. The non-linear behaviour of the complex system should be investigated at this stage so that the catastrophic failure conditions caused by the

FIGURE 8.3 Interrelation between real and virtual entities during the lifecycle of products.

combination of factors or discontinuities are tentatively identified. Once the construction of the virtual system is completed and validated, the information is used in real space to create a physical version as the twin of the digital entity. It is important to understand that if the above described process is carried out correctly, the number of UUs can be significantly reduced. It is obvious that the exponential advances in computing capability will permit expanding the possibilities that can be examined. Moreover, testing the virtual models by qualified experts is extremely important and at the same time it may be desirable to make use of the least qualified personnel to check the sturdiness of the system. The subsequent production phase is for starting to build the physical systems as DTIs with specific configuration. The information and data flow is reversed as compared to the previous stage. In other words, the data about the physical build is sent to the virtual space and the virtual version will correspond to the physical version in a more exact representation.

In the third stage of support and sustainment the real and virtual systems should maintain their linkage both ways. Since changes occur in the real system due to parts replacement or state modifications, this operational phase might be helpful to find out if the PDs actually happen and PUs are eliminated. This is the stage to detect if UUs are minor, major or catastrophic. Changes in the physical systems are captured in the virtual systems. On the other hand, information from the virtual systems can be used to predict performance and failures of the physical systems. The last stage of disposal or decommissioning is important to assess the impact of the system on the environment on disposal. In addition, this stage may provide lessons for future so that the next generation of systems can be built, avoiding the past problems of performance.

8.4 ADOPTING DIGITAL TWIN TECHNOLOGY IN MANUFACTURING

There is hardly any difference in view among the experts about the wide variety of role that the digital twin technology can play in manufacturing industry at large. It is progressively recognized that the manufacturing industry will specifically gain from adopting the technology by visualizing, designing, monitoring, and maintaining their systems and equipment. The gains will emanate from the fact that digital twins are designed to model complicated assets or processes that interact in many ways with their environments, the outcomes of which are difficult to predict otherwise over an entire lifecycle of the assets or the process equipment.

Broadly speaking, there are four essential hardware and software components such as sensors, datasets, analytics and actuators at the core and six steps of actions such as creating, communicating, aggregating, analysing, forming insights, and acting further on the basis of such insights. For illustration purposes, this is depicted in Figure 8.4 [5].

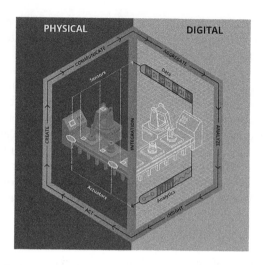

FIGURE 8.4 Principal components of implementing the digital technology [5].

The sensor signals enable the twin to capture operational and environmental data pertaining to the physical process or system. As a result, if any change action is warranted in the physical entity, the digital twin prompts the action and the actuator triggers the physical process. It is obvious that the actions are dependent on the virtual model of the physical system and the state of the virtual model, the inputs for which are diverse in character and may consist of the real-world operational and environmental data, bill of materials, enterprise systems, design specifications, engineering drawings, customer feedback, and so on. The dataset is generated from external data feeds and sensors, which communicate to the digital world through technologies such as edge and communication interfaces between the physical-to-digital systems and vice versa. The dataset is subjected to analytics through algorithmic simulations and visualization processes, which produce insights to act upon.

It must be admitted at this stage that the twinning of a physical system and its digital analogue is more complex than the framework described above. Figure 8.4 is just an illustration to emphasize the integrated, holistic, and interactive approach of the technology.

8.4.1 STRATEGIC APPROACH FOR ADOPTION

The approach for adoption of the digital twin technology cannot be the same for all situations and for all organizations. The virtual models can be at different levels of detail. If the exercise is oversimplified in terms of detailing, the digital twin may not serve the purpose. If it is overdone, the exercise may be lost in the complexity of sensors and data feed sources and the extent of processing technology required. A balanced approach has been proposed in [5] and it is presented in Figure 8.5.

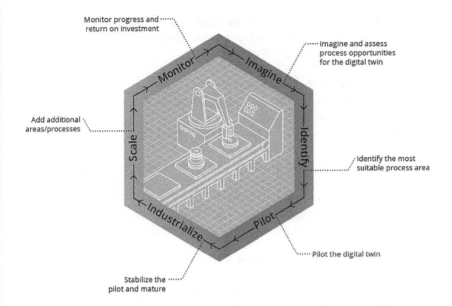

Monitor progress and return on investment

Imagine and assess process opportunities for the digital twin

Add additional areas/processes

Monitor

Imagine

Scale

Identify

Industrialize

Pilot

Identify the most suitable process area

Pilot the digital twin

Stabilize the pilot and mature

FIGURE 8.5 Illustration of a practical approach for application of digital twins [5].

The approach has to start from exploring the process opportunities that may derive benefits from digital twin construct. It is desirable to shortlist a set of scenarios that could make those identified opportunities more productive, efficient, and reliably predictive. The process segments that could respond to the digital twin approach rather quickly may be identified. Having gone through the exploration stage, it is logical to narrow down on the physical system that could be subjected to digital configuration at pilot scale. A preliminary assessment of the best chance of success may underpin such selection. It may also be borne in mind that the planned pilot configuration may be small in scope but it should have enough possibility of displaying its impact on the total business or the enterprise. The next step is to expand the scope of the pilot configuration after obtaining its positive results and potential impact. Before implementing such full-fledged expansion in the digital twin configuration, enough consideration will have to be directed towards developing the data base, data standards, and organizational changes necessary for the project and the post-project scenarios. In scaling up the twin, it is extremely important to monitor and measure the relevant parameters and introduce changes to the digital version iteratively.

8.5 FUNCTIONALITY AND STRUCTURAL CONFIGURATION OF DIGITAL TWINS IN MANUFACTURING

It is pertinent to recall that a manufacturing digital twin generally offers an opportunity to simulate and optimize the production system including its logistical aspects, and enables a detailed visualization of the manufacturing process from single components to the whole assembly. Targeted to increase competitiveness, productivity and efficiency, this activity range of digital twins can be broken down as follows:

1. Analysis and prediction of asset condition or state: systemic anomaly, wear and tear evaluation, reliability assessment, preventive and predictive maintenance. etc.
2. Digital reflection of assets: integration of lifecycle data of a machine or process in the form of a data thread
3. Management of assets as components and as aggregated factories
4. System and process optimization in current operation
5. Production planning and control: order planning on the basis of statistical assumptions, execution planning by the production units
6. Improved decision support with the help of visualized diagrams at different levels
7. Providing training and learning platforms

In order to execute the above range of functions the digital twins must have an appropriate structure. In this context, a question has often been raised whether a digital twin should be structured as a visualization system only or it should include the control function in production or manufacturing. If the latter is included, the structure should also change as discussed below. Furthermore, the structural built-up should necessarily cover the complex network of manufacturing, which include heterogenic information sources such as those of measurements on one hand and data gathering from the process, execution elements and decision-making organs on the other, hierarchical and stand-alone functions, various types and forms of information, and different methods and models of information processing. The structure should also have the following attributes: interfacing with process hierarchies, scalability to accommodate increase in hierarchy and increased addition of details at every stage of the process, ability to represent different types of processes in the same display format, ability to view different scales of visualization with different amount of details. These attributes are additional to the basic framework of a digital twin to connect different modelling methods and to function with a communication layer to maintain uninterrupted data exchange with the physical counterpart.

The structural configuration of a digital twin for diagnostics, optimization, and prediction has been discussed comprehensively in [6]. The diagnostics are classified as static and dynamic. The static diagnostic tasks relate to situations, where all the key performance indicators (KPIs) of the processes are known, stored, and set beforehand without any likelihood of changes over a period of time. Conventionally, such tasks can be carried out via a synergetic system involving sensors, a control application, a knowledge base (KB), and an OPC (Object linking and embedding Process Control) server. The sensors feed data to the OPC server, which is connected to the knowledge base management system (KBMS) forming a part of the knowledge base (KB) to produce control signals. The KB is the source of rule sets emanating from mathematical models and empirical relations. Signals from KBMS are transmitted to controlled objects and operator terminals.

Unlike the static diagnostics, the dynamic diagnostics refer to situations, where the KPIs are unknown and need to be computed before control. Such situations occur mostly when the production plans are modified, source materials change, or the facilities are modernized. Under those circumstances the digital twin components such as Digital Model, Execution Environment, DBMS (Data Base Management System) and Control System must work in a cooperative mode (Figure 8.6).

In order to obtain data from the control plant, DCS (Distributed Control Systems) or ICS (Improved Control Systems) are installed in direct link with the control plant. For secure operations, a PSS (Protection Safety System) is also provided. Further, it is proposed to use CPAMs (Compositional Programme Analytics Models), which are essentially the model relations of KPIs at different functional levels, expressing how KPIs of different levels depend on one another or on control parameters. The CPAMs can be of different types. They can be simply formula-derived models or analytical models or neural networks based models, or even files of ChemCAD, MATLAB, MathCAD, or similar other sources. The pattern of information flow from block to block is displayed in Figure 8.6.

The 'Execution Environment' is a crucial subsystem, which computes the relevant KPI from the current data and CPAM, and generates reference data that are sent back to DBMS in order to check if the current data meet with

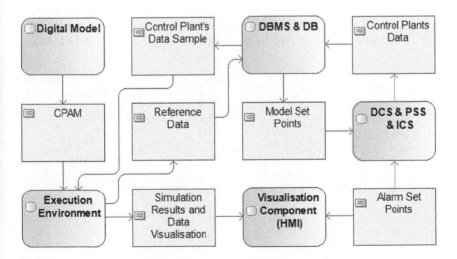

FIGURE 8.6 Functional scheme of the dynamic diagnostics [6].

the defined KPI limitations. After the comparative evaluation, the set points go back to the DCS to provide control signals, based on modelling values. At the same time, the data are displayed in HMI (Human Machine Interface) to show the difference between the set points and current values. The alarm set points have the links with both HMI and DCS. The entire scheme is also based on the flow of information and data amongst DCS, Control Plants Data, DBMS, Model Data Points and Reference Data blocks.

8.5.1 PROCESS AND SYSTEM OPTIMIZATION

The main difficulty in process optimization is caused by the hierarchical nature of the processes, as a result of which the control parameters and process parameters under optimization may belong to different levels. For example, if the control parameter is the volume of goods produced and the process parameter under optimization is a manufacturing parameter like the process temperature, they will belong to different levels and may not apparently display any direct dependence of functions. This type of relational discordance of two functions is schematically shown in Figure 8.7.

As depicted in the figure, the transformation of a source material from State 0 to State 2 via an intermediate State 1 goes through a vector U_1. The KPIs of State 1 and State 2 are unlikely to be the same and the relation of vector U_1 with KPI_2 cannot be simple. If these two parameters have to be related as dependent functions, the processes may have to be decomposed as illustrated in Figure 8.8.

In this decomposition, vector U_1 is may be connected to KPI_1 through the progressive decomposed states, and using the same approach, KPI_1 may be related to KPI_2. In other words, $KPI\_0.1 = f(U\_1)$; $KPI\_1 = f(KPI\_0.1)$; $KPI\_2 = f(KPI\_1)$. In these exercises, what is important is the step of identification, which may require complicated models that are based on neural networks. For this purpose, historical data are necessary as stored in DB. After

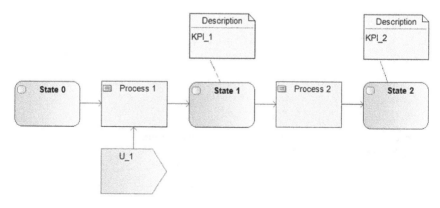

FIGURE 8.7   Schematic diagram of the production process [6].

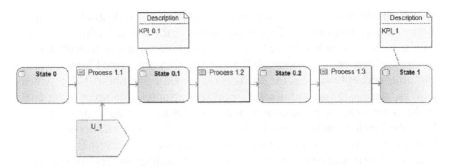

FIGURE 8.8   Block diagram showing the decomposition of Process 1 [6].

the analytical relation of KPI_2 and vector U_1 are composed, any of the multi-object optimization algorithms can be applied. The prediction task can be solved using the CPAM and considering 'time' as one of the parameters.

8.5.2 Tentative Structural Configuration of Digital Twins

Following [6] and as discussed above, the total functional scheme of a digital twin should be structured to implement static and dynamic diagnostics, systemic optimization, and prediction tasks. The structure should be capable of using the digital models of control plants and the configuration should be able to identify models if they are not given. The executive environment has to be the repository of the models. The system identification and chosen algorithms are particularly significant for optimization. The reference data, generated from the executive environment, must be sent to the data base controlled by the DBMS. In this context, a differentiation between a data base and a KB can be considered. While the data base may store current data, the KB may be the repository of historical data and information about the performance indicators. The interactions between the main digital twin components are the key features of the total system. After the simulation and visualization, the results of such processes are visible at the human-machine interfaces. It is important to note that the digital twins need to create the set points based on modelling data. Such set points, combined with the data in the KB, may be sent in the form of alarm set points and deviation signals to the control system. Finally, the DCS or its improved variants provide the real control signals to reach the control objects.

8.6 DIGITAL TWINS DRIVING THE PILOT PRODUCTION ENVIRONMENT

In manufacturing industry, for new products and processes, the setting up of small trial facilities has been a very common practice. The purpose of such plants is proving of concepts and collection of design parameters for scaling up the production process to plants of appropriate industrial capacities requiring large investments. Experience shows that the trial facilities for continuous processes such as those for petroleum refineries may start with micro-reactors of capacities less than 1000 mL, when they are termed 'bench scale', and after concept proving, the reactors are scaled up to about 100 L, when they are designated as 'pilot scale'. Further capacity scale up of such trial facilities for technology demonstration purposes may go up to around 1000 L. In the batch manufacturing industry, the bench-scale and pilot-scale trial facilities are often in the range up to about 20 kg and 100 kg, respectively. The strategy of setting up such pre-production pilot facilities suffers from the lack of process flexibility to establish the evolving trend of any new production system, inherent potential of scalability, spending of resources in reconfiguring the facility even where such opportunity exists, absence of any scope of defining the systemic limitations, and difficulties of identifying reasons for unpredictable performance of the system.

In this context it may be relevant to mention that integrated pilot plants are rarely set up in the cement industry due to the complexity of the end-to-end process and envisaged difficulties of applying the pilot plant data to designing plants of industrial capacities. The traditional practice is to test raw materials for crushing and grinding in small facilities that are available with the original equipment suppliers, who also have their individual data banks for design purposes. Similarly, for testing the thermal behaviour of the raw materials, more empirical tests and approaches are adopted. The maturity of the cement manufacturing processes and the accumulated data banks available with the equipment and technology providers help in avoiding catastrophic failures in most cases, but non-achievement of the process guarantees and desired performance is not uncommon in the industry. Furthermore, the new product technologies are almost always tried out at the plant production facilities involving production losses and incurring substantial expenditure on setting up make-shift arrangements.

In order to mitigate the above short-comings of physical pilot production facilities in manufacturing industries in general and in the cement industry in particular, integrated pilot installations having both physical and virtual components, based on the digital twin concept, may provide significant advantages. As discussed in [7], the concept relies on in situ and ex situ measurements along with modelling and simulation of the existing and evolving resources/processes at operational, tactical, and strategic levels. Virtual dashboards supplement the pilot plant to provide insights for decision-making for all perspectives and by all stake-holders involved. The conceptual elementary blocks of such a pilot plant is shown in Figure 8.9.

It is presumed that an asset, either physical or virtual, is related to other assets by one or more relations, which is expressed by 'perspective' in the above figure. One of the relevant types of relations is the 'consist of' relation, allowing subsystems to be built in the system, enabling a pilot plant to consist

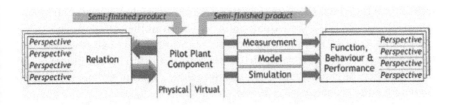

FIGURE 8.9 Elementary building blocks of a pilot plant [7].

of subservient pilot plants. The pilot plant components will express their individual behaviour, which will be different for different perspectives. When the component is physical, its behaviour can either be measured or derived from models or simulations. If the component is of virtual type, the model can be based on the real-life conditions from the previously stored information or from measurements at comparable assets at different locations.

8.6.1 Modelling Approach in the Pilot Facility

It is known that partial models, theories, best practices, laboratory experience, etc. are only available at the starting point to describe the behaviour of a process or asset. Since such data or models do not capture the asset's behaviour for all perspectives, the approach given in Figure 8.10 can be conveniently adopted in pilot plants in order to develop the predictive model. As shown in the figure, the offline and online simulations as well as the in-line optimization can be made to converge for robust optimization.

While many prevailing approaches start from the available data and information and try to convert that to a feed-forward predictive model, the pilot plant concept can change the path. Once it is established what result is required, and with what sensitivity and accuracy, the available knowledge, theory, and existing models can all be combined to determine a rudimentary structure to capture process parameters and machine learning settings, as well as their known or projected relations. Based on sensitivity analysis, the relations are clustered according to their anticipated impact. This leads to an overview of the data/information that should be available to render a

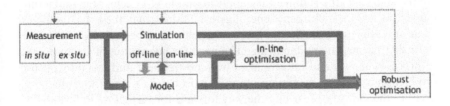

FIGURE 8.10 Modelling approach for pilot plant [7].

meaningful outcome. However, it is known in practice that models and simulations in essence rely on adequate information from the actual process. The pilot plants can be instrumented with sensors on the physical assets on one hand and can be provided with virtual sensors on the virtual components on the other. Sensor can also be used on external resources to provide meaningful data for the process model development.

Thus, the virtual or integrated pilot plants can serve as the backbone of process development for a new production environment. It is not only able to capture and integrate a pool of data and information to demonstrate possibilities, variants, and alternatives but also indications of their performance.

8.7 RELEVANCE OF DIGITAL TWINS FOR CEMENT MANUFACTURING

During the last four decades, the cement industry has undoubtedly progressed well into instrumenting and automating the manufacturing process. Way back in the late nineteen seventies, the practice of parametric measurements, both online and offline, and operation of programmable logic controllers (PLC) were common features of the production lines. For viewing of the processes, mimic display panels were used on the front panel of the control desk. The mimic panels were subsequently furnished with indicator lamps and push button switches for emergency shutdowns. This kind of display helped in monitoring the whole plant and also facilitated identification and exact location of faults and breakdown. From the nineteen eighties, the second-level controllers were in use above the PLC level. This was the period when the optimization strategies, based on advanced data processing technology, slowly penetrated into the operational practice in the cement industry. At the turn of the century, the application of the plant controller was attempted to be made more effective with fuzzy logic and the resultant basic software, termed as 'expert systems', was customized to suit different operational situations at the time of installation and used. Subsequently, the focus moved from the control function to manoeuvring the performance trends of production lines. This change of strategy brought into play the new technology of model predictive control (MPC), based on artificial neural networks (ANN) and gave rise to 'production optimizers', which are in use in some plants. These developments are recognized as the rudimentary application of AI to cement manufacturing. However, the field experience has established that these technologies could not effectively tackle the process variables and perturbations that are intrinsic to cement manufacturing. The need for adaptation of technologies to process changes and uncertainties was too frequent and too difficult for production optimizers to remain in use in most cases. A big concern has surfaced in the industry regarding the way forward for further digitalization of cement manufacturing. Can the digital twin technology provide a robust solution?

The application potential of digital twins has been discussed at length in [8]. The authors observed that since the digital twin technology provides an extended insight into the production process as a replica in real time, the management of the process can be more effectively optimized. The digital twin is completely accessible for any information that is not available in the real plant or not available in time. In principle, this approach can therefore be considered as a continuation of MPC technology but it goes far beyond the earlier model because of its comprehensibility and imaging.

8.7.1 INFORMATION DENSITY CONSIDERATION

It is estimated that the automation system in a cement plant now has about 10,000 to 20,000 measuring channels that accept and process analogue and digital signals at a frequency of about 1 Hz [8]. Because of the complexity of the cement production process and also considering the heterogeneity and variability of the measured data, the authors doubt if the quantity of data is sufficient for automatic production control. At the same time it must be understood that the normally measured dataset in cement manufacturing represents only a fraction of the total information that is generated in the course of the process. The combination of 3D space, processing time, size-reduction, and thermo-chemical process components, varying temperature in different process segments, high-temperature solid-gas reactions, liquid-phase sintering, flow of particulate suspensions in gas streams, emission of polluting and non-polluting gases, and several other consequential parameters make the present level of information density grossly inadequate for automatic control and operation of the complete process. Hence, for information density, much larger variety and volume of data and more closely measured and acquired data will be necessary. If a fully autonomously driven car generates a data volume of 4000 gigabytes per operating hour, it is not difficult to comprehend the magnitude of data that will be essential for autonomous operation of a cement plant having numerous drives and controls. It is evident that under the real industrial conditions it may not feasible to increase the information acquisition density by that order of magnitude. The digital twin may seem to be a solution for this impasse. Since the digital twin is capable of utilizing real-time mathematical problems as well as soft sensors that are operated in parallel with the real plant, the data acquisition can be increased many folds.

A newly introduced system, called aixPert Optimizer, offered by aixergee GmbH, has been reported in [8]. The system combines the real-time Big Data analytical functions, separate sensors, and laboratory data in a model format, which combines statistical and AI algorithms with deterministic models like process and CFD simulations. It is claimed that the combination of the models facilitates adaptive prediction of multidimensional condition variables. The design architecture of the system is schematically shown in Figure 8.11. The digital twin operated in parallel in real time consists of a data-driven

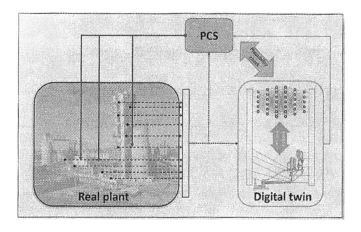

FIGURE 8.11   Basic design diagram of aixPert Optimizer [8].

ANN-based model, a real-time CFD model, and some other mathematical process models. The advantages of different modelling strategies are utilized and interlinked. For example, the predictive capability of ANN-based models is combined with the strength of deterministic process models. Furthermore, the digital twin also works together with the real plant. The deviations in process parameters can be detected and the cause can be found from the historical data and scrutiny of process models. The system is modular in character and can help optimization of subsystems and through reinforcement of the modular controls the entire production line can be optimized.

8.7.2   DIGITAL TWIN OPTIONS IN CEMENT MANUFACTURING

The general approach towards developing digital twins for cement manufacturing will essentially remain the same as discussed earlier. The initial step will be envisaging the need primarily based on either production efficiency or customer demands, which should be followed by the second step of narrowing down what is envisaged to be a selected part of the total process that will have the potential of demonstrating the impact of the digital twin concept all across the organization. Considering this as a pilot project, the implementation step may be undertaken after procuring appropriate tools and resources. Once implemented and the outcome assessed, the scope and boundaries of the pilot project may be progressively expanded to spread through the organization in order to realize the total benefits in real terms. What is important is to identify a few critical subsystems within the total cement manufacturing process that can be taken up on priority for introducing the digital twin concept. Priorities will differ from situation to situation but the following subsystems may be kept in view.

A. **Limestone mining**: There is no need for reinstating that the cement manufacture begins with limestone mining. The important parameters of mining and the crucial issues of quality control of run-of-mine limestone have been discussed in [9]. The salient aspects are as follows:

1. processing of exploration data
2. 3D modelling of the limestone deposit
3. quarry design and opening up of mine faces
4. determining the mining option
5. deploying the fleet of mining and hauling equipment
6. optimizing operations for grade control, blasting pattern, blending, etc.
7. planning advance drilling programmes for quality control
8. transporting limestone to crushers
9. long-term planning including mine closure
10. observing safety and environmental norms
11. restoring the landscape of the mine site

It is obvious that the open-pit mining operation for limestone looks simple but consists of multifarious inter-connected functions that begin before the cement plant is set up and end after the plant life is over. Present-day mining software packages allow rapid iterative investigations of mining scenarios to determine the most appropriate course of action. The basic features of these software packages include a relational geological database, 3D deposit modelling for quality and quantity, blending of raw materials from the quarry faces, long-range quarry scheduling, blast design and layout, production planning from the working benches, discounting the value of mining blocks and locating the optimal limit of mining, and several other aspects.

The software packages, however, are mostly segmental, focusing on specific aspects of geology and mining. The mining simulators typically use static empirical models as inputs. In most cases with complex geology, there is gross mismatch between what is predicted and what is actually observed in terms of quality, quantity, and plant performance over the life of the mine. Hence, the digital twin concept for the limestone mine value chain optimization appears to be a worthwhile strategy to adopt in cement manufacture. This can provide the right platform for integration of 3D geological models with plant operational and performance data, for backward reconciliation of mine blocks, and most importantly, for analytics and forward prediction. The conviction comes from the fact that the world's first digital twin for mine value chain optimization came into operation in 2018 [10]. Although it was for metal mining, the concept and approach in cement production would not be much different. The operational silos between the geologists, mining engineers, and plant engineers will be broken down and the mining options and constraints can be visualized by multi-disciplinary teams for proper work plan, based on very large volume of historical performance data of both the mine and the plant in conjunction.

B. Limestone calcination: The maximum heat consumption within the cement manufacturing process is in the limestone-bearing raw meal calcination, occurring at about 900°C, and amounting to about 2000 kJ/kg of clinker. Hence, this sub-process has been in focus for both equipment design and process control, resulting in the application of numerous systems of varying performance. But in all cases the degree of calcination and the corresponding energy efficiency of the special vessels added to the preheater string depend on multiple operating parameters:

1. inside temperature distribution
2. residence time of the raw meal
3. gas/solids separation
4. dust circulation
5. varying concentration of CO, NO$_x$, and CO$_2$
6. kinetic properties limestone decomposition

The CFD models have shown that, depending on the size of the equipment and process fluctuations, the temperature distribution, the solids concentration, and the gas dispersions vary in 3D space and over time. Hence, it can be easily understood that the traditional control arrangement with single-point temperature measurement in the calciner may not always serve the purpose. It is, therefore, highly pertinent to identify the complete management of the calciner operation including its control regime for application of digital twins. The scope of the digital twins can be progressively enhanced for complete combustion management of the kiln systems.

C. **Alternative fuels combustion management:** From the perspectives of waste utilization, energy conservation and greenhouse gas reduction the use of alternative fuels through co-processing in clinker making has gained significant importance in recent years. The global use of alternative fuels, primarily biogenic and non-biogenic residues, which was just above 5% in the year 2014, is projected to increase to 30% under two-degree scenario by 2050. It is essential, therefore, to organize the upstream and downstream facilities and processes for effective and efficient combustion of alternative fuels. The kiln management, which is at the heart of cement production, becomes more crucial in burning alternative fuels. Various process effects such as increase in waste gas volume, shift in kiln temperature profile, blockage and build-ups in various parts of the system, creation of reducing environment, modification of clinker microstructure, higher emissions of SO_2, NO_x, CO, and other volatile elements, have been either experienced or foreseen. The higher the substitution of conventional fuels by the alternative varieties, the more intense will be the process effects. Further, depending on the composition, calorific value, feeding systems, etc. the alternative fuels are fed to calciners, or kilns, or both. Thus, in enhancing the use of alternative fuels, it is important to be able to predict the changes that are likely to occur in the burning process. The studies made on simulation of clinker processes with alternative fuels have been discussed in [11]. Most of the previous studies had been carried out using commercially available software packages like Aspen plus, Chemapp, etc. Another simulation tool known as Particle Swarm Optimization has also been attempted.

Considering the wide variety of alternative fuels and their compositional and granular features, physical state of supply, hazardous or non-hazardous nature, pre-processing requirements, etc., it is obvious that there are many unknowns to deal with in using alternative fuels. Certain process effects such as preheater clogging, clinker melt properties, clinker phase and microstructure modifications, occurrence of volatile cycles in kiln-preheater section, etc. cannot be measured with the help of sensors for predictive purposes. Hence, this is another area in the contemporary cement manufacturing that deserves to be considered for application of AI tools.

D. **Grinding of multi-component blended cements:** It is known that the grinding operation is the most energy-intensive part of the cement manufacturing process. Consequently in many plants the ball mills have been replaced with energy-saving systems like vertical roller mills, hydraulic roll presses, horizontal roller mills and also hybrid circuits of ball mills and hydraulic roll presses. It can further be surmised that all the above mill circuits can be provided with predictive operational settings, based on analytics of historical and current operational data, if the necessary AI platform is created. But will that be adequate for assuring the quality parameters of multi-component cements with high clinker substitution? Perhaps not, unless the impact of supplementary cementitious materials to be blended in different combinations on the hydration properties of the finished cements is quantitatively assessed and captured by models. A central element in such models will be the reactivity of the materials to be blended and their compatibility.

The combination of thermodynamic modelling and physical test results of cements in mortar or paste are often used to correlate microstructural parameters to mechanical properties. The hydrate phase assemblage in a hydrating system can be predicted by the thermodynamic modelling approaches. The

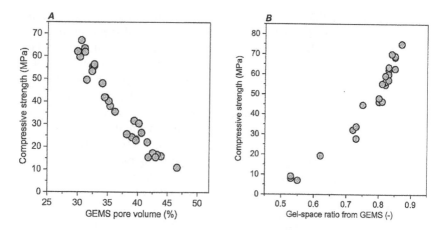

FIGURE 8.12 Relation of measured compressive strength with porosity and gel-space ratio from GEMS [12].

Gibbs Energy Minimization Software (GEMS) and some other tools have been instrumental in recent times to predict the phase assemblage, porosity, and volume stability of blended cements in general. The model findings have in good agreement with the experimental data as shown in Figure 8.12 [12].

The thermodynamic modelling approach may be considered as a powerful practical tool to predict the use of supplementary cementitious materials in composite cements. Hence, the prognostics and predictions in cement grinding operation, particularly for the present-day needs of multi-component blended cements may be effective if the operational data analytics is combined with application of hydration modelling tools such as GEMS and supporting experimental models, based on physical tests. As discussed earlier, this kind of complex approach can be made feasible via digital twins and, therefore, the cement grinding section may be considered for application of digital twin technology.

The above examples are intended to indicate certain possibilities of applying digital twins technology to cement manufacturing. The opportunities are definitely more widespread and diverse. Primarily, it can be in four segments: products, processes, utilities or production facilities. The product related digital twins will have focus on design and operation for making the product based on several operating conditions to increase economic effectiveness, to enhance product attributes, and to predict potential failure in different scenarios well in advance. The process twins are directed towards process simulation solutions such as the clinker quality prediction, volatile cycles, energy conservation, emissions control, and so on. In the third segment of utilities, a specific mention may be made of air compressors, which constitute a relatively high investment area in cement plants both in terms of capacity and distribution systems. Given the high cost of producing and distributing compressed air, several measures are normally taken in the plant to minimize the losses but the measures can be further improved with automatic solutions. The service providers are now able to offer remote service capability to interface with SCADA system and other third-party instrumentation devices, and also to perform data analytics and reporting. Hence, such utility areas may offer opportunities for digitalization. Finally, the plant twins will focus on complete operation simulation to predict and simulate the overall plant operations.

8.8 TECHNOLOGICAL PREPAREDNESS WITH ENABLING TOOLS

Creating and using digital twins in all fields of application, will require, in addition to strong domain knowledge, an enabling environment with appropriate technologies and effective tools. A comprehensive technical survey of enabling tools and technologies has been published recently [13], although other reviews, more commercially oriented, are also available, as in [14].

The compilation and review of technologies that has been reported in [13] were based on five dimensions or interconnected modules of digital twins: physical entities, virtual models, services, connections, and application fields. The physical entities, which may consist of a product, a system, an activity related process or even an organization, can be divided into three levels: unit level, system level, and system of systems level. The virtual models, which in form are 3D and geometric, represent physical phenomena such as deformation, fracture, corrosion, etc.; the behavioural patterns such as state transition, performance degradation, response mechanisms against changes in the external environment; rules extracted from historical data or provided by domain experts for reasoning and decision-making. The services refer to those that are provided by digital twins such as optimization, diagnosis, prognosis, etc., on one hand, and those that are required by digital twins for dealing with data, knowledge, and algorithms, on the other. The connections refer to the triangular two-way linkages from the physical world to virtual world, from the virtual world to services, and from the services to the physical world with digital data connectivity at the centre of the triangle. The application field is of course defines the start and end points of developing the enabling environment with the aforesaid interconnected modules.

8.8.1 SALIENT TECHNOLOGY REQUIREMENTS

The specific technological requirements are different for each of the above modules. For example, a precise profiling of the physical entity is essential for preparing a precise virtual model. It can only be done, when there is suitable hardware for sensing and measurement, capability of tracing the process dynamics and control, and understanding the material science and engineering. Once the profiling of the physical entity has been accurately done,

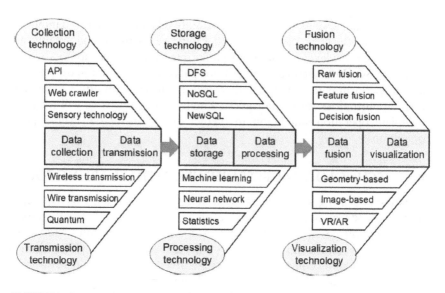

FIGURE 8.13   Important steps and technologies in data management [13].

generating large volume of data and information, the technologies for data management comes to the fore. For illustration purposes, the sequence and components of data management and technology requirements are shown in Figure 8.13 [13].

Generally speaking, any hardware or software or a network can be the source for data. The hardware data includes the sensor-based information about the static attributes and the dynamic status, identification through codes and other IoT technologies, and real-time status. The software data can be collected through application programming interfaces (APIs) and open data base interfaces. The network data can be collected from the internet through web crawlers, search engines, and public APIs. Both wired and wireless systems are adopted for data transmission, which, of course, are dependent on transmission protocols, access methods, multi-access schemes, coding, etc. For data transmission it is necessary to explore the applicability of high-speed, low-latency, high-performance, and highly secure data transmission protocols and corresponding devices. It is relevant to mention that the quantum transmission technology is potentially applicable, though it is futuristic for the present status of digital twin concepts for manufacturing. Since the collected data is necessarily processed and analysed further, the data storage technology is as critical as the data transmission itself. The data storage technologies are inseparable from the database technologies. Due to increasing volume and heterogeneity of data, Big Data storage technologies such as distributed files system (DFS), NoSQL, and NewSQL are becoming more relevant. An important trend is to store all this in the cloud with managed services for on-demand scalability and compute. Data processing refers to the extraction of useful information by data cleansing, data compression, data smoothing, data reduction, data transformation and the associated activities. Thereafter, the pre-processed data is analysed with the help of numerous statistical and machine learning methods. The statistical methods are applied for studying parametric correlations, clustering, discriminant analysis, dimension reduction, time-series analysis, and so on. The machine learning methods include regression, decision trees, random forests and methods based on neural networks. Data fusion and data visualization are the subsequent steps in data management. Data fusion deals with synthesis, filtering, correlation and integration of the multi-source data and the measure can be taken at the raw data level, at the feature level, and at decision level. The methods adopted for fusion are numerous and include classical reasoning, weighted average, Kalman filtering, Bayesian estimation, Dempster–Shafer evidence reasoning, fuzzy set theory, rough set theory, neural networks, wavelet theory, support vector machine, etc. Some of the methods are suitable for all the three levels and some are more appropriate for the feature-level and decision-level data fusion. Data visualization in the form of graphics and dashboards is well known.

In addition to data management, the other crucial technological component in the implementation of digital twin concept is modelling and simulation, the complexities of which are reflected in Figure 8.14 [13].

The modelling functions, processes and resources are captured in the figure for an overall comprehension. It is evident that this activity depends on a highly interdisciplinary approach. It has to reach a high level of reliability,

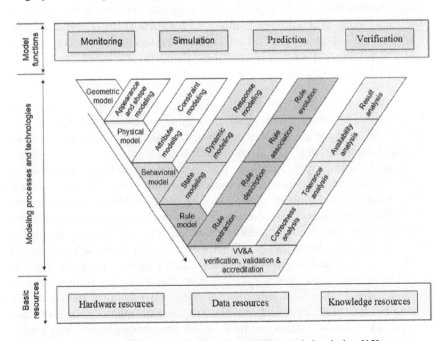

FIGURE 8.14 The technological spread of modelling and simulation [13].

reproducibility and multi-objectivity to be effective for operational purposes. For the digital twins to be efficaciously workable, the predictive models and failure models need to be optimized through machine learning.

8.8.2   IoT Platforms and Enabling Tools for Digital Twins

In the manufacturing context, the industrial internet of things (IIoT) is the key strategic consideration for realizing the full potential of digital twin concept as products, processes and people behind machines are captured through devices and sensors for an effective digital twin to be built and applied. The selection of the right IoT platform for a specific manufacturing business or a plant is a primary task, particularly in view of the fact that in this new and growing field already there is a wide choice in the following categories:

- IoT infrastructural platforms (such as AWS IoT core, Azure IoT Hub, IBM IoT, Google IoT core, etc.)
- IoT network providing platforms (e.g., AT&T, Orange Business Services, Telefonica, Verizon, Vodafone)
- Iot business platforms for vertical applications and different industrial profiles (more than a few hundred at present).

Since the first category consists of large platforms, primarily focused on infrastructure, and the second category platforms are specialized in communication, customization of these platforms for specific business applications by the users is difficult and time-consuming. Consequently, several IoT platforms have been and are being built on top of the infrastructure providers, offering additional tools and services so that these platforms can be adapted quickly for specific businesses. Such platforms are receiving more attention for digital twin applications.

As far as the digital twin-specific software development is concerned, a large number of comprehensive tools are commercially available. They can be applied for multiple activities aimed towards creating digital twins. An illustration, adapted from [13], is given in Table 8.2.

It would be improper to presume that the software packages mentioned in Table 8.2 are the only ones of relevance for digital twins. There are several other packages and tools of varying capabilities available for use. For example, 'aPriori' is a digital manufacturing simulation package used for insights such as 'design for manufacturability' and 'design to cost', helping the manufacturers to make better designs in shorter time. Another package, named 'IoTiFY' is intended to work as a visualization platform, while 'IoTSYS' is capable of providing communication protocol stack for communication between smart devices. It may also be relevant to mention that tools for changing the physical world mostly are control related. For example, 'TwinCAT' software system can turn almost any compatible PC into a real-time controller with multiple PLC systems. Autodesk Tandem is another tool

TABLE 8.2
Functional Comparison of a Few Selected Digital Twin Tools

Functional capability	PTC's Predix Thingworx	Siemen's Mindsphere	ANSYS	Dassault's 3D Experience	Foxconn's Beacon
Knowing the physical world			#	#	
Changing the physical world	#	#			
Geometric modelling				#	
Physical modelling			#	#	
Behavioural modelling			#		
Rule-based modelling	#				
Data collection	# #	#			#
Data transmission	#	#			
Data storage	#			#	#
Data processing	#			#	#
Data fusion	#			#	#
Data visualization				#	#
Simulation	#	#			#
Optimization	#	#			#
Diagnosis & prognosis	# #	#			#
Platform	# #	#		#	#
Connections in digital world	#	#		#	
Connections between digital and physical world	# #	#		#	#

that brings project data from multiple sources and formats to create a digital twin and can track the asset data from design to operations. It can also connect the digital world with the real world.

From the above discourse it is obvious that the sphere of enabling tools and technologies is already large and is expanding further, keeping the multifarious user needs in view. Due to different formats, protocols and standards, there are problems of integration and application for a particular objective. Evaluation and selection of tools with respect to the specific fields and objects will be the most crucial issues for application.

8.9   DIGITAL TWINS FOR LEARNING AND TRAINING

In the manufacturing industries including cement, the use of computer-based training simulators has been prevalent since the nineteen eighties. In most cases, such facilities have been set up to train operators in plant processes. They are generally at the functional level and not at the enterprise level. The real plant data and models, mostly one-dimensional, form the basis of such simulators. The training sessions are executed through interaction of trainers and trainees in offline facilities, often provided by the original equipment manufacturers, and sometimes by the users in the plant premises. Visualization of equipment, process, graphics, alarm systems, etc. makes the training more realistic and effective than mere classroom lessons. Introduction of faults and abnormal conditions by the trainer and rectification of faults and bringing back the operation to a steady state by the trainee generally serve as the proof of training success.

Despite the merits of computer based simulators, they still fall short of exposing plant operators and line managers to real world situations and real-time problems. Tackling the problems and issues of the real world continues to be in the hands of those, who are experienced. In this context, digital twins show a strong potential for learning and training. It is understandable that existing production environments have only limited opportunities for learning – and especially for making mistakes. A digital twin, being a fully working system or subsystem of a production environment, can serve as an excellent training facility or a learning factory. This is due to the fact that the trainees do not endanger or disturb the actual production environment. They can repeat and vary the activities, observing simultaneously the repercussions of their actions on different perspectives involved. All this can be done without any major investment for physical modifications.

It is pertinent to mention here that for new products or processes the pilot project development with digital twins, as discussed in Section 8.6 of this chapter, can have several advantages as a learning platform. In addition to having the new operators trained, the development engineers get exposed to the new system during the course of its creation and may come across process inconsistencies and unsafe operations, which may modify the course of project development without major investments in physical alterations at a later period. Digital twins, thus, unfold a new horizon of learning and training.

8.10   CONCLUDING OBSERVATIONS

The concept of digital twins is just about two decades old and in this short span of time it has caught up with the manufacturing sector to a notable extent. The successful adoption has been possible due to the key enabling IoT technologies that include reliable sensors, high-speed networks, low-cost data storage, and Big Data analytics. In addition, several other technologies such as PLM, CAD, VR/AR, support the integration of digital twins with the manufacturing systems. It is predicted that the adoption will increase significantly in the next five years.

In the industrial IoT environment, the major advantage of digital twins is foreseen in evaluating the manufacturing decisions, not only on the basis of analytics, but also due to visualization of information with live and historical data. Furthermore, with digital twins it is possible to achieve a wider spread of collaborative involvement in decision-making of a broader set of professionals than what is possible on the shop floor. Another notable advantage of digital twins is that different versions of a given situation can exist side by side, which can be divided by iteration per department so that ideas can be tested against specific requirements. Benefits of remote service centres for commissioning of machines or troubleshooting of equipment can be derived with the help of digital twins, which may also provide opportunities for connecting separate systems/processes for improved tracking and monitoring. In effect, digital twins have immense potential to gain control over complex processes and system-of-systems.

Like a few other bulk manufacturing industries, an integrated cement production plant is a system of interconnected unit processes, machinery and equipment having a large number of motors, drives and distributed control systems. In its modern configuration, a cement plant is one of the most appropriate instances of industrial IoT. While a large number of material properties and operational parameters are regularly measured with sensors, there are several other features and attributes of the process that cannot be directly determined in a quantitative manner. Hence, there is a specific thrust in the industry towards developing soft sensors. In continuation of this effort, the application of digital twin technology is also being explored, more at the functional level than at the enterprise level.

Presently, examples of application of digital twins to cement manufacturing are few in the public domain as the development efforts are mostly in the trial stage. A particular digital twin based system that has been reported is 'aixPert Optimizer' that combines the real-time Big Data analytics, separate sensors, laboratory data in a model format that synthesizes statistical and AI algorithms with deterministic models.

The potential opportunities for introducing digital twins in cement manufacturing are large. In fact, pilot projects in limestone mining, precalcining, alternative fuel firing, grinding of multi-component blended cements, and utilities, can be identified, developed, tried out, assessed and progressively scaled up in scope for deriving the actual benefits. Broadly speaking, the future of digital twins will lie in creating the requisite infrastructure and expertise to utilize the emerging technologies of machine learning, object recognition, acoustic analytics, advanced signal processing, and many other developments in digital transformation.

REFERENCES

1. Michael Grieves, *Digital twin: manufacturing excellence through virtual factory replication*, Florida Institute of Technology, March 2015. Available at https://researchgate.net/publication/275211047
2. Michael Grieves, John Vickers, Digital twin: mitigating unpredictable, undesirable emergent behaviour in complex systems (Working paper excerpts, uploaded by Michael Grieves on 31 August 2016). Available at https://researchgate.net/publication/307599727
3. https://ge.com/research/offering/digital-ghost-real-time-active-cyber-defense
4. W. Kritzinger, M. Karner, G. Traar, J. Henjes, W. Sihn, Digital twin in manufacturing: a categorical literature review and classification, *IFAC PaperonLine*, 51–11, 2018, 1016–1022.

5. Aaron Parrott, Lane Warshaw, *Industry 4.0 and the digital twin*, Deloitte University Press, 2017.

6. D. Kostenko, N. Kudryasov, M. Maystrishin, V. Onufriev, V. Potekhin, A. Vasiliev, *Digital twin application: diagnostics, optimization and prediction, Proceedings of the 29th DAAAM International Symposium*, B. Katalinic (Ed.), Vienna, Austria, 2018, 574–581.

7. E. Lutters, Pilot production environments driven by digital twins, *South African Journal of Industrial Engineering*, Vol. 29, n. 3, Pretoria, November 2018.

8. M. Mersmann, M. Weng, Optimizing cement production with the aid of digital twins, *Cement International*, Vol. 27, No. 5, 2019, 20–27.

9. Anjan Kumar Chatterjee, *Cement Production Technology: Principles and Practice*, Chapter 1, Basics of mineral resources for cement production, CRC Press, Taylor & Francis Group, Florida, USA, 2018, 1–20.

10. https://petrodatascience.com/casestudy/worlds-first-digital-twin-for-mine-value-chain-optimization/

11. Anjan Chatterjee, Tongbo Sui, Alternative fuels – effects on clinker process and properties, *Cement and Concrete Research*, Vol. 123, 2019, 105777.

12. B. Lothenbach, M. Zajac, Application of thermodynamic modelling to hydrated cement, *Cement and Concrete Research*, Vol. 123, 2019, 105779.

13. Qinglin Qi, Fei Tao, Tianliang Hu, Nabil Anwer, Ang Liu, Yongli Wei, Lihui Wang, A. Y. C. Nee, Enabling technologies and tools for digital twin, *Journal of Manufacturing Systems*, October 2019. Available at https://researchgate.net/publication/336870688

14. https://g2.com/categories/digital-twin

9 Developments in Application of Sensors to Sustainable Manufacturing of Cement

Kamal Kumar
Holtec Consulting Private Limited, Gurgaon, India

Anupam
National Council of Cement and Building Materials, Ballabgarh, India

Anjan Kumar Chatterjee
Conmat Technologies Private Limited, Kolkata, India

CONTENTS

9.1   Introduction 294
9.2   Overview of Sensors Applications 294
    9.2.1   Data Processing and Communication..................................296
    9.2.2   Smart Sensors 296
9.3   Sensors Application in Cement Manufacturing.............................297
    9.3.1   Location of Temperature Sensors...299
    9.3.2   Clinker Cooler..302
    9.3.3   Analysis and Monitoring of Gas Emissions 303
    9.3.4   On-Stream Analysis of In-Process Solids 304
    9.3.5   Sensors in Mining, Crushing and Pre-blending 307
9.4   Soft Sensors in Process Industry..308
    9.4.1   Basic Design Approach for Soft Sensors 308
9.5   Soft Sensors for Cement Manufacture..310
9.6   Soft Sensor Development in the AI Environment..........................314
    9.6.1   Data Collection and Processing..317
    9.6.2   Telemetry Endpoint..318
9.7   Future Role of Soft Sensors in Intelligent Cement Production.........319

DOI: 10.1201/9781003106791-9

9.8   Concluding Observations 319
References..320

9.1   INTRODUCTION

It is highly probable that the basic concept of sensors application to industry might have evolved from the five sensing organs of the human body, the
eyes, ears, nose, tongue, and skin to process information about the world
outside in order to react, communicate and to keep the body healthy and
safe with the help of sight, sound, smell, taste, and touch, respectively.
Thermostat, capable of sensing and maintaining temperature, was introduced for industrial purposes in the late nineteenth century, and the infrared sensors have been around since the late 1940s. Progressively, sensors
emerged as a class of devices to detect and respond to changes in environment caused by the intensity and spectra of light, heat and temperature,
motion, pressure, and some other parameters. Since the output of such
devices can be stored and shared through connected devices for monitoring
and controlling purposes, the manufacturing industry gradually expanded
the use of sensors to a significant extent. The metal fabrication sector
emerged in the forefront of incorporating sensors, particularly in the
computer-integrated manufacturing. The large-scale chemical plants
including the production of cement found the use of sensors highly advantageous for automation.

In recent times, the internet of things (IoT), which include the industrial
internet of things (IIoT), are bringing the application of sensors to a new
level. It is known that the IoT is a complex technology having several layers in
its architecture and a network of connected and interacting devices. Although
the components of the IoT systems would vary, depending on the scope and
scale of application, most of these devices share data collection mechanism
and rely on sensors. The IoT systems have already made an entry into the
cement manufacturing sector and consequently, the sensors in use now and to
be used in near future are advancing from traditional to smart and intelligent
types, Further, with the advent of artificial intelligence, the development of
'soft sensors' has become imminent.

In this background, this chapter attempts to summarize the status of the
sensor technology and to trace the journey of the cement manufacturing sector from the use of conventional sensors to smart and soft sensors.

9.2   OVERVIEW OF SENSORS APPLICATIONS

As already stated, sensors are simple devices that detect, measure, and convert
physical parameters such as light, heat, motion, pressure, and similar entities
into electrical signals or changes in electrical properties. They are physically
small, low power-consuming, mostly wireless, and operationally robust
devices that are used viably in large numbers even in harsh environments.

Some illustrations of types, operational principles, and common applications of sensors are given in Table 9.1.

From Table 9.1 it is evident that a large number of sensors are designed and manufactured, based primarily on electromagnetism, optics, and electronics. Since the sensors are hardware based, the criteria for selection of sensors for

TABLE 9.1
Types and Applications of Commonly Used Sensors

Categories of Sensors	Types Based on Operational Principles	Illustrative Applications
Temperature (with differing ranges)	Thermocouples	$-200°C$ to $+1750°C$
	Resister temperature detectors (RTD)	$-200°C$ to $+600°C$
	Thermistors	$-50°C$ to $+250°C$
	IC semiconductor	$-70°C$ to $+150°C$
	Infrared	
Pressure (gases and liquid)	Potentiometric	Industrial filters, process flows,
	Inductive	HVAC, automobile, medical,
	Capacitive	barometric, wearable devices, etc.
	Piezoelectric	
	Strain gauge	
	Magnetic (variable reluctance)	
Level	Ultrasonic	Point level and continuous level for
	Capacitance	liquids, powders and granular
	Optical interface (including Laser)	materials
	Microwave radar	
Humidity	Capacitive	Absolute and relative humidity
	Resistive	measurement for industrial safety,
	Thermal conductivity	HVAC, indoor air quality, etc.
Proximity (non-contact detection of objects)	Inductive	Metallic and non-metallic detection
	Capacitive	
	Photoelectric	
	Ultrasonic	
Motion detection	Passive infrared	Security systems, privacy invasion,
	Ultrasonic	and wearable devices such as
	Microwave	smart watches
Vibration detection (accelerometer)	Hall effect	Condition monitoring including
	Capacitive	tilting and acceleration
	Piezoelectric	
Angular velocity (gyroscopic)	Ring laser	Speed of rotation
	Fibre-optic	Three-dimensional object location
	MEMS	
Noise	Acoustic probe	Compliance of industrial, healthcare, and social standards
Image	Charge-couple device	Digitization of images
	Complementary metal oxide semiconductor (CMOS) imagers	

different applications are numerous and include output type, power, sensing range, operational mechanism, and environmental conditions. In reality, there is no single sensor that provides more benefit to a manufacturer than the other. There are different ways to connect to a system. Hence, each manufacturing set-up will need a specific plan of action. The programmable logic controllers (PLC) are often compared with sensors. Although PLCs are capable of collecting data from equipment, considerable time and efforts are needed to interact with them essentially through trial and error. Further, it is often more difficult to pull out data from PLCs. It is therefore operationally more convenient to make use of sensors, particularly if a single data point is desired in a set-up. This is one of the prime factors for the expansion of sensors application.

9.2.1 Data Processing and Communication

When using sensors of any kind, there is a need for signal processing techniques to reduce or enhance certain aspects of the signal. Digital filters can be used to reduce noise and to get a usable signal. This is a prerequisite for the sensors to accurately estimate the key parameters and enable a reliable control. The signal may need digital-to-analogue and analogue-to-digital conversions so that data can be easily transferred from one software to another.

In an industrial process, the data communication is usually integrated into a commercial process control system that handles all the information generated by the measurement instrumentation in the process. Many commercial sensors are standalone equipment and can be connected to a number of applications.

Sensors usually come with their own software where signal processing and algorithms are implemented. The information given by the software can be useful for monitoring purposes of the process but in order to implement the signals from the sensors into a soft sensor or a hybrid model to achieve an automated control, this information needs to be accessible in real time.

Open Platform Communications (OPC) is a technology used for easy exchange of information between platforms and is widely used in manufacturing industry. One powerful tool to achieve this is soft sensors that use online signals from the process to derive new information

9.2.2 Smart Sensors

Smart sensors are microprocessor driven and include features such as the communication capability, self-calibration, and self-diagnosis. Often they have data pre-processing capability so that the load to gateways, PLCs or cloud resources is reduced. The building blocks of smart sensors are schematically shown in Figure 9.1 [1]. The application algorithms are performed by a built-in microprocessor unit (MPU). An important component is the execution of process-specific signal conditioning tasks, which include filtering

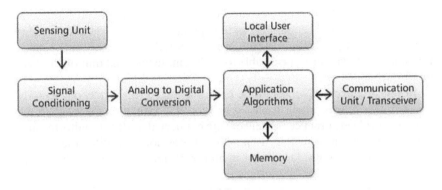

FIGURE 9.1 Building blocks of smart sensors [1].

and compensation. The step helps in preserving integrity and isolation of signals from the harsh industrial environment.

The MPU's intelligence can be used for certain other functions such as the handling of the calibration data so that the sensors may automatically respond to production changes. The MPUs can also detect production deviations and generate warning alerts. The smart sensors generally have built-in capability of complying with different communication standards such as Wi-Fi, Bluetooth, etc. These sensors are mostly capable of detecting and measuring two or more parameters simultaneously. For example, combining humidity and temperature or current, power and temperature is common in practice. In brownfield environment, an important pre-requisite for sensors is the capability of working with digital and analogue inputs so that the data from the older sensors can be read.

With significant improvement in the sensors technology itself and with increasing demand in IoT and IIoT in different economic and industrial segments, the sensors market is predicted to expand at a compounded annual growth rate of over 11% [2]. In addition to manufacturing, the major drivers for growth of the sensors market will be the smart cities, healthcare, automotive sector, and wearable devices.

9.3 SENSORS APPLICATION IN CEMENT MANUFACTURING

Since cement manufacturing is a complex, resource-intensive and energy-consuming chemical process, its competitiveness and viability depend on cost economics, plant productivity, equipment availability and product quality. These four dimensions of the manufacturing units require real-time information on key process variables and product attributes. Generally speaking, the entire process control function in the manufacturing units is structured with Distributed Control Systems (DCS) in which (PLC) are dispersed throughout the process link, each PLC controlling a certain section of the process. The fully developed DCS comprises supervising control, data acquisition software, and the personal computers as the operator interface. The process conditions for each production section are monitored and the resulting signals transmitted by wire or optical cable to each remote terminal unit or PLC via input/output devices.

In this architecture, sensors have been playing a critical role, particularly for several key tasks, which are listed in Table 9.2. Generally speaking, in the present worldwide practice the sensors are connected to PLC monitoring and control devices in sealed local junction boxes for providing the sought information and data for effective and real-time control of the entire

TABLE 9.2
Sensors Generally Used in Cement Production Lines

Sensor Type	Location & Spread in the Production Line	Purpose
Temperature sensors of different ranges	Multiple locations such as motor bearings, conditioning tower, flue gases, preheater cyclones, kiln shell etc.	Continuous monitoring at critical locations
Vibration sensors (accelerometers)	A network of accelerometers positioned close to machinery parts	Condition monitoring and wear rate assessment
Level sensors	Multiple locations such as crusher, stacker, conveyor transfer stations, raw meal silos, conveyor belts, clinker cooler, clinker silo, cement silos, solid fuels storage, liquid tanks, packaging machines, trucks, and others	Point level detection, continuous level detection,
Pressure sensors of different ranges	Various locations in the entire cement plant such as cyclones in preheater, grinding circuits, flue gas duct, across bag filters, grinding mills (raw, coal and cement), compressors, process fans, clinker cooler, kiln hood	Monitoring the operation behaviour like cyclone build-up, location of null point, compressors and Process fans performance, etc.
Mass & volume flow sensors including IR and Laser types	Crusher hoppers, belt conveyors at different sections, fans and blowers including the cooler fans and induced draft fan of the kiln	Controlling solid materials balance, avoiding solids spillage, monitoring air flow through the kiln system, etc.
Gas sensors	Kiln inlet, preheater outlet, coal mill systems, electrostatic precipitators, etc.	Measuring CO, NO, O_2, SO_2, CxHy as necessary. Monitoring of coal mills for detecting air ingress and smouldering fires.
Proximity sensors	Mobile mining equipment and transport vehicles, stacker arms, conveyor belts, valve discharge, etc.	Ensuring driving safety of mobile vehicles, positioning of stacker arms, belt drift correction, monitoring belt tension, etc.
Chemical sensors	In-process granular and powdery materials; phase composition of clinker and cement	Process and quality monitoring

production line. It is important to note that in the production line, the key components to monitor are the motor and gearbox assemblies, particularly in crushers, conveyor belts, grinding mills (raw, coal, cement), elevators, separators, process fans and blowers, kiln and cooler drives. It is observed that despite the deployment of efficient dust collectors, the ingress of fine dust into the machinery with moving parts affects the efficiency of lubricants, leading to increased wear of bearings, shafts and seals. Detecting the manifestation of small vibrations and tiny changes with the help of sensors becomes crucial in the plant operation.

The cement plants operate a large number of mobile and conveying facilities, any malfunctioning of which can cause serious disruption of the entire line of production. Hence, the proximity sensors have proved to be very useful in the system. Since the handling of the production line involves the flow of solids varying from lumps to fine powder, liquids of differing composition and functionality, combustible materials and gases, and finally air flows of both compressed and induced draft streams, the real-time data collection needs a variety of sensors as illustrated in Table 9.2.

It is important to ensure that the temperature profile of the kiln system is controlled with the help of temperature measurements at all critical locations in the production line and the progress of all chemical process reactions are monitored with the help of monitoring the composition of exhaust gases on one side and the in-process materials, clinker and cement on the other.

At the same time, the environmental compulsions and performance of pollution control equipment call for the use of different types of sensors. Finally, the condition monitoring of the main machinery and the essential auxiliaries ensure the operational efficiency of cement manufacturing. Hence, the correct selection and application of various types of sensors as illustrated in Table 9.2 is extremely crucial for cement manufacturing. A few specific uses of sensors in some segments of the production line are discussed further to illustrate the criticality of sensors in cement manufacturing.

9.3.1 Location of Temperature Sensors

It goes without saying that the entire production line requires a variety of temperature sensors of different objects having widely varying temperature ranges. Although initially contact thermocouples were the only technology available and required frequent replacement due to mechanical damage or corrosion by process gases, in many susceptible locations in the system non-contact pyrometers are now in use. A typical spread of locations of sensors of different types is given in Figure 9.2 [3]. The maximum temperatures often specified for the sensors located at the above points are as follows:

- Motor bearings:
  Located near the bearing greasers and resistant to vibrations, this sensor checks the temperature of the bearing reliably and

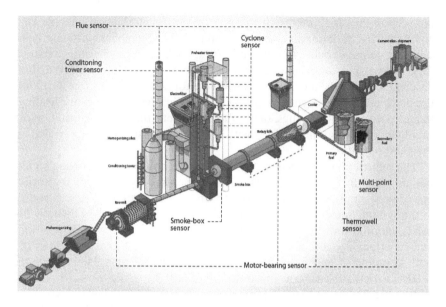

FIGURE 9.2 A typical illustration of the locations of temperature sensors in the cement production line [3].

instantaneously. It can detect at an early stage any overheating which may lead to malfunctions. Maximum temperature up to **200°C**

*   Conditioning tower:
    Installed to check the temperature of the gases, its rugged, compact design makes it highly resistant to shocks and vibrations. Its sensing element is protected by a stainless-steel sheath capable of withstanding corrosive gases. Maximum temperature up to **450°C**

*   Exhaust gases:
    Used in discharge chimneys, this sensor measures and controls the temperature of the gases discharged into the atmosphere. In this way, it can be used to ensure correct operation of the installation and environmentally-friendly production. Maximum temperature up to **350°C**

*   Preheater Cyclone:
    Installed upstream and downstream (discharge chute) of each cyclone, this sensor measures the gas and material temperatures. This monitoring helps to control and prevent any obstructions due to clogging of the fines. For the lower levels, the sensor is equipped with anti-abrasion protection. This rugged sensor withstands severe mechanical shocks.

*   Upper preheater cyclones: Maximum temperature up to **700°C**
*   Lower preheater cyclones: Maximum temperature up to **1000°C**
*   Smoke chamber:

This sensor is capable of withstanding high temperatures and high concentrations of clinker dust. It monitors the temperature of the hot gases recovered from the rotary kiln and from the preheater tower. Sensing element: S thermocouple with high-temperature protection/Measurement temperature up to **1300°C**

In addition to the above, the temperature measurement of the burning zone in the rotary kiln is essential but the direct application of sensors is fraught with problems of obscure dusty environment, continuous movement of clinker granules, and presence of unstable coating layer over the refractory lining inside the kiln. Even the video cameras installed in many kiln systems lose their effectiveness due to heavy dust curtains in the kilns.

While single-spot measurements from a ratio pyrometer are often effective, more comprehensive information is obtained from thermal imaging using an infrared system. A relatively simple system of this category, called Near Infrared Borescope (NIR-B), can be installed under the burner, sighting onto the clinker bed as shown in Figure 9.3 [4]. It has the capability of measuring the flame temperature, clinker temperature, and observing the overall dynamic condition of the kiln. With the help of a line scanner, the condition of the refractory brickwork can also be monitored.

The infrared thermography has now become more useful for monitoring the rotary kiln section in the form of kiln shell scanner. A simplified schematic diagram of a kiln shell scanner is shown in Figure 9.4 [5]. The instrument system scans the shell along the longitudinal central line with high speed to yield the temperature profile of the kiln surface.

In the modern sophisticated design of the system, an operator has access to the thermal image of the entire flattened shell surface with temperature-indicating colour codes. The thermal profile is constructed by superimposing and merging multiple curves so as to maximize the kiln coverage and to

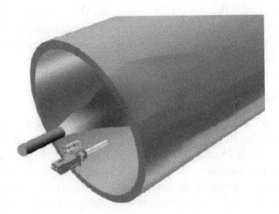

FIGURE 9.3 Location of Near-Infrared Borescope in the kiln [4].

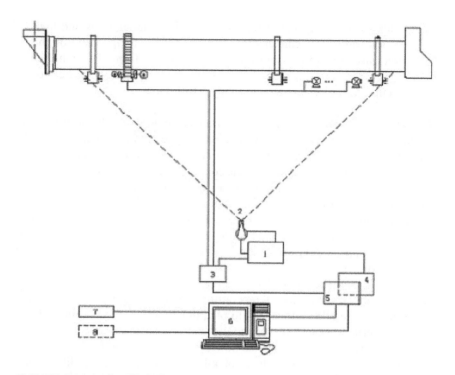

FIGURE 9.4 A simplified diagram of a kiln shell scanner showing the high-speed scanner (1), temperature measuring sensor (2), control cabinet for the IR sensor (3), intelligent signal processing device (4), I/O port device (5), computer (6), printer (7), and communication Ethernet card (8) [5].

eliminate the shadow zones. The kiln positions along with the kiln and tyre rotation pulses are considered in data processing. The transmission of processed data to computer is effected either through a fibre-optic interface or via industrial wireless Ethernet cards. Some of the commercial systems are reported to have the capability of three-dimensional walk-in-the-kiln facility from selected view position, which provides a realistic view of the coating condition inside the kiln without any stoppage [6, 7].

9.3.2 CLINKER COOLER

The IR scanners are used over clinker conveying belts for detecting hot or under-burnt clinker lumps. These types of scanners are capable of detecting hot spots as small as 25 mm, based on their suitable resolution and frequency as well as the measurement of a very large number of temperature spots. However, the measurement of temperature distribution inside the clinker cooler is still an area of study, although it is highly important for cooler operation from the following perspectives:

- Uniform and fast cooling
- Avoiding snowman formation
- Preventing the occurrence of red rivers
- Boosting of energy recovery from the secondary and tertiary air to the kiln

In a recent study, using a highly specialized mid-infrared (10.6 µm) camera, a data set has been obtained, capturing the radiation emissions of a clinker cooler [8]. The necessity of image pre-processing and calculating reflectance-corrected thermal images for temperature estimation without the use of reference marking or additional instrumentation has been emphasized in the paper. While further studies are in progress, it has been observed that IR-based thermography could become an indispensable tool for clinker cooler operation.

9.3.3 ANALYSIS AND MONITORING OF GAS EMISSIONS

Measuring the composition and volume of process gases such as O_2, CO, NO, and C_xH_y at certain control points in the production line is as essential as the monitoring of emitted gases such as SO_2, NO_x, HCl, HF, and volatile Hg, in addition to dust and water vapour. Depending on the process requirements and environmental regulations the standard operating practices are adopted. The monitoring practice can be online or offline, continuous or periodic with central or portable analysers.

Traditionally speaking, the gas analysis is mostly carried out, using the extractive sampling method. A representative portion of the gas is extracted from the process stream with the help of a sampling probe and then it is conditioned before it is fed to the analyser. A variety of measurement techniques are utilized to determine the concentration of different gases in a gas mixture. Some illustrations are given in Table 9.3. The subject has been dealt with in detail in [9].

TABLE 9.3
Measurement Techniques for Different Gases

Techniques	Different Gases						
	O_2	CO	CO_2	NOx	SO_2	H_2S	Hydrocarbon (CxHy)
Non-dispersive IR spectra		#	#	#	#		#
Non-dispersive UV spectra				#	#		
Paramagnetic property	#						
Chemoluminiscence				#			
Flame ionization							#
Ion-specific potentiometry	#	#		#	#	#	
Solid-state electrolyte	#						

From the above table, it is evident that the infrared spectrum is a very effective chemical sensor for multiple gases. A large number of gas molecules can be identified by their characteristic frequencies. Depending on the scope and intent of application, the infrared spectra can be used in dispersive, non-dispersive, or Fourier Transform mode. The noble gases and the divalent gases such as O_2 do not absorb IR spectra. For measuring O_2, however, other sensitive chemical sensors, based on its paramagnetic property or its electrical behaviour on passing through a platinum-coated ZrO_2 (solid-state electrolyte) are available. It is relevant to mention here that in cement plants continuous monitoring is the preferred option. In fact, CO_2, NO_x, and SO_2 have been continuously monitored for several years in some plants. However, in recent times due to more frequent use of alternative fuels and raw materials, monitoring other gases such as HCl, HF, and the Hg vapour has become necessary, in addition to measuring total organic carbon (TOC), dioxins, furans and other persistent organic pollutants (POPs). Consequently, the online continuous monitoring of gases and organics has become more complexly instrumented with chemical sensors of high sensitivity. The effective modern monitoring systems are designed comprehensively, consisting of gas sampling assembly, Fourier Transform IR Spectrometer for multiple gases, ZrO_2-based Oxygen Analyser, Flame Ionization Detector for organics, Cold-Vapour Atomic Absorption or Fluorescent spectrometer for mercury, and the appropriate software system for data collection, interpretation and reporting [10].

9.3.4 On-Stream Analysis of In-Process Solids

The in-process materials in the production line, varying from lumps to fine powder, are subjected to analysis with instruments that are based on the following types of sensors:

- Gamma radiation isotope
- Near infrared spectra
- X-rays

It is widely known that the popular practice to-day for monitoring the composition of crushed stone is to use the technique of Prompt Gamma Neutron Activation Analysis (PGNAA), an online method, in which the material passes without any further size reduction through the neutron bombardment of radioactive californium 252 [8]. The PGNAA analyser is installed on a belt conveyor structure going to the pre-blending stockpile after the crusher. It consists of two units – a shield block assembly and an electronics enclosure. The shield block assembly houses the detectors and neutron sources, provides shielding and supports the conveyor belt. The electronics enclosure contains the components for signal processing, computerized data analysis, modem for remote diagnostics and connection with the operator console, which can be located at a distance of more than a kilometre. The instrument analyses the material every minute for all major oxides, alkalis, sulfate and chloride.

Another development in the on-stream analysis, apart from the widely used bulk analyser based on γ-radiation, is the application of infrared spectra that are provided by the stabilized white light source. The light illuminates the target bulk material to be analysed as it passes the unit on an existing conveyor belt. The infrared radiation excites vibrational oscillations of the molecular bonds in the material under test, which results in reflection and absorption spectra that are characteristic of minerals being analysed. The Near Infrared (NIR) ranges are applied for analysing limestone materials. It is claimed that the IR based online bulk analyser shows better performance for the cement raw material constituents than the traditional γ-ray equipment. One additional advantage in this new development is the avoidance of potentially hazardous excitation sources [10].

The x-rays are used in two modes for two purposes: Fluorescence Spectrometry for determining the elemental composition of materials and Diffractometry for phase analysis and both the systems have been in offline use for decades. A specific mention may be made of the practice of analysing the hot meal sample as entering into the kiln on a regular basis in order to monitor the degree of calcination achieved in the precalciner and also the occurrence of volatile cycles in the kiln system. The sample is taken from the hot meal chute between the bottom cyclone and the kiln inlet. The hot meal is rapidly cooled in a special sampler without allowing it to be in contact with the ambient air. The sample is transported either manually or by a pneumatic conveying system to the central laboratory equipped with the x-ray fluorescence spectrometer. The analysis results are obviously crucial for monitoring the volatiles cycle by taking timely action to prevent high concentration of alkalis and chlorides in the pyro system, and also to adjust the kiln bypass, if installed. Similarly, the x-ray diffractometry has continued to be a highly reliable offline tool for raw materials characterization, pyro-processing, clinker mineralogy, diagnostic studies on kiln coating formation, lump formation in cement silos and bags. Determination of free CaO content in the cement clinker is still regarded as an index of clinker quality. The quantitative x-ray diffraction measurement of the clinker phases and their grain growth features are used to identify process disturbance and to predict quality.

Recent developments in the use of x-ray diffraction are changing the traditional methods of quality and process control, as they have the ability to measure mineral phases or compounds formed directly in real time. Cement and clinker production involves chemical reactions to produce precisely controlled blends of phases with specific properties. So far there has been overwhelming dependence on either offline or online oxide or elemental analysis of raw or in-process materials for QC. Methods and equipment are now available for continuous quantitative on-stream analysis of the mineral or phase composition of cement and clinker. The instrument is a standalone piece of equipment, which is installed at the sampling point. A sample for analysis is extracted from the process stream and after due preparation online the sample passes through the x-ray beam. The diffracted x-rays are collected over 0 to 120° by a detector. The Rietveld structural refinement technique is applied to analyse the resulting diffraction pattern. The analysis of the moving stream is done in close frequency of, say, once every minute. All analysis results are communicated directly to the plant PLC system. The real-time measurement of the mineral composition of cement and clinker for process control is a paradigm shift for the cement industry. The discernible benefits of using on-stream x-ray diffraction are the following:

- Control of kiln burner based on free lime, clinker reactivity, alkali, and sulfur contents
- Control of cement mill separators and feed rates and proportions to achieve consistent cement strength at minimum power consumption
- Control of gypsum dehydration through cement mill temperature to give consistent setting times
- Control of mill weigh-feeders for different feed materials

A typical illustration of how the increase in free lime in cement had affected the formation of the silicate phases in an operating online installation is shown in Figure 9.5 [11]. Broadly speaking, the net benefits of implementation of such online QC systems are the optimum performance and cost, reduced risk of product failure, and consequent marketing benefits.

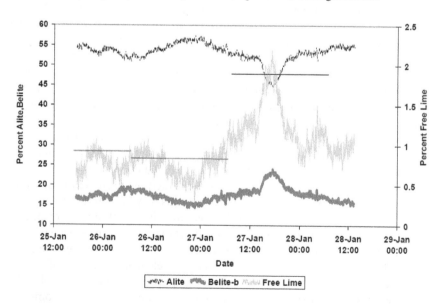

FIGURE 9.5   Increase in free lime in cement affecting the formation C_3S and enhancing the amount of C_2S as observed in a plant record [11].

9.3.5 Sensors in Mining, Crushing and Pre-blending

At the mining stage in the limestone quarry, there is always a serious concern for safety, as there is simultaneous movement of a large number of vehicles such as shovels, bucket extractors, wheel loaders, haul trucks, etc. Avoiding close proximity of these vehicles at the time of backward and forward movements in limited space, generating of alert signals for drivers, ensuring continuous unloading into the crusher hopper, avoiding interruptions due to the presence of over-sized stones, and proper flow of crushed stone to pre-blending stockpile are extremely crucial operational requirements. Now-a-days, proximity sensors and Laser volume flow meters are used to provide solutions to the above problems (Figure 9.6) [12]. The sensors are built with rugged construction, high reliability, and wear-free operation which allow

FIGURE 9.6 Detecting over-sized stone in the crusher hopper (1) and monitoring the crusher discharge flow on the belt (2) along with ensuring the feed flow into the crusher hopper (3) with the help of Laser flow meters [12].

continuous quantity control at the crusher or at the transfer points. By integrating the loading centre-of-gravity detection feature, the maximum transport performance and correction of one-sided conveyor belt loads are ensured. Conveyor belt control permits the detection of belt slippage, thus minimizing the maintenance costs. Further, correct positioning of the vehicles for loading and unloading is as important as ensuring the positioning of the stacker for bed building in the stockpile. Hence, the entire section of raw materials extraction and preparation is highly dependent on new and smart sensors.

The strong role of different sensors in the overall process and quality monitoring functions as well as in ensuring plant health and operational safety, discussed above and as summarized in Table 9.2, is extensively recognized, in the cement manufacturing industry. Nevertheless, it is observed that since the traditional and even the smart sensors are hardware based, they are not capable of continuous monitoring of a host of physical and chemical transformational properties of in-process materials and finished products. There is a discernible gap between the performance of and demands from the sensors technology. For example, despite the adoption of effective mill control systems and installation of online Laser particle size analysers, the required level of uniformity of the cement properties is seldom achieved without any significant operational intervention in controlling the product fineness. Despite the adoption of expert systems in kiln control, the complexity of clinker formation chemistry cannot be tracked or predicted. Because of such shortfalls the sensors technology is moving to a hybrid character of both software and hardware orientation, which is discussed in Section 9.4.

9.4 SOFT SENSORS IN PROCESS INDUSTRY

Soft sensors are data-driven inferential sensors which try to estimate the difficult-to-measure properties or parameters from the knowledge and data of easy-to-measure process variables. Thus, soft sensing is a modelling approach and apparently the name is derived by adding 'soft' from software to a sensor. Soft sensors are developed from the historical and real-time data provided by the installed sensors, spot measurements of process variables, and offline laboratory tests performed at certain time intervals. In the last two decades, soft sensors established themselves as an effective alternative to the traditional means for acquisition of critical process variables, process monitoring and other tasks related to process control. Soft sensors are not only augmenting online process measurements; they are also used as backup sensors, when the hardware sensor is in fault or removed due to maintenance or replacement.

9.4.1 BASIC DESIGN APPROACH FOR SOFT SENSORS

The basic steps in any data-driven soft sensors design method are data collection, data pre-processing, model selection, parameters identification, and

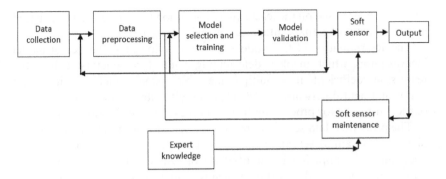

FIGURE 9.7   Basic steps in designing soft sensors [13].

finally model validation, which are shown schematically in Figure 9.7 [13]. The raw data from plants often suffer from several drawbacks such as redundancy, inaccuracy or low accuracy, presence of outliers, multidimensionality, measurement noise, and others. The unprocessed flawed data sets are not used for model development as the resulting control system may lead to suboptimal operation, off-specification products, higher operational cost, etc. Hence, pre-processing or cleaning of data collected from any industrial base turns out to be an indispensable step. Generally, statistical or heuristic techniques are performed for prepossessing the data set. The 'principal component analysis' and 'clustering analysis' are the two frequently used methods. Further, various univariant and distance-based multivariate outlier detection techniques are applied for data cleaning.

The processed data sets are subsequently used for developing the model, which has two distinct dimensions: selection of the model structure and identification of the specific model. In practice, the process models are not always available in ready-to-use forms; they are usually built up. Based on the basic structure, two broad types are classified:

- first principles models (mechanistic)
- data-based models (empirical)

The first principles models describe the physical and chemical background of the process and hence require detailed understanding of the process. The data-based models are developed from the actual process input–output data. The development of the first principles models is difficult as it involves solving highly complex process model equations. On the other hand, availability of large amount of plant historical data has shifted the attention towards the development of data-driven soft sensors. This category of models is closer to the ground realities and reflects the true conditions of the process in a better way, although the acceptability depends on the data set used and processing of data. For all practical purposes, the model structures may range from the statistical ones based on the least square regression to complicated ones based on neural networks, fuzzy inferences or their combinations.

The determination of the model structure is followed by the step of model identification, which involves determining a set of parameters such as the regression coefficients in a multiple linear regression model or weights and biases in a neural network model. This step ultimately helps in identifying a particular model for proceeding further in the course of development.

The processed data sets that are used for soft sensor modelling are usually divided into training sets and validation sets. It can be done either by random selection or by applying Kennard-Stone or Duplex algorithm. The model errors are determined statistically before undertaking the validation step which is complex and more so for non-linear models. Finally, the acceptance of a model depends on high prediction accuracy for both the training set and the validation set of data. A comprehensive description of the basics of model development is given in [13].

9.5   SOFT SENSORS FOR CEMENT MANUFACTURE

The need for soft sensors, specific for the cement manufacturing process, was felt more strongly since the beginning of the present century and in the last decade or so the developmental efforts have been quite extensive. A pioneering study was reported in 2007 [14]. In this study, the principal component regression and partial least square models were derived for predicting free lime content in clinker and NO_x emission from the kiln. Another study for estimating the free lime content in clinker by the least square support vector regression technique was also reported later [15]. Modelling of the grate cooler has received attention as it is a very difficult part of the process to control due to its non-linearity and hysteresis. Although some approximation models of the grate coolers, based on energy and mass balance, are known, they are not effective in designing a cooler controller. A study was carried out to build up non-linear predictor models by MLP neural network and LM training function [16].

A specific mention may be made of the study carried out on cement grinding operation as reported in [13]. The work involved the processing of 158 input–output data of a vertical roller mill (VRM) with grinding capacity of 235 tons per hour. Three process inputs such as the clinker flow, hot air flow, and classifier speed were considered to predict the output of cement fineness. The collected data were divided into training and validation subsets, using Kennard-Stone maximal intra-distance criterion. The training set of data was used for development of different kinds of data-driven soft sensors, which included linear and support vector regressions, artificial neural network, fuzzy inference, and neuro-fuzzy models. Performance of the developed models was assessed with the validation data set by using six different statistical indicators. The neuro-fuzzy and the back propagation neural network models showed better performance than

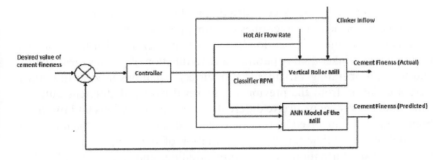

FIGURE 9.8 Closed loop monitoring and control of cement fineness [13].

the others. It was observed that between the two disturbance variables, the clinker flow rate had stronger impact than the hot air flow rate. A closed loop structure of monitoring and control of cement fineness is illustrated in Figure 9.8.

The development of soft sensors for the clinkering process and quality parameters was more data-intensive and involved more complex modelling. The study was carried out for a kiln system having clinker production capacity of 10,000 tons per day and it was based on nine inputs and eight outputs. The inputs included four parameters of raw meal quality and five kiln operational variables. The outputs included free lime, C_3S and C_3A, in addition to the oxide constituents. After processing, the data set consisted of 223 data set pairs of input–output. Subsequently, Kennard-Stone algorithm was used to divide the training and validation data subsets. Three kinds of feed-forward neural network models and two types of fuzzy-inference models were worked upon. The performances of the developed soft sensors were assessed with the validation data set by evaluating six statistical model evaluation parameters. For illustration purposes the fuzzy-inference model structure is shown in

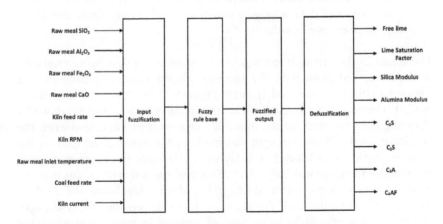

FIGURE 9.9 Fuzzy inference structure of the rotary kiln variables [13].

Figure 9.9. Of the models considered, the Takagi-Sugeno fuzzy-inference model outperformed the other ones in estimating the outputs.

It may be pertinent to mention here about another model study approach for predicting clinker quality parameters as a function of the raw meal quality and the kiln operating variables [17]. Fourteen equations used in this study were a collation from the previously published works of different authors. This model, a system of twelve differential-algebraic equations and two algebraic equations, involved appropriate mass balance, energy balance, and material residence time for the axial evolution of the components that take part in clinker formation, temperature profiling, and movement of solids inside the kiln. The model reportedly had mean square error, coefficient of determination, worst case relative error, and variance account for unknown values of 8.96×10^{-7}, 0.9999, 2.17%, and above 97%, respectively. As claimed, this first principles-based model is designed to provide real-time estimates of clinker quality parameters with high accuracy and capture wider ranges of real plant operating conditions.

Prediction of cement strength with soft sensors is another important aspect in cement manufacturing. A review of salient past studies is available in [18], which shows that various attempts have been made in this direction, starting from the use of simple control charts to linear and non-linear regression of existing data, application of fuzzy logic and neural network. A specific study reported by the author was the model development from more than 1700 data sets of cement fineness, cement composition, and cement strength obtained from a single mill of a single plant. Two classes of models were developed:

a. Static models, wherein for a given data set the parameter values were computed through non-linear regression and then the values were utilized to predict future strength;

b. Movable horizons models, where the parameters were estimated from a moving set of data belonging to a predetermined past time interval, e.g., three or four months.

The actual 28-day strength test results of a standard cement being produced in the concerned plant with the model-predicted values are compared in Figure 9.10 (for the static model) and in Figure 9.11 (for the movable horizons model). The investigation revealed that the model error was brought down by changing from the static to the movable horizons model and also when the prediction of the 28-day strength was made by the model developed on the 7-day strength results instead of 1-day strength results. In the given instance the lowest error value was 3.0%. The efficacy of the soft sensors was demonstrated by using a proportional-integral controller regulating the 28-days strength. Although the study defines a way for predicting the 28-day strength of cement, it continues to be an area of concern in manufacturing as the

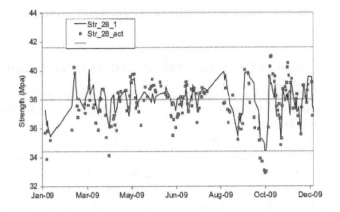

FIGURE 9.10 Comparison of actual 28-day strength values with the values predicted from the static model developed on 1-day strength results [18].

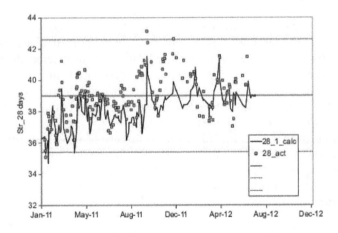

FIGURE 9.11 Comparison of the actual 28-day strength values with the values predicted from the movable horizons model on 1-day strength results [18].

essential step of reducing the holding time for the produce before dispatch is yet to be reached.

In fact, soft sensing of difficult-to-measure process variables in clinker making and clinker quality parameters continues to be an area of extensive research. Interestingly, numerous studies in this field have yielded valuable results, but they are applicable within the range of data collected for model development, This is also observed from the fact that the expert systems widely in use for process control always require tuning and adjustment for specific plants for their effectiveness. Since the available systems are based on the steady-state operation of the manufacturing process, which in reality is not so, the intensity of development needs to be significantly enhanced.

9.6 SOFT SENSOR DEVELOPMENT IN THE AI ENVIRONMENT

In the past, technological advancements in process control and automation have been driven by the need to have sequential operation, interlocking, quality control, higher throughputs, and higher efficiency in terms of power and heat. In future, it is envisaged that technological advancements in process and automation control will be influenced by unmanned operations, artificial intelligence-driven controls, a requirement for consistent quality, and sustained optimized efficiency. Cement production operations are increasingly becoming more technology oriented. Since the industrial high-volume data services focusing on performance and monitoring of plant operations are now available, the current technology trend is likely to be driven more and more by continuous raw data acquisition from the plant DCS systems and transfer to the Cloud environment for processing. The data retrieved from the plant DCS can be subjected to rigorous deep data analysis by knowledgeable experts using proprietary plant analytics software. The software can be fully integrated with domain specific algorithms providing diagnostic content to facilitate process intervention for operational improvement. A typical system architecture is schematically presented in Figure 9.12 [19]. Multiplicity of process variables for section wise analytics is shown in Figures 9.13 and 9.14.

Control is further complicated by the occurrence of process disturbances. These include substantial thermal cycles, the build-up of rings of clinker on the refractory lining of the kiln and the falling away of clinker attached to the lining surface. Furthermore, poor kiln operating practice and changes in feed and fuel chemistry aggravate these disturbances. To tackle these

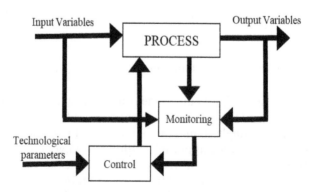

FIGURE 9.12 Typical control system architecture [19].

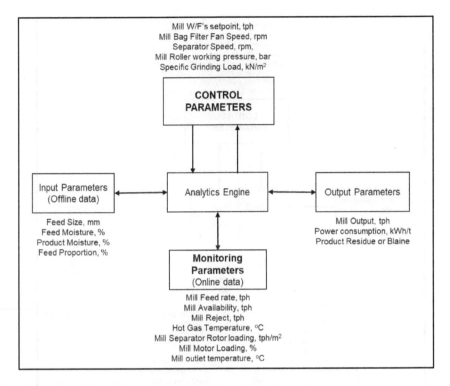

FIGURE 9.13 Process analytics for raw mill (VRM) system.

complications, and proactively solve them, it is important for the analytics engine to capture all the parameters that are necessary to perform real-time diagnostics.

It is important for the analytics engine to be universal, capable of transferring anonymous learning from one cement plant to another. Analytics engine allows for systematic computational analysis of data and can facilitate fast discovery of meaningful patterns that can help improve throughput of a cement plant. Furthermore, the analytics engine should also be able to transfer anonymous learning from one cement plant to another and utilize the power of AI to address common issues and evolve collectively. The architecture of universal analytics engine is presented in Figure 9.15 [20], which shows in detail how the information flows within a single unit plant. For analytics engine to benefit from learning across various plants, it is important for the design to scale to multiple unit plants while providing effective data isolation among various cement plants to mitigate any security risks.

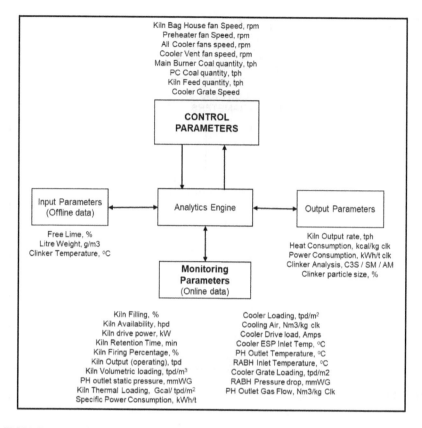

FIGURE 9.14 Plant analytics for the kiln section.

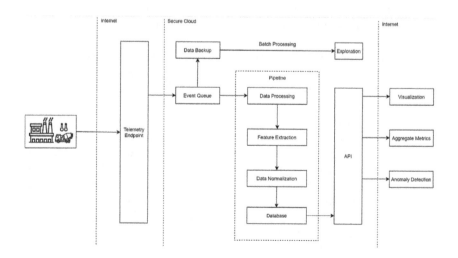

FIGURE 9.15 Schematic diagram for information flow in a single unit plant [20].

9.6.1 DATA COLLECTION AND PROCESSING

The first step to building a scalable universal analytics engine is to capture telemetry from the machinery and upload it to cloud-hosted web service. Sensors are installed to monitor and control the operating parameters for both online data and offline data, and send them over to telemetry endpoint for real-time analysis. The design has to be fault tolerant and therefore provides for backing up a copy of metrics on a local server, in addition to uploading metrics to telemetry endpoint. Local back up improves availability of the system and enables disaster recovery in an event online analytics system is inaccessible or down. Another major consideration while collecting data is to ensure that measurements do not themselves have any adverse impact on the performance.

Data pipeline is a set of data processing components connected in a way where output form one component feeds into the next component as an input. This is a key part of the analytics engine and involves multiple steps on data processing before it is ready to be ingested into the database:

- Data validation: Check for any formatting errors. If the data is invalid, log it as an exception and avoid processing it to corrupt rest of the analysis.
- Feature extraction: To save bytes over the wire, the data often comes in compressed and delta encoded format. This is where data is turned into features that are used as an input to machine learning algorithms as well as statistical analysis.
- Data normalization: The data is transformed into normalized distribution such that principal component analysis (PCA) can help remove redundancy in the data.
- Metric computation: Finally, from the normalized data, the metrics are computed and stored in a database for visualization and charting later.
- Data modelling: Finally, the data is modelled by adding structure and relationships among various parameters. As part of modelling, each row is uniquely identified by cement plant ID, date, and time.

Visualization is a critical step in an analytics engine. Given the scale at which the data is collected in the pipeline, it is impossible to manually analyse the data. Visual elements like charts and graphs provide an accessible way to see and understand trends at scale.

A data model determines the structure of the data and establishes relationship among various elements of the data. In this system, each data row is uniquely identified by the identification number of a cement plant and is sent along with any information originating from a particular cement plant. This also acts as a primary key and helps partition data when stored in the cloud.

TABLE 9.4
Measures and Dimensions of Data Collected for Developing Analytics Engine

Metric	Unit	Aggregation	Description
Clinker quality C_3S	Percentage	95th percentile	Measures quality of clinker
Gypsum moisture	Percentage	95th Percentile	Measures moisture in gypsum
Clinker size	Millimetres	Mean	Measures the particle size of clinker
Clinker temperature	Degree centigrade	Mean	Measures temperature of clinker coming out of the clinker cooler
Cyclone outlet pressure	mm WG	Mean	Measures the static pressure at the outlet of any cyclone
Kiln inlet oxygen	Percentage	95th Percentile	Measures oxygen concentration at the kiln inlet
Fan power	Kilowatt	Mean	Measures power drawn by a fan

Other important information to capture are the date and time, which help visualize the measures and dimensions on a time series. Some examples of measures and dimensions of data collected for the analytics engine are given in Table 9.4.

9.6.2 TELEMETRY ENDPOINT

Telemetry is the collection of measurements at remote plant locations and their automatic transmission to the endpoint hosted on the cloud. This endpoint is exposed to the internet and is built with high availability in mind to receive incoming data from various cement plants. It is recommended to secure this endpoint with Transport Layer Security (TLS) to fight against any possible man-in-the-middle-attacks to steal sensitive data collected from the plants. The endpoint is built with a singular purpose of processing high throughput incoming events and moves the data to an event queue as fast as possible. Introducing an event queue in the architecture leads to two main benefits: capturing data offline in a persistent storage for ad-hoc offline analysis, and in an event of an outage, it helps in data recovery and in maintaining continuity of activities.

It is obvious that the course of sensor development now and in near future will be in compliance with the information processing and data analytics environment as briefly outlined above. The combination of process models and sensor-based online control will go hand-in-hand to effect a course of feed-forward control for optimizing the individual processing steps resulting in intelligent manufacturing with high product quality and high-yield production rate.

9.7 FUTURE ROLE OF SOFT SENSORS IN INTELLIGENT CEMENT PRODUCTION

It is understood that the intelligent cement manufacturing will depart from predefined process models and statistics by using sensed information to self-direct the processing chain on event-based control strategy. The implication of event-based control strategy is to forecast events for which adaptive control of some process variables will be imperative. In other words, intelligent manufacturing of cement will employ process control by objectives rather than control with the prescribed parameters.

It is widely accepted that the smart sensors are mostly able to perform in noisy and adverse manufacturing environment. However, particularly in cement production the following constraints will have to be overcome more comprehensively:

- Chemical changes because of the non-linear behaviour of several simultaneous reactions and formation of intermediate products
- Sensing very rapid changes and large gradients in temperature profiles
- Point measurements, sensing critical points, rates, and changes in rates in combustion and production
- Identifying parameters with constitutive relationships of the measured variables for inferred measurement

In fact, the objective of sensing the in situ processes with the help of measured material properties at different stages of conversion and the corresponding non-linear process variables will have to be achieved.

9.8 CONCLUDING OBSERVATIONS

It is evident from the discourses in the chapter that, like most other process industries, manufacturing of cement strongly relies on sensors technology. Achieving optimum performance of the unit processes and of the complete manufacturing system is only possible if timely and accurate process information is available with the help of a wide variety of sensors employed in the plant. The sensors are designed and manufactured, based primarily on electromagnetism, optics and electronics. Since the sensors are hardware based, the criteria for selection of sensors for different applications are numerous and include output type, power, sensing range, operational mechanism, and environmental conditions. Sensors for measuring material and gas temperatures, solids and gas flows, solids and liquid levels, machine vibration, conveyor and feeder flows, belt drifts, gas emissions and composition, safe movement of transport vehicles, and several other process and operational parameters have played a significant role for many years in cement plants. However, until recently, the application of sensors has often been constrained by issues such as the system noise, signal attenuation, and response dynamics. The constraints have been overcome, to a great extent, by the integration of computing power and IoT technology. The ordinary sensors have been transformed into smart sensors, which have the capability of carrying out complex calculations locally on measured data within a sensor module. Such enhanced capability coupled with miniaturization and operational flexibility has made the smart sensors highly potential for the cement process. The smart sensors with digital architecture can readily integrate with the machine controllers and improve the system performance. A further development is the adoption of fusion technology for combining different parametric measurements into a single sensor. The availability of current-power-temperature-vibration sensor is an example of such advancement, which can be of immense significance in condition monitoring of cement plant machinery.

Notwithstanding the progress made in sensors technology, sensors continue to be the weak link in making cement manufacturing intelligent and autonomous. The major short-comings are as follows:

- Monitoring of chemical changes with non-linear behaviour of sequential and simultaneous reactions and
- Inferred measurements by sensing parameters that are not directly measurable.

These shortcomings may, in all probability, be overcome with the development of soft sensors or inferential sensors. The basic steps in any data-driven soft sensor design method are data collection, data pre-processing, model structure selection and model identification, control parameter identification, and finally model validation. The model structure may range from simple statistical models to more complicated ones such as neural networks, fuzzy inference systems, or combinations. In the last two decades soft sensors have established themselves as a valuable alternative to the traditional means for data acquisition for critical process variables and process monitoring tasks. The soft sensor technology is perceived as a landmark development towards making the cement manufacturing intelligent and sustainable with the use of sensed information to self-direct the process via an event-based control strategy, a control path in response to changes in material behaviour that are deemed as process events.

REFERENCES

1. https://si.farnell.com/smart-sensors-overview-and-latest-technology
2. https://www.i.scoop.eu/global-sensor-market-forecast-2022
3. https://www.pyrocontrole.com/applications/cement-manufacturing
4. https://insights.globalspec.com/article/10752/temperature-measurements-in-cement-manufacture

5. Gold Star M & E Technical Development Co. Ltd., Technical Manual, GS-HG Kiln Shell Scanner, Hefei, China
6. F.L. Smidth Co., ECS/Cemscanner, Kiln Shell Monitoring System (www. flsmidth.com).
7. https://www.zkg.de/eu/artikel/2210270.html
8. R. Gabriel, S. Keller, J. Matthes, P. Weibel, H.B. Keller, S. Hinz, *Infrared measurements and estimation of temperature in the restrictive scope of an industrial cement plant, ISPRS Annals of Photogrammetry, Remote Sensing & Spatial Information Sciences*, Karlsruhe, Germany, Vol. IV-1, 2018.
9. *Emissions monitoring handbook*, Gasnet Technologies, Vantaa, Finland (www. gasmet.com).
10. Anjan Kumar Chatterjee, *Cement production technology: principles and practice*, Chapter 8: Advances in plant-based quality control practice, CRC Press, 2018.
11. Don Summit, G. Duene Crutchfield, A. L. Gadek, Con Manias, Peter Storer, Experiences with continuous on-stream XRD at the Ash Grove Leamington Plant, available at https://www.researchgate.net/publication/4018377
12. Cement Industry, No. 88014937, SICK AG, Germany (www.sick.com)
13. Ajaya Kumar Pani, Design of soft sensors for monitoring and control manufacturing processes, Thesis for PhD, Birla Institute of Technology and Science, Pilani, India, 2013.
14. B. Lin, B. Recke, J.K. Knudsen, S.B. Jorgensen, A systematic approach for soft sensor development. *Computer and Chemical Engineering*, 31(5), 2007, 419–425.
15. J. Qiao, Z. Fang, T. Chai, LS-SVR based soft sensors model for cement clinker calcination process *Proceedings of International Congress on Measuring Technology and Mechanotronics Automation (ICMTMA)*, Vol. 2, 2010, 591–594.
16. M. Seraj, M.A. Shooredeli, Data-driven predictor and soft sensors models of a cement grate cooler based on neural network and effective dynamics, *Iranian Conference on Electrical Engineering (ICEE)*, Tehran, 2017, 726–731 (doi:10.1109/IranianCEE.2017.7985134).
17. N.E. Moses, S.B. Alabi, Predictive model for cement quality parameters, *Journal of Materials Science and Chemical Engineering*, 4, 2016, 84–100 (http://dx.doi.org/10.4236/misce.2016.47012).
18. Dmitris Tsamatsoulis, Prediction of cement strength: analysis and implementation in process quality control, *Journal of the Mechanical Behavior of Materials*, 21(3–4), 2012, 81–93.
19. Dražen Slišković, R. Grbic, Ž. Hocenski Methods for plant data-based process modeling in soft-sensor development: *Automatika*, 52, 4. 2011.
20. Microsoft Azure IoT Reference Architecture Version 2.1.

10 Integrated Enterprise Resource Planning in Sustainable Cement Production

Rameshwar Dubey
Liverpool John Moores University, Liverpool, UK

CONTENTS

10.1 Introduction 323
10.2 Underpinning Theories 325
    10.2.1 Sustainable Business Development..325
    10.2.2 Enterprise Resource Planning 327
    10.2.3 Big Data and Predictive Analytics..327
    10.2.4 Integrated Enterprise Resource Planning 328
10.3 Resources for Building BDPA Capability 329
    10.3.1 Tangible Resources..329
    10.3.2 Human Resources 331
    10.3.3 Technical Skills 332
    10.3.4 Management Skills...333
    10.3.5 Intangible Resources 333
    10.3.6 Data-Driven Culture 334
10.4 Developing IERP in Cement Industry..335
    10.4.1 Sampling Design and Data Collection 336
    10.4.2 Structural Self-Interaction Matrix...336
    10.4.3 Final Reachability Matrix 337
    10.4.4 Level Partitioning..338
    10.4.5 Fuzzy MICMAC Analysis 339
    10.4.6 Theoretical Model for IERP in the Cement Industry 340
10.5 Concluding Recommendations...341
References...342

10.1 INTRODUCTION

The cement industry has a long history and the demand for cement has risen rapidly over the last decades. The industry is resource-intensive and highly energy-consuming with fuel accounting for 30%–40% of the production costs.

DOI: 10.1201/9781003106791-10

According to Cembureau (2008), the cement industry is capital and energy-intensive, but not labour-intensive. According to Ghemawat and Thomas (2004), the minimum scale that is efficient for a cement plant is approximately 1 million tons of annual capacity. Labour usage in the cement industry is relatively low because it is a continuous process with a high level of automation. Cement plants are normally located near the quarries which are the source of their main raw materials. The main reason for their location is that approximately 1.6 tons of main raw materials are required to produce 1 ton of cement (Cembureau, 2008). The massive growth of the global cement industry during the last three decades and the predicted demand growth for cement in the next two decades (Schneider et al. 2011; Singh & Dubey, 2013) make it imperative for the industry to adopt a sustainable production framework (Luo et al., 2017).

The downstream component of the cement supply chain varies from country to country (Dubey et al., 2012). Concrete (and therefore cement) demand is created primarily by residential, non-residential, and infrastructural construction. Cement sales are normally related to economic growth, macroeconomic factors, and climatic conditions. These issues have local and regional cycles.

The cement industry is an ideal example of a continuous manufacturing process where the traditional mass production system is adopted in order to produce, accumulate, and move thousands of tons of materials between the work areas. Nowadays the challenge is to change the cement industry from traditional mass production into a more effective production system aiming to increase productivity, overall performance, and capacity utilization to meet high market demand. The cement industry is forced to reduce the production costs and delay times in order to take advantage of the global competition environments. Since the maintenance cost turns out to be as high as 20–25% of the production cost (Al Mohaisen etal., 2002), there is a continuous focus of the plant management to bring down the maintenance cost (Eti et al., 2006). In addition, the strategy of adopting the Total Quality Management (TQM) has also been emphasized in the industry (Dubey & Singh, 2013). For achieving the sustainability objectives, the researchers have delved into the development of green concrete products (Berry et al., 2009).

Broadly speaking, the problems of the industry can be minimized through enhancing coordination among various partners involved in the cement production supply chain. The poor coordination has been attributed to the lack of information sharing and poor visibility. Thus the cement industry needs to invest in emerging technologies like advanced ERP, big data and predictive analytics, blockchain technology, and IoT (internet of things).

This chapter intends to present an exploratory study carried out by the author on developing a model for integrated enterprise resource planning (IERP) in cement production plants. It first focuses on the underpinning theories before providing an overview of the study and the development of the theoretical framework of the implementation of IERP and then concludes with a set of pertinent recommendations.

10.2 UNDERPINNING THEORIES

10.2.1 SUSTAINABLE BUSINESS DEVELOPMENT

The United Nations (UN) has given a much-needed impetus at political level for sustainable development. The concept has evolved over a period of time to blend and balance environmental, economic, and social goals (Jeble et al., 2018). There is no consensus on meaning of sustainability. Different organizations interpret it in different ways. Some organizations may be striving for financial self-sustainability, whereas another may be committed to financial–social objectives or yet another may be focusing entirely on environmental sustainability (Swanson and Zhang, 2012). Nevertheless, sustainability has become a part of common business nomenclature in recent years. It is gaining acceptance as a right measure of firm's overall performance. Internationally recognized standard ISO 26000 provides guidelines for businesses on conducting business while caring for their social duties. This means firms have responsibility towards ensuring well-being of the society. The industry has significant carbon footprint. Thus, it has a significant impact on both environment and society.

There are several definitions of sustainability. Jha et al. (2014) consider green manufacturing with linkages to corporate social responsibility and corporate governance as three components of sustainability (Figure 10.1). Elkington (1994) defines sustainability as a triple bottom line (TBL) which occurs at the intersection of environment, social, and economic objectives (Figure 10.2).

Moving one step further Caradonna (2014) suggests economy as a subset of society and society as a subset of environment (Figure 10.3).

Thinking beyond quarterly profits, businesses will not be able to sustain on the planet that fails (Winston, 2014). Incremental investments into sustainability are no longer sufficient (Dubey et al., 2017). As the climate is changing, a different long-term approach spanning decades and generations will be needed (Winston, 2014). While designing facilities, business processes, or

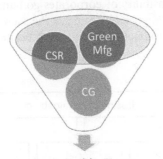

Sustainable Firms

FIGURE 10.1 Sustainable firms [Jha et al., 2014].

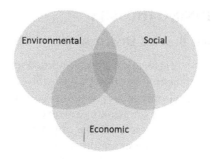

FIGURE 10.2   Triple bottom line [Elkington, 1994].

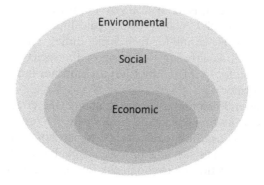

FIGURE 10.3   Caradonna model of sustainability [Caradonna, 2014].

products, all three aspects of sustainability need to be considered for long-term sustenance.

TBL provides a good framework for firms to practice sustainable business development. This ensures that firms can be economically sustainable while ensuring the protection of environment and society. The goal of profit alone may lead to the exploitation of nature as well as society. Sustainable development and social responsibility of corporates go hand in hand (Dubey et al., 2017). Svensson and Wagner (2015) echo the view (Figure 10.4) that economic, social, and environmental objectives of business sustainability need to be addressed together.

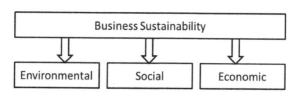

FIGURE 10.4   Business sustainability based on TBL [Svensson and Wagner, 2015].

10.2.2 ENTERPRISE RESOURCE PLANNING

"Enterprise resource planning (ERP) systems integrate the data and processes of organizations into single system and single database. This database functions as a hub that stores, shares, and circulates data from within the different departments and business functions. ERP systems are one of the most adopted IT systems by organizations" (Elragal, 2014, p. 242).

The cement manufacturing firms adopt ERP systems to manage the everyday large volume of operations and information, which are created from within the organization (Dutta and Bose, 2015). ERP systems are "tightly coupled" in comparison to the traditional systems that usually stay with the organizations prior to the ERP implementation (Liang et al., 2007; Elragal, 2014). The cost reduction is one of the main drivers for an ERP implementation. Secondly, ERP help integrate the technical and operation business functions to streamline the flow of information with the material flow of goods or services (Koh et al., 2006). The ERP has been a source of competitive advantage for many organizations. On the other hand, in order for the organization to utilize the ERP capability, they must have a thorough understanding of the ERP tools so that it can be used to the maximum potential (Elragal, 2014; Zhenyu and Prashant, 2001). Elragal (2014, p. 243) argue that *"Acquisitions, mergers, and joint ventures could be drivers of organizations to adopt ERP systems, in order to unify, utilize and manage the huge information and work flow among them"*. ERP systems require many organizational changes which could impose high risks if the implementations are not thoroughly planned, executed, and managed, as statistics from literature and practice show high rates of implementation failures (Zhenyu and Prashant, 2001).

10.2.3 BIG DATA AND PREDICTIVE ANALYTICS

Predictive analytics is the most useful technique for getting insights from data about what can happen in future from available big data (Jeble et al., 2018). It is defined as the process of discovering meaningful patterns of data using pattern recognition techniques, statistics, machine learning, artificial intelligence, and data mining (Gupta and George, 2016; Srinivasan and Swink, 2018). It includes statistical and empirical models to create empirical predictions, as well as methods for assessing the predictive power of the models (Jeble et al., 2018; Srinivasan and Swink, 2018). Also, referred to as advanced analytics, it simply means application of data analytics techniques to answer questions or solve problems (Dutta and Bose, 2015). It is a further progression of business intelligence (BI) and data mining combined with statistical techniques. BI processes help analysis of internal and external data to enable business executives to make intelligent decisions. The questions and variables are developed by experts in the field of study whereas in case of predictive analytics, selection of model and relationship are data driven (Jeble et al., 2018).

It is a systematic analytical process, wherein a computer algorithm finds out patterns and underlying relationships of dependent and independent variables. It is designed to find the optimum regression coefficients of relationship to minimize the errors in the model. The process uses advanced information systems to go through several iterative steps to find out the optimum outcomes to the problem. Process mining has emerged as a new research avenue for the analysis of process based on event logs. It opens opportunities for conformance and discovering new processes in various fields such as cement industry (Dutta and Bose, 2015). The Hadoop framework provides a solution for dealing with these analytics requirements. Based on the source and nature of different data, there are various analytical methods which support data mining and statistical analysis techniques.

Text analytics techniques derive real-time and meaningful information from unstructured data sources such as documents, emails, web pages, and social media. It is being pursued in some of the emerging areas such as sentiment analysis, opinion mining, or for extracting information from text sources (Giannakis et al., 2020). In recent years, soon after a product launch, sentiment analysis with social media data provides early indicators of consumer feedback about products.

As the data on social media is growing, it contains valuable information for business firms, governments as well as NGOs; it's being tapped for deriving value. It requires a different process of data collection and analysis due to its large volume, continuous flow, and variety of data to arrive at meaningful information.

10.2.4 Integrated Enterprise Resource Planning

Information systems evolution has occurred over several decades (Jeble et al., 2018). During early phase of IT revolution, information systems were used to improve productivity by automating several operational functions and were a source of competitive advantage. During this phase well-developed IT capability helped firms to achieve better customer coordination, customer service, and improved economic performance (Chen and Tsou, 2012). However, in recent years, as technology has moved forward, a good information system no longer provides competitive advantage. Firms need data analytics capabilities on top of information systems to remain competitive. Firms are collecting hordes of data coming from internet, smart phones, cloud computing, and IoT. They need to develop soft and hard factors to IERP.

For developing such a capability, a firm first needs to recognize strategic significance of big data resources for competitive advantage, develop competencies in big data technologies, acquire knowledge on tapping value from big data, and transform itself towards data-driven culture (Dubey et al., 2019). According to Akter and Wamba (2016), firms need to focus on cutting edge technology, quality analytics resources, and analytics-driven management culture for developing big data analytics capabilities. Another recent study identifies various resources that can be integrated to build a big data and predictive analytics (BDPA) capabilities (Gupta and George, 2016), and it validates the relationship between BDPA and firm's market and operational performance. Current study goes one step further in developing an instrument to measure BDPA capability and its impact on supply chain sustainability on social, environmental, and economic parameters. This is unique contribution of this study.

Several firms have developed infrastructure to gather large datasets, analyse them, and use them either for making operational decisions or predictions. This additional information helps them to gain market share or improve profitability. This ability to gather, integrate, and utilize a firm's big data specific resources is defined as BDPA capability (Gupta and George, 2016). Drawing on the resource-based view (RBV) logic, Gupta and George (2016) have identified tangible (data, technology and other basic resources), human (management skills, technology skills), and intangible (organizational learning and culture of making decisions based on data) resources as building blocks of BDPA capabilities.

Each of these resources is critical for the firm for superior BDPA capabilities. For example, a firm may have a good infrastructure and quality data available from internal and external data sources. However, without good analytics team firm may not be able to derive value from these data assets. Good analytics team will be able to define problem statement, analyse datasets and apply right technique which will result in actionable insights. Similarly firms need good management team that can motivate and rally the team towards business goals. Building BDPA capabilities is a continuous process. With constant growth of data and development in techniques, the field of analytics continues to evolve. Business environment is changing at unprecedented rate. Therefore, firms need to keep pace with these changes by encouraging continuous learning. In the next section, various resources that contribute towards building BDPA capabilities are discussed.

10.3   RESOURCES FOR BUILDING BDPA CAPABILITY

10.3.1   TANGIBLE RESOURCES

According to Gupta and George (2016), tangible resources include capital, buildings, IT infrastructure, networks, connectivity, data sources, etc. These resources are necessary for engineers to develop analytics solutions. IT infrastructure provides much required foundation for firms to provide business applications and services, share information across different organizational functions, to enable the communication and execution of business strategy within the organization (Chen and Tsou, 2012). It includes hardware (such as application and database servers), software (such as ERP, CRM, SRM), database systems (such as Oracle, DB2, Microsoft SQL servers, etc.). Information systems provide the backbone for operational transactions in accounting, inventory, procurement, sales, logistics, etc. Besides these enterprise systems, other applications such as emails, mobile technologies, enterprise integration applications facilitate communication within organization as well as with business partners to ensure continuity of business. Most of the firms have had this infrastructure in place over last few decades. This IT capability is a prerequisite for using big data applications which require high-speed servers to quickly process and store large volumes of different datasets. Thus big data applications are the logical extension of information technology capability (Jeble et al., 2018). There is a recent trend of investments into big data and relevant technologies. However, investments alone may not provide the competitive advantage from big data. It is important that in addition to these investments, firms devote enough time to their big data analytics projects to accomplish their objectives (Dubey et al., 2019; Gupta and George, 2016). These resources will not provide competitive advantage on their own, but these are required as a foundation for building capabilities. Thus, availability of data, technology, time, and money are some of the basic resources towards the BDPA objectives (Dubey et al., 2019; Gupta and George, 2016; Jeble et al., 2018). Data is one of the most critical resources required for modelling and analysis (Dutta and Bose, 2015). It can be classified as internal data from organizational systems or external data from business partners, social media, internet, or other external sources.

Big data can be further classified as a combination of structured, semi-structured, and unstructured data. Structured data comes from transaction processing systems such as ERP or CRM; this data is stored in well-defined relational database management (RDBMS) system. The primary purpose of internal "small data" was to support internal business decisions related to tracking stock-outs, accounts payables, receivables, product pricing decisions, etc. Semi-structured data comes from XML files, system logs, or emails. Unstructured data originates from social media, text files, emails, word documents, and a variety of other files. Companies are able to get a complete view of their customers and operations by combining structured and unstructured data. To understand a complete map of customer journeys, data is gathered from semi-structured or unstructured sources such as weblogs. Unstructured data helps in getting insights into those aspects of customer relationships that cannot be obtained from internal data. For example, data originating from call centre logs or social media helps in getting an understanding of customer experiences and how that affects customer attrition. External sources of data about suppliers can provide information of supplier's technical capabilities, financial health, quality management, delivery reliability, weather, and political risks. This information helps to manage sourcing risks (Jeble et al., 2018).

Most of the firms have implemented information systems to record transactions and maintain master files in the last few decades. Now, realizing the value of big data, there is a recent trend of investments into big data and relevant technologies. However, investments alone may not provide the competitive advantage from big data. It is important that in addition to these investments, firms devote enough time to their big data analytics projects to accomplish their objectives (Gupta and George, 2016). These resources will not provide competitive advantage on their own, but these are required as a foundation for building capabilities. Once the mix of data is collected, a firm needs to develop data mining, text mining, and web mining capability (Dutta and Bose, 2015). By applying statistical techniques intelligence can be derived from data to make actionable decisions. Thus the availability of data, technology, time, and money are some of the basic resources towards the BDPA objectives (Gupta and George, 2016).

10.3.2 HUMAN RESOURCES

Wright et al. (1994) define human resources as people who are employed directly by the organization and include procedures adopted towards fulfilling organizational goals. Human resources are one of the most important resources that firms possess (Kazlauskaité and Bučiūnienė, 2008). Skilled human resources are important for the success of any IT project. In case of analytics projects which have its roots in information technology, data originating from a variety of sources can be considered as raw material. Large datasets are analysed and converted into meaningful information by analytics professionals. Knowledge, skills, and training in the technology coupled with analytical capabilities of resources play an important role in success of analytics projects. There are several studies which suggest human resources as a source of sustained competitive advantage because they meet the RBV criteria of being valuable, rare, inimitable, and non-substitutable (Kazlauskaité and Bučiūnienė, 2008). Chen and Tsou (2012) consider technical skills and managerial skills essential for the success of information technology projects. In order to provide value to the firm, human resources must possess skills and superior competencies. Human resources can add value to the firm only when they have required competencies such as knowledge, skills and abilities (Gupta and George, 2016) and they are motivated to perform. It is not sufficient just to possess highly skilled human resources, but a firm must have the ability to manage and deploy them towards strategic goals of the organization. However, skilled resources required for analytics projects are rare due to complex qualitative and quantitative skills resource for success of such projects. It is a challenge to find the analytics resources with technical, analytical, and governance skills required for big data analytics projects (Akter and Wamba, 2016). A large volume of data needs to be captured, transformed into meaningful information, analysed, and presented. Therefore, technical and analytical skills of data scientists are critical for the firm to tap value from big data (McAfee et al., 2012). According to Dubey and Gunasekaran (2015), big data analytics professionals are in great demand due to an acute shortage of skilled professionals in this field. They have identified set of hard and soft skills for developing analytics resources which are not easy to find among technology professionals. As a requirement for analytics project resources are firm specific and analytics capability of a firm is developed over a period of time, it's hard for competitors to replicate the same. Experience gained from past analytics projects helps the firm to take analytics capability to the next level and beyond their competitors. Technical skills and knowledge in analytics can be generally transferred from one resource to another experienced resource through knowledge sharing and training. However, domain-specific knowledge and analytics skills built over a period of time from experienced resource in a particular organization are hard to transfer to another resource. A particular resource can be replaced by another resource in the short run, but every time such a substitution happens, project performance gets adversely affected and competitive advantage for the firm is lost. A similar impact is likely to happen when resource substitution is carried out in any knowledge intensive industry. From the above discussion it is clear that human resources are critical input for building BDPA capabilities and meet the criteria for being a source of competitive advantage.

10.3.3 Technical Skills

Traditional information systems analyst has a strong background in business analysis, understanding of functional domains, and possesses database management skills. They can grasp business requirements and convert those into functional designs. They understand business processes can enable business functions using information systems and can have good problem-solving skills. Big data analyst requires somewhat different skillsets. Commonly referred as data scientist, needs to possess specific skills and knowledge in statistical analysis, machine learning and business acumen to understand business problems, articulate research problems, problem-solving skills, strong communication, and people skills (Davenport, 2014). Dubey and Gunasekaran (2015) have identified hard skills such as knowledge of statistics, quantitative techniques, multivariate data analysis, research method, domain knowledge in finance, marketing, etc. for a successful career in the big data analytics field. Data scientists need additional skills to manage variety of large datasets and also ability to get insights from patterns from data. According to Davenport (2014), many large firms are augmenting their existing analytical staff with data scientists who possess a higher order of IT capabilities and ability to manipulate big data technologies. Developing these skills for data scientist takes time, and hence there is acute shortage of skilled data scientists in the industry. Several educational institutions have introduced data science as part of their advanced courses.

10.3.4 Management Skills

As discussed in previous section, data science skills are rare and it takes time to develop them. Industry and academic institutions have geared towards developing data scientists to fulfil the needs of industry and academia. Another challenge industry faces is shortage of management skills for analytics projects. These skills are equally important for analytics projects as the managers play an important role in a leading and culture building role (Davenport, 2014). Managers have an overall responsibility towards motivating the team member to get the best out of them. Success of analytics projects depends on how well managers can assemble a team with the right skills and aligns team members towards common goals. Managerial skills include several soft skills such as ability to identify right analytics resources, developing resources with technical skills, leadership, team player, listening skills, learning capabilities, communication, and relationship building skills (Dubey and Gunasekaran, 2015) as they need to deal with the internal and external stakeholders of the project. Success of analytics projects depends on a good mix of both soft and hard skills within the project team. Thus a firm will possess valuable, rare, inimitable, non-substitutable (VRIN) human resources with hard skills as well as soft skills in analytics. Technical resources on their own will not provide competitive advantage to the firm. That will depend on the skill of the managers to recognize, develop, and exploit these resources within the firm towards goals of organization. Thus, in addition to investments in basic resources required for big data analytics projects, a firm needs human resources with skills in big data analytics technology as well as management skills to run the projects effectively (Gupta and George, 2016). A firm's human resources bring their experience, knowledge, business skills, management skills, their associations to the firm. These help in executing business strategies of the firm effectively. Thus big data analytics field requires mix of technical and management skills to execute projects successfully and to provide valuable actionable insights to business functions. This being an interdisciplinary field, resources need skills in computer science, data management, and statistical techniques to execute projects.

10.3.5 Intangible Resources

Intangible resources possessed by the organization are more difficult to measure and not reported on the firm's balance sheet or any other reports (Gupta and George, 2016). There are several subtle factors such as concern for quality, business processes, and firm's unique culture that defines intangible resources for the firm. Prior studies have identified these as a source of sustained firm performance (Dubey et al., 2019). It plays an important role in business process management, implementation of innovations, and superior firm performance. Organizational culture determines the extent of success for key strategic initiatives such as business process reengineering (Dubey et al.,

2019). Such strategic initiatives require extensive support from every layer in the organization. Culture of embracing change and adaptability of organization facilitate speed of execution. Organizational culture consists of several factors specific to a particular firm, and it gets built over a period of time. It includes formal and informal practices evolved within an organization, various beliefs, and values followed by the people associated with the firm. Schein (2010) defines organization culture as the overall environment, various formal and informal practices that firm develop around their people, values, and beliefs of the organization. Dubey et al. (2019) postulate that organizational culture gives organizations a sense of identity and determines, through the organization's legends, rituals, beliefs, meanings, values, norms, and language, the way in which "things are done around here". Culture is built over a period of time, differs from company to company and it's hard to replicate. Leadership team of the organization as well as its employee contributes in developing a unique culture over period of time. It's hard for competitors to replicate the Toyota culture of trust-based relationship across supply chains or McDonald's functional integration capabilities. Similarly, recent work in big data analytics has confirmed organizational culture as an important success factor for analytics projects (Gupta and George, 2016). Different constructs in organizational culture that contribute to success of IT and analytics initiatives are discussed in the next two sections.

10.3.6 DATA-DRIVEN CULTURE

Organizations are driven by a set of decisions taken by senior managers in response to challenges in the business environment. These decisions are made based either on available information or the gut feeling of the leadership team. Research suggests that firms with data-driven decision process perform far better on financial and operational results (Ikemoto and Marsh, 2007; McAfee et al., 2012), higher level of productivity and market value, and certain measures of profitability (Brynjolfsson et al., 2011). New generation companies such as Google and Amazon are known to use data for decision-making. They collect, store, and analyse data generated from every customer interaction. This data includes social media data, online searches, weblogs, call centre logs, clickstreams, etc. This was not possible in the traditional brick and mortar business model. Analysis of this kind of data leads to an online or offline product recommendation to customers leading to improved business opportunities. Usage of big data for decision making has great potential for traditional businesses as well. The culture of data driven decision-making can be considered as an intangible asset for the firms (Brynjolfsson et al., 2011). In the framework of data-driven decision-making (Ikemoto and Marsh, 2007), information is extracted from large volumes of data, which leads to gaining knowledge and better decision-making (see Figure 10.5). Thus, for realizing the full potential of big data owned by firms, senior decision-makers in the company embrace evidence based

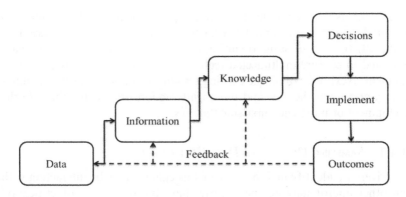

FIGURE 10.5   Data-driven decision-making process (Ikemoto and Marsh, 2007).

decision-making and firms develop data-driven culture (Brynjolfsson et al., 2011; Gupta and George, 2016; McAfee et al., 2012).

10.4   DEVELOPING IERP IN CEMENT INDUSTRY

Following Whetten (1989), we use two guiding principles for developing the IERP: first, the constructs or factors or drivers identified for our study must be comprehensive, and second, they are parsimonious in nature. To ensure that the drivers are comprehensive in nature, we adopt a two-stage process. Firstly, we have undertaken an extensive review of existing literature, and secondly, we have attempted to avoid overlapping drivers or constructs to ensure that the drivers are parsimonious in nature (Luo et al., 2018). However, this is a critical stage of theory development (Dubin, 1978; Whetten, 1989), and thus we have taken utmost care to ensure parsimony of the drivers.

Table 10.1 indicates the drivers of IERP in the cement industry. However, in the absence of adequate literature the nature of the associations between

TABLE 10.1
Drivers of IERP in Cement Industry

Drivers	References
Institutional pressures	Liang et al. (2007); Dubey et al. (2019)
Waste reduction	Dubey et al. (2012)
Manufacturing flow management	Elragal (2014); Srinivasan and Swink (2018)
Reverse Logistics	Mishra and Singh (2020)
Customer focus	Dubey et al. (2012); Srinivasan and Swink (2018)
Suppliers involvement	Srinivasan and Swink (2018)
Information sharing	Gupta and George (2016); Jeble et al. (2018)
ICTs	Gupta and George (2016)
Competitive advantage	Liang et al. (2007); Gupta and George (2016); Srinivasan and Swink (2018); Dubey et al. (2019)

these drivers is not well understood. In the past, researchers have attempted to ground their theoretical model in existing organizational theories (see Luo et al., 2018). The existing organizational theories in recent years have attempted to capture multifaceted reality with a finite, internally consistent statement. Next, we exploit the existing tensions or oppositions in surrounding organizational theories and use an interaction process to stimulate the development of more encompassing theories.

10.4.1 Sampling Design and Data Collection

The drivers we identify in Table 10.1 of this chapter may be interacting with each other and/or may not necessarily share the same level of criticality/importance in practice. Hence, we need to capture direct relationships among them. To understand the interaction among these drivers, a survey instrument was developed in which each possible connection between two driver's two questions was asked: for example, "*ICTs leads to competitive advantage*" and next "*Competitive advantage leads to ICTs*". Each relationship between two drivers was measured on a five-point Likert scale with anchors ranging from strongly disagree (1) to strongly agree (5). In prior research, scholars have used dichotomous scales (Yes/No) (see Luo et al., 2018). However, the Likert scale provides greater statistical variability among survey responses in comparison to the dichotomous scale. Before using a questionnaire for data collection, we pre-tested our survey instrument in two stages. In the first stage, five experienced researchers were asked to critique the questionnaire for ambiguity, clarity, and appropriateness of the drivers used (Luo et al., 2018). Based on the feedback, minor modifications in language were made to enhance clarity of the questions. Following Dillman's (2011) total design test method, the survey then was emailed with a cover letter to senior members of the Cement Manufacturers Association of India (CMA). The selection of India for the study was quite relevant as it continues to be the second largest producer of cement after China (IBEF, 2021). From the CMA membership directory, 45 senior members were selected with more than 10 years' experience in IT department. We received 36 usable responses representing an 80% response rate. We believe that we owe such a high response rate to the use of email for the survey and follow-up telephone calls to each respondent. The data analysis and the results are discussed below.

10.4.2 Structural Self-Interaction Matrix

Based on the 36 survey responses, we calculated a mean score for each direct relationship between two drivers. In our study, we assume that a driver does not impact another if the mean score is less than three. The bidirectional relationship $(i \rightarrow j, j \rightarrow i)$ is represented with mean scores as $(\bar{w}_{ij}, \bar{w}_{ji})$. We capture the bidirectional relationship between two drivers using the letters V, A, X, and O.

TABLE 10.2
SSIM

Drivers Numeric Code	Drivers	Drivers								
		IX	VIII	VII	VI	V	IV	III	II	I
I	Institutional pressures	V	A	A	V	V	V	V	V	
II	Waste reduction	V	A	A	V	V	X	V	–	
III	Manufacturing flow management	V	A	A	O	V	O	–	–	
IV	Reverse logistics	V	A	A	O	V	–	–	–	
V	Customer Focus	V	A	A	O	–	–	–	–	
VI	Suppliers involvement	V	A	A	–	–	–	–	–	
VII	Information sharing	V	A	–	–	–	–	–	–	
VIII	ICTs	V	–	–	–	–	–	–	–	
IX	Competitive advantage	–	–	–	–	–	–	–	–	

The letter V denotes a relationship in which node i leads to node j ($\overline{w}_{ij} > 3$), but the connection is not reciprocal (i.e. j does not lead to i or $\overline{w}_{ji} \leq 3$). Letter A denotes a relationship in which driver node j helps to achieve node i ($\overline{w}_{ji} > 3$), but the reverse is not true (i.e. i does not lead to j or $\overline{w}_{ij} < 3$). Hence, A is the opposite of V. The letter X denotes a relationship in which both nodes impact each other (i.e. i impacts j, but also j impacts i ($\overline{w}_{ij} > 3$ and $\overline{w}_{ji} > 3$)). Similarly, the letter O represents a relationship in which neither node is associated with one another (i.e. there is no connection between i and j ($\overline{w}_{ij} < 3$ and $\overline{w}_{ji} \leq 3$)). Based on 36 survey responses, we developed the total interpretive logic matrix (see Table 10.2).

10.4.3 FINAL REACHABILITY MATRIX

To obtain a final reachability matrix from Table 10.1, we followed two steps (Luo et al., 2018). In the first step, the variables V, A, X, and O are converted into 0s and 1s. This matrix is often referred to as the initial reachability matrix (see Table 10.3) using the following rules:

TABLE 10.3
Initial Reachability Matrix

	I	II	III	IV	V	VI	VII	VIII	IX
I	1	1	1	1	1	1	0	0	1
II	0	1	1	1	1	1	0	0	1
III	0	0	1	0	1	0	0	0	1
IV	0	1	0	1	1	0	0	0	1
V	0	0	0	0	1	0	0	0	1
VI	0	0	0	0	0	1	0	0	1
VII	1	1	1	1	1	1	1	0	1
VIII	1	1	1	1	1	1	1	1	1
IX	0	0	0	0	0	0	0	0	1

TABLE 10.4
Final Reachability Matrix

	I	II	III	IV	V	VI	VII	VIII	IX	Driving Power
					Drivers					
I	1	1	1	1	1	1	1*	0	1	8
II	0	1	1	1	1	1	0	0	1	6
III	0	0	1	1*	1	1*	0	0	1	5
IV	0	1	1*	1	1	0	0	0	1	5
V	0	0	0	0	1	1*	0	0	1	3
VI	0	0	0	0	1*	1	0	0	1	3
VII	1	1	1	1	1	1	1	0	1	8
VIII	1	1	1	1	1	1	1	1	1	9
IX	0	0	0	0	0	0	0	0	1	1
Dependence	3	5	6	6	8	7	3	1	9	

First, if the entry (i, j) is "V", then the corresponding entry in the reachability matrix (i, j) is replaced with 1 and the (j, i) entry is 0;

Second, if the entry (i, j) is "A", then the corresponding entry in the reachability matrix (i, j) is 0 and the (j, i) entry is 1;

Third, if the entry (i, j) is "X", then the corresponding entry in the reachability matrix (i, j) is 1 and the (j, i) entry is also 1;

Fourth, if the entry (i, j) is "O", then the corresponding entry in the reachability matrix (i, j) is 0 and the (j, i) entry is also 0.

In the next step, we checked the transitivity between the links. The transitivity uses triads as the unit of analysis. The transitivity principle is used in interpretive links to check the consistency of the model developed (Luo et al., 2018). As per the transitivity principle, if i leads to j and j leads to k, then the supposition that i leads to k must hold true. The transitivity property also helps to remove any possible gaps in the realized relationships among the variables. The final reachability matrix for drivers shown in Table 10.4 is prepared by adopting the criteria and the transitivity principle.

10.4.4 LEVEL PARTITIONING

The process of ranking different drivers into hierarchical levels is called level partitioning. To obtain the levels of drivers, the first step is the calculation of reachability and the antecedent sets from Table 10.3 (Luo et al., 2018). In any iteration, if the reachability set intersection with the antecedent set is the reachability set itself, then that variable will be placed in the top level of the hierarchy. The final output of level partitioning is shown in Table 10.5 and the conceptual framework of drivers of IERP is shown in Figure 10.6.

TABLE 10.5
Level Partitioning

Drivers	Level
IX	Level 1
V, VI	Level 2
II, III, and IV	Level 3
I, VII	Level 4
VIII	Level 5

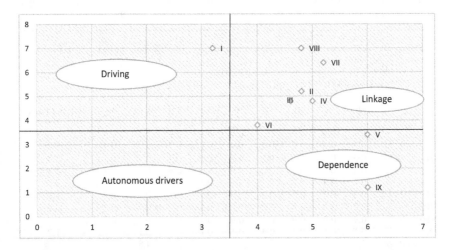

FIGURE 10.6   Classification of drivers.

10.4.5   Fuzzy MICMAC Analysis

Prior studies have used conventional MICMAC analysis (see Dubey and Ali, 2014; Dubey et al., 2015a, 2015b). Luo et al. (2018) noted that conventional MICMAC analysis only considers 0 or 1. Hence, the binary interaction between two drivers may not adequately reflect reality in practice. The judgement has some degree of fuzziness, and thus the fuzziness element of decision-making must be reflected in the final reachability matrix. In our case, we have considered that the interaction between any two drivers may acquire any value ranging between 0 and 1 (where no interaction=0; very low=0.1; low=0.3; medium=0.5; high=0.7; very high=0.9; complex=1) based on expert judgement. We have obtained fuzzy direct reachability matrix (FDRM) (see Table 10.6).

Next, to obtain a fuzzy stabilized MICMAC matrix, we have taken the FDRM as the base to begin the process. The matrix is multiplied repeatedly until the hierarchies of the driver power and dependence stabilize (Dubey and Ali, 2014; Luo et al., 2018). We have obtained the stabilized matrix as shown in Table 10.7.

TABLE 10.6
Fuzzy Direct Reachability Matrix (FDRM)

	I	II	III	IV	V	VI	VII	VIII	IX
I	0	0.9	0.9	0.5	0.5	0.7	0.5	0.3	0.7
II	0.3	0	0.5	0.7	0.9	0.7	0.1	0.1	0.7
III	0.3	0	0	0.1	0.9	0.1	0.1	0.3	0.7
IV	0.1	0.7	0.1	0	0.9	0.1	0.3	0.3	0.7
V	0.3	0.1	0.1	0.1	0	0.1	0.5	0.5	0.7
VI	0.1	0.3	0.3	0.1	0.5	0	0.7	0.7	0.7
VII	0.7	0.9	0.7	0.9	0.7	0.7	0	0.7	0.7
VIII	0.9	0.7	0.9	0.9	0.7	0.7	0.9	0	0.9
IX	0.1	0.1	0.3	0.3	0.1	0.1	0.1	0.1	0

TABLE 10.7
Fuzzy Stabilized MICMAC Matrix

	I	II	III	IV	V	VI	VII	VIII	IX	Driving Power
I	0	0.9	0.9	0.9	0.9	0.9	0.9	0.7	0.9	7
II	0.3	0	0.5	0.7	0.9	0.7	0.7	0.7	0.7	5.2
III	0.3	0.7	0	0.7	0.9	0.1	0.7	0.7	0.7	4.8
IV	0.3	0.7	0.5	0	0.9	0.7	0.5	0.5	0.7	4.8
V	0.3	0.3	0.3	0.3	0	0.1	0.7	0.7	0.7	3.4
VI	0.3	0.3	0.3	0.3	0.5	0	0.7	0.7	0.7	3.8
VII	0.7	0.9	0.9	0.9	0.9	0.7	0	0.7	0.7	6.4
VIII	0.9	0.9	0.9	0.9	0.9	0.7	0.9	0	0.9	7
IX	0.1	0.1	0.3	0.3	0.1	0.1	0.1	0.1	0	1.2
Dependence	3.2	4.8	4.6	5	6	4	5.2	4.8	6	

Using the data shown in Table 10.7, the co-ordinates of each driver are noted in the Table 10.8, which are used as input for generating the MICMAC structure (see Figure 10.6).

From Figure 10.6 we can see that nine drivers of the IERP are classified into four clusters based on their driving power and dependent nature. Prior studies (see Dubey and Ali, 2014; Luo et al., 2018) have explained each of the clusters.

10.4.6 Theoretical Model for IERP in the Cement Industry

Following a synthesis of the TISM and MICMAC analyses, we develop a theoretical model (see Figure 10.8). The model shows regulatory pressures (regulatory pressures are also termed external pressures or institutional pressures resulting from government or regulatory bodies) have a direct influence on the adoption of ICTs and information sharing between partners. This observation supports earlier work (see Jeble et al., 2018).

TABLE 10.8
Co-ordinates of the Drivers

Driver	X	Y
I	3.2	7
II	4.8	5.2
III	4.6	4.8
IV	5	4.8
V	6	3.4
VI	4	3.8
VII	5.2	6.4
VIII	4.8	7
IX	6	1.2

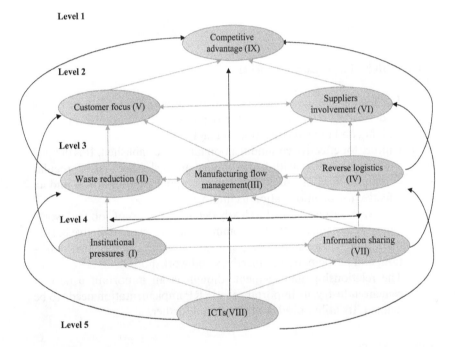

FIGURE 10.7   IERP model.

10.5 CONCLUDING RECOMMENDATIONS

A number of recommendations can be made out of the present study. These recommendations would be insightful for the cement industry, not only to those who have successfully implemented IERP but also to those who are in the process of implementation. Companies may address soft and hard dimensions in effective implementation of IERP which are discussed as:

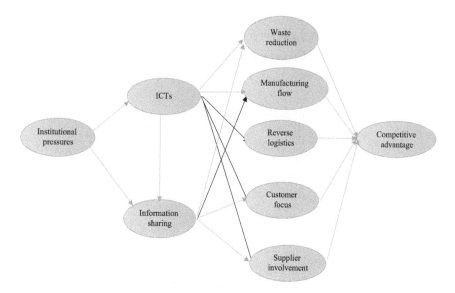

FIGURE 10.8   Final theoretical model.

i. Leadership should assume a supportive and motivational role of the leader for its team members for IERP initiatives and also facilitate training and related education in IERP.
ii. Culture for effective communication must be established. It is defined as an organization with formal information sharing system, regular communication of quality performance, goal and initiatives, and an effective use of information technology.
iii. HR Work systems must be clearly defined in terms of job and work, a flexible system, an effective communication system, effective participation of partners and associates in managing systems, cost reduction, and participation in meetings and workshops.
iv. The relationship management, though is an important aspect in cement industry, its importance in IERP implementation needs to be defined. Its utility needs to be explored further.

REFERENCES

Akter, S., & Wamba, S. F. (2016). Big data analytics in E-commerce: a systematic review and agenda for future research. *Electronic Markets*, *26*(2), 173–194.

Al-Muhaisen, M., & Santarisi, N. (2002). Auditing of the maintenance system of Fuhais plant/Jordan Cement Factories Co. *Journal of Quality in Maintenance Engineering*, 8(1), 62–76.

Berry, M., Cross, D., & Stephens, J. (2009, May). Changing the environment: an alternative "Green" concrete produced without Portland cement. In *Proceedings World of Coal Ash Conference*, Lexington, KY, USA.

Brynjolfsson, E., Hu, Y., & Simester, D. (2011). Goodbye pareto principle, hello long tail: the effect of search costs on the concentration of product sales. *Management Science, 57*(8), 1373–1386.

Caradonna, J. L. (2014). *Sustainability: a history*. Oxford University Press.

Cembureau (2008). 2007 world cement production by region.

Chen, J. S., & Tsou, H. T. (2012). Performance effects of IT capability, service process innovation, and the mediating role of customer service. *Journal of Engineering and Technology Management, 29*(1), 71–94.

Davenport, T. H. (2014). How strategists use "big data" to support internal business decisions, discovery and production. *Strategy & Leadership, 42*(4), 45–50.

Dillman, D. A. (2011). *Mail and Internet surveys: the tailored design method—2007 Update with new Internet, visual, and mixed-mode guide*. John Wiley & Sons.

Dubey, R., Singh, T., & Tiwari, S. (2012). Supply chain innovation is a key to superior firm performance an insight from Indian cement manufacturing. *International Journal of Innovation Science, 4*(4), 217–230.

Dubey, R., & Singh, T. (2013). Soft TQM for sustainability: an empirical study on Indian cement industry and its impact on organizational performance. In *Mechanism Design for Sustainability* (pp. 77–104). Springer, Dordrecht.

Dubey, R., & Ali, S. S. (2014). Identification of flexible manufacturing system dimensions and their interrelationship using total interpretive structural modelling and fuzzy MICMAC analysis. *Global Journal of Flexible Systems Management, 15*(2), 131–143.

Dubey, R., & Gunasekaran, A. (2015). Education and training for successful career in Big Data and Business Analytics. *Industrial and Commercial Training, 47*(4), 174–181.

Dubey, R., Gunasekaran, A., & Sushil, Singh T. (2015a). Building theory of sustainable manufacturing using total interpretive structural modelling. *International Journal of Systems Science: Operations & Logistics, 2*(4), 231–247.

Dubey, R., Gunasekaran, A., Papadopoulos, T., & Childe, S. J. (2015b). Green supply chain management enablers: mixed methods research. *Sustainable Production and Consumption, 4*, 72–88.

Dubey, R., Gunasekaran, A., Childe, S. J., Papadopoulos, T., & Fosso-Wamba, S. (2017). World Class Sustainable Supply Chain Management: critical review and further research directions. *International Journal of Logistics Management, 28*(2), 332–362.

Dubey, R., Gunasekaran, A., Childe, S. J., Blome, C., & Papadopoulos, T. (2019). Big data and predictive analytics and manufacturing performance: integrating institutional theory, resource-based view and big data culture. *British Journal of Management, 30*(2), 341–361.

Dubin, R. (1978). *Theory development*. New York: Free Press.

Dutta, D., & Bose, I. (2015). Managing a big data project: the case of Ramco Cements Limited. *International Journal of Production Economics, 165*, 293–306.

Elragal, A. (2014). ERP and big data: the inept couple. *Procedia Technology, 16*, 242–249.

Elkington, J. (1994). Towards the sustainable corporation: win-win-win business strategies for sustainable development. *California Management Review, 36*(2), 90–100.

Eti, M. C., Ogaji, S. O. T., & Probert, S. D. (2006). Reducing the cost of preventive maintenance (PM) through adopting a proactive reliability-focused culture. *Applied Energy, 83*(11), 1235–1248.

Ghemawat, P., & Thomas, C. (2004). *Identifying the sources of sustained performance differences: a study of the international cement industry.* Mimeo, Harvard Business School.

Giannakis, M., Dubey, R., Yan, S., Spanaki, K., & Papadopoulos, T. (2020). Social media and sensemaking patterns in new product development: demystifying the customer sentiment. *Annals of Operations Research*, 1–31.

Gupta, M., & George, J. F. (2016). Toward the development of a big data analytics capability. *Information & Management*, *53*(8), 1049–1064.

IBEF(2021). Indian cement industry analysis. https://www.ibef.org/industry/cement-presentation (Date of Access: February 11, 2021).

Ikemoto, G. S., & Marsh, J. A. (2007). Chapter 5 Cutting through the "data-driven" mantra: different conceptions of data-driven decision making. *Yearbook of the National Society for the Study of Education*, *106*(1), 105–131.

Jeble, S., Dubey, R., Childe, S. J., Papadopoulos, T., Roubaud, D., & Prakash, A. (2018). Impact of big data & predictive analytics capability on supply chain sustainability. *International Journal of Logistics Management*, *29*(2), 513–538.

Jha, S. K., Singh, A. K., & Prakash, A. (2014). Understanding green manufacturing (GM). *Journal of Production Research & Management*, *4*(1), 33–45.

Kazlauskaitė, R., & Bučiūnienė, I. (2008). The role of human resources and their management in the establishment of sustainable competitive advantage. *Inžinerinė ekonomika*, (5), 78–84.

Koh, S. C., Saad, S., & Arunachalam, S. (2006). Competing in the 21st century supply chain through supply chain management and enterprise resource planning integration. *International Journal of Physical Distribution & Logistics Management*, *36*(6), 455–465.

Liang, H., Saraf, N., Hu, Q., & Xue, Y. (2007). Assimilation of enterprise systems: the effect of institutional pressures and the mediating role of top management. *MIS Quarterly*, 31(1), 59–87.

Luo, Z., Dubey, R., Gunasekaran, A., Childe, S. J., Papadopoulos, T., Hazen, B., & Roubaud, D. (2017). Sustainable production framework for cement manufacturing firms: a behavioural perspective. *Renewable and Sustainable Energy Reviews*, *78*, 495–502.

Luo, Z., Dubey, R., Papadopoulos, T., Hazen, B., & Roubaud, D. (2018). Explaining environmental sustainability in supply chains using graph theory. *Computational Economics*, *52*(4), 1257–1275.

McAfee, A., Brynjolfsson, E., Davenport, T. H., Patil, D. J., & Barton, D. (2012). Big data: the management revolution. *Harvard Business Review*, *90*(10), 60–68.

Mishra, S., & Singh, S. P. (2020). A stochastic disaster-resilient and sustainable reverse logistics model in big data environment. *Annals of Operations Research*, 1–32.

Schein, E. H. (2010). *Organizational culture and leadership* (Vol. 2). John Wiley & Sons.

Schneider, M., Romer, M., Tschudin, M., & Bolio, H. (2011). Sustainable cement production—present and future. *Cement and Concrete Research*, *41*(7), 642–650.

Svensson, G., & Wagner, B. (2015). Implementing and managing economic, social and environmental efforts of business sustainability: propositions for measurement and structural models. *Management of Environmental Quality*, *26*(2), 195–213.

Singh, T., & Dubey, R. (2013). Soft TQM practices in Indian cement industry–an empirical study. *International Journal of Productivity and Quality Management*, *11*(1), 1–28.

Srinivasan, R., & Swink, M. (2018). An investigation of visibility and flexibility as complements to supply chain analytics: an organizational information processing theory perspective. *Production and Operations Management*, *27*(10), 1849–1867.

Swanson, L. A., & Zhang, D. D. (2012). Perspectives on corporate responsibility and sustainable development. *Management of Environmental Quality: An International Journal*, *23*(6), 630–639.

Whetten, D. A. (1989). What constitutes a theoretical contribution?. *Academy of Management Review*, *14*(4), 490–495.

Winston, A. (2014). Resilience in a hotter world. *Harvard Business Review*, *92*(4), 56–64.

Wright, P. M., McMahan, G. C., & McWilliams, A. (1994). Human resources and sustained competitive advantage: a resource-based perspective. *International Journal of Human Resource Management*, *5*(2), 301–326.

Zhenyu, H., & Prashant, P. (2001). ERP implementation issues in advanced and developing countries. *Business Process Management Journal*, *7*(3), 276–284.

11 Implementation of Digital Solutions in Cement Process and Plants

Sriram Seshadri
F. L. Smidth & Co., Copenhagen, Denmark

Jeyamurugan Kandasamy and Manikandan Rajendran
F. L. Smidth India, Chennai, India

CONTENTS

11.1 Introduction 349
11.2 Digitalization Approach for Cement Plant – Mine to Packer 349
    11.2.1 Smart Machines 350
    11.2.2 Plant Control System 350
    11.2.3 Process Optimization 351
    11.2.4 Quality Management Systems..351
    11.2.5 Plant/Enterprise Management Systems.............................351
    11.2.6 IoT Platform Foundations 351
    11.2.7 Connected Asset Insights 351
    11.2.8 Connected Asset Health...352
    11.2.9 Connected Operation 352
    11.2.10 Connected People 352
    11.2.11 Connected (Business) Process 352
    11.2.12 Connected Innovation...352
11.3 Smart Machines and Digitalization of Plant Control Systems 353
    11.3.1 Smart Machines 353
    11.3.2 Essentials of Plant Control Systems.................................354
    11.3.3 Upgrading the Existing Plant Control Systems................356
11.4 Process Optimization Solutions 358
    11.4.1 Ball Mill Application 360
    11.4.2 Multi-Fuel Application 360
    11.4.3 Kiln and Cooler Application...361

DOI: 10.1201/9781003106791-11

347

   11.4.4 Vertical Roller Mill Application 361
 11.5 Quality Management Systems 363
   11.5.1 Sampling and Transport...363
   11.5.2 Sample Preparation..364
   11.5.3 Sample Analysis...365
   11.5.4 Laboratory/Quality Control Software...........................365
   11.5.5 Quality Optimization Software 365
      11.5.5.1 BlendExpert – Pile Control 366
      11.5.5.2 BlendExpert – Mill......................................367
 11.6 Automated Cement Laboratories 368
 11.7 Plant/Enterprise Management Systems 369
   11.7.1 Enterprise Asset Management System 369
   11.7.2 Enterprise Resource Planning and Its Components 371
   11.7.3 Management Information Systems 371
 11.8 Foundation of IoT Platform..372
   11.8.1 Edge Gateway or Field Agent 373
   11.8.2 Cloud Platform 373
 11.9 Connected Asset Insights 374
   11.9.1 Assets Performance Insights..375
   11.9.2 Exploratory Workspace...375
   11.9.3 Connected Asset Health...376
      11.9.3.1 Smart kiln – Online Condition Monitoring.....380
      11.9.3.2 Condition Monitoring Services for
           Vertical Roller Mills 381
      11.9.3.3 Mill Case Studies 382
      11.9.3.4 Filter Bag – Online Condition Monitoring......384
 11.10 Connected Operation – Pyroprocessing Section 385
   11.10.1 AI in Process Optimization 386
   11.10.2 Predicting the Kiln Red Spot by Using
      Temperature Anomaly 386
   11.10.3 Monitoring Kiln Coating Stability Index 388
 11.11 AI-Enabled Intelligent Production Management System...............388
 11.12 Connected People..391
   11.12.1 Remote Operations-Control Room 391
   11.12.2 Remote Troubleshooting Support 392
 11.13 Application of Drone Technology 394
 11.14 Connected Business Process 397
   11.14.1 Automated Diagnostic Solutions 398
   11.14.2 Asset Performance Management..................................399
   11.14.3 Remote Plant Operations...399
 11.15 Connected Innovation 399
   11.15.1 Digital Twin 400
   11.15.2 Augmented Reality/Virtual Reality 401
 11.16 Concluding Observations 402
Acknowledgments...403

11.1   INTRODUCTION

"Increase output and lower the production cost" – this could be the motto for most cement producers around the world who have left no stone unturned in their quest to increase production and reducing cost of operations in a sustainable way. The only hope for any real further gains is step change in the approach. In short, to optimize further, a revolution is needed. And now, the revolution is Industry 4.0, often referred to as the fourth Industrial Revolution which will take us to the next level.

Unlike the previous three industrial revolutions, Industry 4.0 is not about replacing existing machines with new machines or assembly lines but more about exploiting new digital technologies like Internet of Things and Cloud Computing to create powerful connections between existing physical assets and digital systems. This opens possibilities and new solutions for further optimization. With the enormous energy consumption, rising cost and sustainability challenges, and the overall complexity, cement production will benefit immensely from Industry 4.0.

Compared to other industries, cement producers are already late to the party, and only few players in the cement industry have been implementing Industry 4.0 initiatives systematically. The good news is that this gives the first movers big advantage over the rest not only with the productivity gains but also most importantly building the digital transformation culture within their organization.

The digital transformation is not about the additional sensor or data or tool or change in process but deals with all aspects including organization, activities, processes, competencies, operating models to fully leverage the changes with digital technologies. This is the ability to transform current way of work, for example, moving from scheduled maintenance to predictive or prescriptive maintenance or working from remote operation centers or engaging new business models (pay by use, performance-based contracts, etc.).

If business leaders think they can digitize a business or digitalize enough processes to digitally transform, they are misunderstanding the terms and missing out on opportunities to evolve, gain competitive advantage, respond to consumer and employee expectations and demands, and become efficient agile businesses.

Therefore, industry 4.0 is not about implementing a technology or replacement of machines, but a comprehensive industrial revolution in transforming the future cement plant toward an intelligent and sustainable production.

11.2   DIGITALIZATION APPROACH FOR CEMENT PLANT – MINE TO PACKER

So how do you start or where do you start with digitalization in a cement plant? The key is this is not a product or project or a tool, this is an evolution, and as mentioned previously, this has started in most of the plants long before with automation solution.

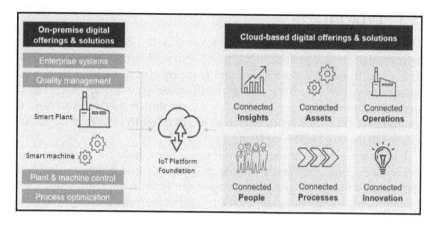

FIGURE 11.1   Key components of digitalization of cement plant.

The new IoT capability now brings transformation in both existing on-premises digital solutions with new capabilities and new possibilities with the Cloud-based digital solutions. These solutions are getting highly interconnected to make them improve continuously with new capability to operate based on the learnings across plants and technologies.

The key components of digitalization of cement plant are shown in Figure 11.1 and include the following:

11.2.1   SMART MACHINES

The first step with digitalization of plants is to make the machines smarter. This includes building machines with more advance instrumentations, control solutions, connectivity capabilities, communication capabilities, more standardized interfaces, ability to provide more insights to operators, self-learning capability, etc. In short "Smart Machines" are self-configuring, self-m onitoring, self-healing, and autonomous machines for achieving exceptional levels of efficiencies and growth in productivity.

11.2.2   PLANT CONTROL SYSTEM

The plant control system is a fundamental and essential part of every machinery in the production line. It can be differentiated into machine control, product control, and process control systems. Process control systems control the overall process and give you complete and reliable operation of your plant. In general, most of the process sections (from Crusher to Packing) are interconnected and the operators in Central Control Room (CCR) get a complete overview of the plant to supervise and operate it. The machine and product control systems are used to control a standalone or part of a small process area.

11.2.3 Process Optimization

Process optimization solutions are the high-level control solutions to further automate and optimize the process in the cement plant. This includes technologies like MPC (Model Predictive Control), AI technologies, simulation technologies, soft sensors, etc., to stabilize the plant operation and optimize production.

11.2.4 Quality Management Systems

Quality plays a vital role in cement production. At every stage in the cement process, you are continuously making decisions that impact final product quality, your plant's productivity and ultimately, profitability. Good decisions rely on high-quality data, so you need sampling, preparation, and analysis systems that you can trust. The quality management system has got the complete overview of all these systems including quality optimization solutions.

11.2.5 Plant/Enterprise Management Systems

The plant enterprise system is about solution which helps you control your operations end-to-end – from raw material procurement to shipping, monitor your inventory in real-time to manage your raw material and fuel requirements. This includes Computerized Maintenance Management System (CMMS), Enterprise Asset Management System (EAM), and Enterprise Resource Planning (ERP) including Finance, HCM, and SCM.

11.2.6 IoT Platform Foundations

The IoT platform is a set of software layers that manage and connects devices, the edge hardware, access points, and data networks to the other end which is usually the end-user application. It helps you in bringing the physical objects online. This platform will provide you with the services to connect the devices for a machine to machine communication. It will easy to get started with generic IoT platforms from the market but take significant cost and effort to develop and maintain the platform with all capability needed to use them in operation, so it would be better to partner with someone who has built these for cement industry applications.

11.2.7 Connected Asset Insights

The focus of connected asset insights is to unlock new learnings from the data, to keep everyone connected and informed to make the right decisions with confidence. It empowers the users to monitor the performance of the plant and equipment on-the-go, track performance trends, optimize operations, and respond to critical events faster and with more accurate information.

11.2.8 Connected Asset Health

The focus of the connected asset health solution is to reduce to unplanned downtime of the assets and plants. The online condition monitoring service is one of the growing solutions that connect, monitor, analyze, and optimize operations in response to the demand by the plant operation to reduce their maintenance burden and optimize maintenance strategy. When combined with advance instrumentation, connectivity and analytics, it will support predicting failure well in advance for the plant operation and maintenance team to plan the maintenance/replacements in advance.

11.2.9 Connected Operation

The connected operation deals with optimizing the plant asset performance and operation through the digital solutions like analytical models, computer vision, or simulations. The data from the plant assets and operations are continuously monitored by the optimization model to provide greater insights into the performance under the given conditions and support the operations by simulating the performance conditions for the future.

11.2.10 Connected People

The aspect of connected people is all about connecting people to the devices – the new digital capability which augments people's ability to do their job better. It is also about the ability to interact with the assets or process easily through their mobile phone or wearable devices to make smart decision using the contextual information provided by these connected assets.

11.2.11 Connected (Business) Process

The digital transformation is not just about devices; it also brings significant changes in the way of work and the business process, including maintenance, operation, and management of the plants, for example, the maintenance process will move from reactive or proactive maintenance to predictive maintenance approach. Plants will be able to focus more and more on carrying out operations remotely, inspite of the difficulty of bringing experts to their remote locations. The technology and the transparency will support new business model like "pay by use" or "Risk and reward"-based engagements, etc.

11.2.12 Connected Innovation

This is another component of the digital platform which permits driving more innovations with unlimited possibilities connecting the products, people, and processes. The volume of data collected across plants allows us building digital twins, which could simulate the operation across the product, process area, and even at plant level to simulate the operation to predict the operation in all aspects. This could only improve as we get more data and data points. Similarly, the virtual reality/augmented reality (VR/AR) technologies allow employees not only to get trained to be more efficient but also to be prepared to handle the real field situations with simulations.

11.3 SMART MACHINES AND DIGITALIZATION OF PLANT CONTROL SYSTEMS

All the above-mentioned key components of digitalization, both Cloud-based and on-premises solutions, are going through continuous technological advancement. This will only accelerate in the coming period with more adoptions across plants. For adoption and implementation, it is important to understand their implications in detail as discussed below.

11.3.1 SMART MACHINES

One of the key foundations for the industrial digital transformation with "Internet of Things" is the smart connected assets ("the things"). There are varying levels of intelligence built on these machines, ranging from adding simple instrumentations and actuating, to controlling, displaying, interconnecting, and self-learning to be a fully autonomous smart machine. The machines are connected through the standard Internet and Cloud technologies that enable secure access to devices and information to leverage big data and analytics, and mobility technologies to drive greater business value.

OEMs and end users can leverage the new digital capabilities to better monitor and control the machines. Industrial IoT applications will include not only machine-to-machine (M2M) communication but also machine-to-people, people-to-machine, machine-to-objects, and people-to-objects communication. Some of the elements that encompass Smart Products do exist with some key machines in cement plants, for example, communications, product control, etc. Therefore it should be viewed as more of an evolution rather than a revolution. New elements like the Cloud, mobility, cybersecurity, big data, and analytics will provide a new way of interacting with these machines efficiently.

To benefit from the potential that now exists for the development of new levels of operational intelligence, industries will need to migrate to the plant asset infrastructure over time that enables to exploit these new capabilities. This is where the next generation of machines – the "smart machines" – enters the picture. Again, this does not necessarily mean replacing the existing machines with new machines but in many cases it may mean enabling the existing machines with additional capabilities to make them "smart".

The term "smart machine" implies a machine that is better connected, easy to operate, more efficient, and safe. A smart machine is also capable to

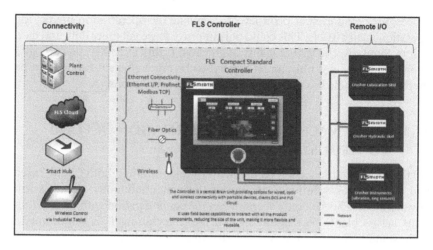

FIGURE 11.2   Characteristics of smart machine.

diagnose itself and support with predictive maintenance practices while mini-mizing its own environmental footprint and total cost of ownership.

The four key characteristics of smart machines are standardized blueprint, efficiency with ease of operation, safety and security, and connectivity (Figure 11.2).

Standardizing blueprint is the important step to enable us to collect asset information in a structured way to be able to compare and benchmark assets within and across the plant.

User experience plays a critical role with adaption of the digital transfor-mation, and it starts with the machine. The smart machines are built with user friendly capabilities to operate and keep the user informed about its operation.

Safety and security are important aspects of the smart machines which ensure both the human and operational safety and security, for example, no distributed stop system across the machine.

Each machine is enabled with the capability to communicate between each other to support the M2M communication but also to other devices or to the IoT Cloud through the right protocol and secured gateway.

11.3.2   ESSENTIALS OF PLANT CONTROL SYSTEMS

We do have different control systems across the plant which include the plant process control system used to operate most part of the process plant, machine, and product control systems which are used to control a machine/product or a small section of the flow sheet. Some of the critical machine/product control systems (e.g., cooler sub-control system, mill hydraulic sys-tem) are connected to the process control systems, but only with necessary information to operate the plant.

FIGURE 11.3 Plant control system pictures.

There are some plant control systems which are tailored for the cement industry that demands a strong domain knowledge and expertise. These systems provide increased flexibility, cut installation times, and ensure safe, reliable, and efficient operations and yet they are user friendly. In effect, the plant control system includes well-known features such as trending, alarm management, and data logging, etc. It uses modern Ethernet-based communication based on uniform engineering standards and optionally provides access on a plant-wide control system level. It is fast, flexible, and offers comprehensive trouble-shooting, yet being a simple-to-use solution. Typical illustrations of the plant control systems are given in Figure 11.3.

With today's tough landscape of rising costs and increasingly stiff competition, the industry needs a reliable control system enabled with remote and timely support from the experts. It is ideal if the plant control solution works equally well with controller hardware from different major brands (e.g., Siemens, ABB, Schneider, Rockwell) which simplifies the task of upgrading the existing systems as they are not dependent on one supplier of hardware.

If we summarize, the plant control system should be able to provide the following benefits:

• Superior automation
• High performance and reliability
• Increased troubleshooting
• Backed by domain/industry know-how
• Simple, yet flexible architecture
• Most importantly, Digital Ready

Another important aspect of plant control system is the system integration. It is understood that discrete (standalone) systems or solutions deployed in a plant might be the best individually. However, Integration of standalone or discrete data sources is envisaged as an important step when the digital transformation is taken on hand. Unless the systems are integrated, deployment of high-level plant-wide digital solutions would become a challenge and even if we deploy, it will not produce major or all the benefits for the investment made against digitalization.

For example, an excellent machine control system is deployed to run the stacker-reclaimer systems in the raw material storage yard. What if this is not integrated with the crusher control system and the raw material hopper filling in the mill feed systems? Will you be able to stop the feed belt to hopper when the hopper is full, or will you be able to stop the crusher when the stacker belt stops?

Energy meters are installed in every substation and for all high-power drives and motors. What if you have not integrated all the meters to the main control system? Will you be able to get a central energy monitoring information and the intelligent load management based on the process priorities?

Raw mill baghouse is installed with the latest control panel with a dedicated display. What if the control panel is not integrated with the main control system? Will you be able to get the operation completely harmonized with respect to the plant process? Even though a mill hydraulic product control system is completely integrated with all the sensors and actuators. What will happen if this is not integrated with the plant control system? Will you be able to get all the key information about hydraulic system from the central control location?

It also needs to be noted that the systems need to have the compatibility for integration, and if it is obsolete or outdated, then we might not be able to integrate the whole system and to get the required data to enable deployment of plant-wide digital solutions. Hence, before deciding the integration of discrete or standalone data sources, we need to have the right instruments or control systems which are compatible for integration.

11.3.3   Upgrading the Existing Plant Control Systems

In this section, an attempt will be made to explore how an upgrade of an existing control system may be carried out and to what extent. For illustration purposes the older or out of date systems are shown in five different levels in Figure 11.4.

Level 1 of upgrading involves human–machine interface (HMI)/SCADA/ control System software, where you could keep all the existing systems, and as a first step, the upgrading exercise could be undertaken without long stoppages. However, in this approach, it is necessary to make frequent updates of the software to keep the systems more secured. If any of the systems is observed to be approaching the "end-of-life" or "end-of-support" notification from the supplier, it needs to be considered for upgrade on topmost priority to avoid any unexpected stoppages and production loss because of the antiquated systems.*

Level 2 of the upgrading exercise involves the controller layer, where the PLC/DCS controllers could be changed faster, and more efficient control actions, security, and more particularly, easy integration with the HMI/ SCADA layer. It will be possible at this stage to make complex control logics with the latest controller hardware. In general, Level 2 upgrading lasts for more years than Level 1.

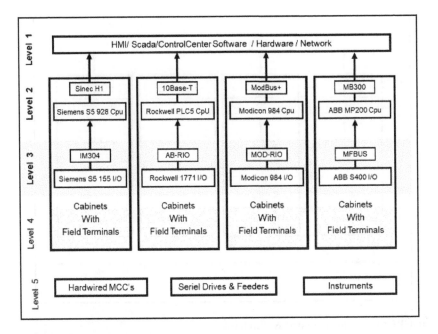

FIGURE 11.4   Different levels of the past systems.

Level 3 of the upgrading exercise involves the Input/Output (I/O) modules. The input modules receive the inputs from the field sensors and the output modules are used to operate the actuators with reference to the control logics built on the controller's I/O modules, either a digital or an analog type. I/O modules are mounted on the control cabinets and are kept in different remote locations. In this level, what is intended is to change only the I/O modules, keeping the same control cabinet and field wiring. Level 3 upgrading is expected to last longer than Level 2.

Further, in hierarchy, the Level 4 upgrading exercise is undertaken, when it is assessed that the field wiring and the field terminals are required to be changed because of the age of the cables and the condition of the cabinets. It is a bigger and more time-consuming activity than the previous ones; this level of upgrading must be executed very carefully with all the old documentation remaining intact.

Level 5 upgrading involves changing the motor control cabinets, drive panels, and field instrumentation. This is a major modification that involves the change of complete field cabling with I/O cabinets. This is carried out in case the field devices/actuators are antiquated, and in such cases, where there are changes in plant equipment and process.

Figure 11.5 is an illustration of typical upgraded systems at each level. The model details are only for illustration purposes and may not reflect the latest

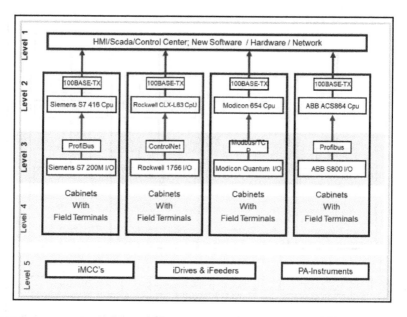

FIGURE 11.5 The upgraded system levels.

scenario. It must be noted that the technology in the concerned fields are fast changing and dynamic with new innovations.

11.4 PROCESS OPTIMIZATION SOLUTIONS

Process optimization solutions are the high-level control solutions to further automate and optimize the process in the cement plant. Process optimization solutions are called by different names in the industry today, namely Fuzzy Control, Process Expert, APC, high-level control, MPC technology, expert optimizer.

The key benefits of the process optimization solutions are the following:

- Stabilizing the plant operation and reducing frequent process disturbances
- Improving consistency between operators
- Reducing the impacts of variations in fuel quality and increasing the use of alternative fuels
- Optimizing the production
- Reducing power consumption
- Ensuring consistency in product quality
- Reducing equipment disruptions and minimizing wear

Process optimization solution is an advanced process control and optimization solution for complex and non-linear process such as finish grinding or pyroprocesses. Control optimization is performed using the advanced capabilities that are specifically tailored to meet the individual user requirements.

Depending on the application type, advanced expert system techniques such as Model-based Predictive Control (MPC), fuzzy-logic, and AI Technologies are used in different applications to enable hybrid control schemes to meet the requirements of the given process control issues. These modules will perform complex and steady evaluations of process conditions and execute adequate control actions on a more frequent and reliable basis than human operators. These systems are built to assist the operators on their control tasks on a 24-hour basis.

The technologies for process control can be broadly defined as follows:

- MPC: This is a multi-input, multi-output controller that handles process dynamics and interactions. MPC performs optimization calculations to drive the measurements to predefined targets or ensure they remain within a set band limit.
- Symbolic AI technologies: This is focused on high-level "symbolic" (human-readable) representation of problems, logic, and search (e.g., Expert systems, Fuzzy logic, delivering rule-based, intelligent fuzzy control and inference rules).
- Non-symbolic AI technologies: This involves providing raw data to the machine and leaving it to recognize patterns like Machine Learning (ML) or deep learning.
- Kalman filter: This is a soft sensor that generates readings where signals are unavailable or unreliable.

In a cement plant, process optimization solution finds extensive applications in four areas: Vertical roller mill application, multi-fuel application, kiln and cooler application, and ball mill application as shown in Figure 11.6.

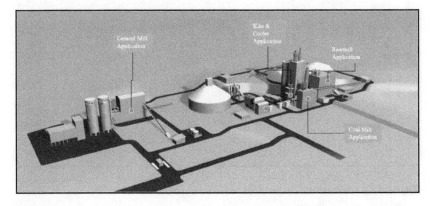

FIGURE 11.6 Typical applications where process optimization solutions are used.

11.4.1 Ball Mill Application

To obtain lowest possible power consumption, maximum production, and lowest possible quality variation, a conventional control solution with PID loops is insufficient. Process delays (e.g., fineness analysis delay, material transportation) cannot be handled well by PID's. Furthermore, the process contains internal couplings. For example, the separator speed impacts not only the fineness but also the mill filling level through the reject flow. Therefore, a change in one of the PID loops causes disturbance in the other PID loop. The two PID loops are consequently often in conflict to reach their own objective. This lack of coordinated action causes undesired disturbances and operation inefficiency. The process optimization solution aims to achieve the best possible grinding efficiency through an advanced Multi-Input Multi-Output (MIMO) control strategy using a MPC technique.

Some of the parameters controlled in the process include feed and fineness control by fresh feed and separator speed, feeder ratio control for quality, mill draft, water flow, and online process state estimation. The parameters monitored include product quality in terms of Blaine surface area, online particle size analysis, SO_3, loss on ignition (LOI), separator and fan speed, fresh and reject feed, mill Folaphone or elevator power, draft and temperature, and feeder response to a given set point. The ball mill control points are schematically shown in Figure 11.7.

The key benefits include increase in production, reduction in ball mill specific power consumption, reduction in quality variations, and minimal product change-over time between recipes.

11.4.2 Multi-Fuel Application

With increased focus on cost reductions, many cement plants have started using alternative fuels for kiln and/or calciner firing. This has created more

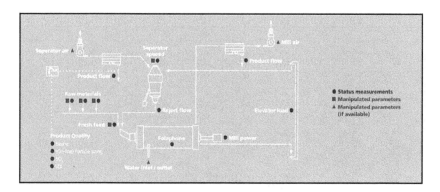

FIGURE 11.7 Schematic diagram of ball mill application.

challenges for the operation of a kiln because of the different characteristics of different alternative fuels such as calorific value, moisture content and chemical composition. The process optimization solution ensures successful management of alternative fuels without compromising stability or quality.

Ratio between fuels is the key parameter controlled in the multi-fuel application and the parameters monitored include fuel costs, fuel feeder, or environmental limitations, current feed rate of different fuels, chemical analysis of fuels, NOx and burning zone temperatures, and emissions data from optional emissions monitoring system.

The key benefits include better transition between fuels, resulting in a stable kiln process, higher utilization of alternative fuels, any number of alternative fuels handled, and better compliance with environmental restrictions.

11.4.3 Kiln and Cooler Application

Controlling a cement kiln has always been a challenging task for cement plant operators. These days, a computer-based pyro control system is not a mere show piece, it is a practical necessity. Both the inherent complexity of the pyroprocess and the far-reaching consequences of off-spec clinker, faulty operating conditions, and production shutdowns make an automated control system indispensable. An unstable kiln and cooler lead to inefficient production and inconsistent clinker quality. The process optimization solution stabilizes the kiln and cooler using advanced process control, resulting in increased production, reduced cyclone blockages, and kiln ring formations, while delivering consistent clinker quality.

Some of the parameters monitored include kiln feed, kiln speed, kiln fuel, calciner fuel, ID fan speed, cooler grate, and fan speed. The parameters monitored include kiln inlet gas analyzer measurements, kiln temperatures and pressure, kiln torque, and cooler pressure and temperatures as illustrated in Figure 11.8.

The key benefits include production increase, reduced cyclone blockages and kiln ring formations, consistent quality with a reduction in standard deviation, and more stable operation.

11.4.4 Vertical Roller Mill Application

Among the operational challenges of a vertical roller mill are the fast dynamics of the process. Compared to ball mills with dynamics changing in 15–20 minutes, vertical roller mills have dynamics changing in 2–4 minutes. The fast dynamics necessitate even closer attention to the process conditions and taking corrective action in time. Other challenges include changes in the grindability of material and water injections causing mill vibrations which may result in mill shutdown (Figure 11.9).

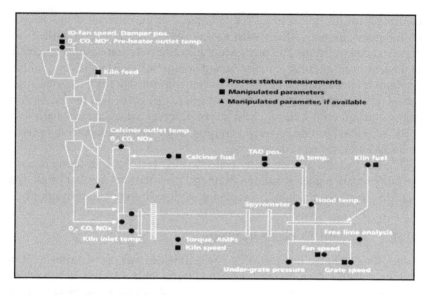

FIGURE 11.8   Schematic diagram of kiln and cooler control.

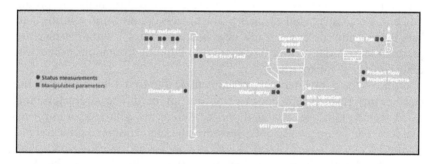

FIGURE 11.9   Schematic diagram of vertical roller mill control points.

Some of the parameters controlled in the process include feed and fineness control by fresh feed and separator speed, feeder ratio control for quality, mill draft, water flow, and online process state estimation. The parameters monitored include product quality in terms of the Blaine surface area or online particle size analysis, SO_3, LoI, separator and fan speed, fresh and reject feed, mill bed level thickness, power and differential pressure, draft and temperature, and feeder response to a given set point.

The key benefits include increase in production, reduction in specific power consumption, reduction in quality variations, control of vibrations to avoid mill shutdowns, minimum product changeover time between recipes and simplified mill start-up.

11.5   QUALITY MANAGEMENT SYSTEMS

The quality control (QC) function in a typical cement plant broadly comprises the following:

- Sampling and transport
- Sample preparation
- Sample analysis
- Laboratory/QC software
- Quality optimization software
- Automated cement laboratories

The above components of the QC function are further described below.

11.5.1   SAMPLING AND TRANSPORT

There is a saying that *"Without correct sampling... preparation and analysis are only a lottery"*. Sampling is the critical first step in the QC chain. Without correct sampling, the preparation and analysis of cement samples will be compromised. It has been calculated that 80% of quality errors can be attributed to poor sampling, making the need for fit-for-purpose samplers paramount.

It is known that collecting samples manually by a lab technician has its own challenges including environmental obstacles which could form a reason for not collecting samples on time, shift changes, availability of resources, and the mindset of the people. If a true representative sample is not available, then all the efforts put on preparation and analysis might go in vain. Further, of course, you need to get your samples to the lab quickly, safely, and with full traceability. So, how do we overcome this most important aspect in having a consistent cement quality or process?

Samplers shall be well performing with proven sampling solutions that support the production process throughout the cement plant – including raw materials, raw mill inlet/outlet, hot meal, hot clinker, cement mill outlet, and finished cement. Further, it is coupled with automatic sample transport where the automated sample transport network links reliable sampling systems to QC laboratory cells. **AutoSampling and transport system** controls automated sampling and pneumatic transport of sampled material from the process areas to the production laboratory (Figure 11.10).

AutoSampling concept offers different levels of automation for the sample taking and transport of samples from the sampling location to a laboratory. The term **"manual sampling solutions"** implies that only sample taking is automated. Sample collection will be manual. The term **"automatic sampling solutions"** implies that the following operations namely sample scheduling, sample call, sample taking, sample collection, and sample transport to the laboratory are fully automated.

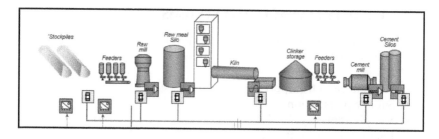

FIGURE 11.10 Typical sampling locations along with automatic sample transport configuration.

Under the fully automated scheme samples will normally be taken and sent from the sampling location to the laboratory according to a defined, fixed sampling scheme, which accommodates the QC/quality assurance (QA) requirements. However, if the laboratory attendant should require an extra sample, it can be "called" from the laboratory at any time when there is material to sample from. The QC software tracks the sample from the process to the final results. Automatically transported samples may in the laboratory arrive in manual receiving stations, from where a laboratory attendant will take the sample carrier, empty out the sample material, and place the carrier for return to the process area. Alternatively, samples may arrive in fully automatic sample receiving stations, from where the incoming sample material automatically is passed on to an automatic sample preparation system.

The benefits of combined sampling and automatic sample transport system include the following:

- Fast turnaround time from sampling to analysis
- Improved product quality and related operational savings
- Sample traceability and improved QC
- Optimized overall sample-taking schedule
- High system availability
- Easy connectivity to automated sample preparation systems
- Less labor requirements, allowing workforce to undertake other important tasks

11.5.2 Sample Preparation

It is widely known that the reliable results from the X-ray fluorescence analysis, extensively adopted in cement plants, are obtained, only when the sample preparation is consistent for the purpose. The sample preparation involves pulverizing, pressing, and fusing equipment to control sample quality at every stage for representativeness and reproducibility. Dosing, cleaning, and fusing units accurately prepare fused bead samples with an automated technology that helps in precisely dosing and weighing both the flux and the sample. A compact unit consisting of an automatic fine grinding mill and pellet press as a single convenient unit is commercially available along with supporting manuals.

11.5.3 SAMPLE ANALYSIS

For precise and reproducible results, various types of advanced analytical instruments are employed in cement plants. These instruments are capable of providing precise and reliable results so that the production can be controlled with confidence. X-ray fluorescence (XRF) and X-ray diffraction (XRD) instruments are widely used in cement laboratories. While XRF is used to analyze the elemental composition of the samples, XRD is used to provide quantitative analysis of the compounds or phases in the samples. Further, particle size analyzers are also widely used in cement industries. In addition, analytical instruments are used for free lime estimation, Blaine surface area determination as well as for determining carbon and sulphur, which help in producing high-quality cement in the most cost-effective way.

11.5.4 LABORATORY/QUALITY CONTROL SOFTWARE

This is an indispensable tool for laboratory management and visualization. This kernel software module includes sample administration, sample tracking, sample preparation recipes, and data import/export facilities. Combining the functionalities of the Laboratory Information System (LIS) and the Laboratory Automation System (LAS) in a unique setup, this software is prepared for any degree of sample automation. In addition, the configuration of user interfaces saves time and ensures consistency when designing and updating sample preparation recipes.

11.5.5 QUALITY OPTIMIZATION SOFTWARE

The quality optimization software is an integral component of the QC package. It takes care of raw material and additive blending, online accounting, and reporting. It also has a range of add-on modules that can help in making the most optimum use of the materials in the quarry. It also serves as a tool to experiment with different blends and production recipes. The quality optimization software modules such as BlendExpert are configured on top of laboratory/QC software and its applications are discussed below.

BlendExpert (BLX) supervises and controls the proportions of material feed to raw mills, cement mills, coal mills, and/or stock piles (or other process blending systems as option) to obtain the desired chemical product quality with respect to chemical constraints, process constraints, material costs, etc. The typical application of BLX is automatic set point control of material feed based on elemental analysis of periodical representative samples of the controlled material stream. The analysis data are received automatically from the

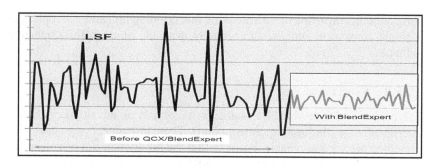

FIGURE 11.11   Variation in LSF of a raw mix with and without BlendExpert.

laboratory XRF spectrometer and/or from an online analysis device such as AtLine (an energy dispersive XRF analyzer) or a PGNAA bulk analyzer. The bands of variation in the Lime Saturation Factor (LSF) of a raw mix with and without the application of BlendExpert are compared in Figure 11.11.

Two major components of BlendExpert software are outlined below.

11.5.5.1   BlendExpert – Pile Control

BlendExpert – Pile Control offers accurate online accounting of raw materials on the stockpile. The software can control the raw material feed so that the stockpiles can be built up as close as possible to raw mill chemistry specifications. Optimizing the feed composition in this way facilitates maximizing the use of materials in your quarry without compromising the product quality and consistency in line with the chemical targets. Some of the key features are accounting, optimizing, sorting, and screening. A schematic overview of the system is given in Figure 11.12. The online accounting features provide comprehensive information on the chemical make-up of completed stockpiles and details within a single stockpile. There will be an alert if the predicted chemistry of a stockpile or layer is out of range. This flexible system can be used for a variety of pile layouts, including both circular and longitudinal stockpiles.

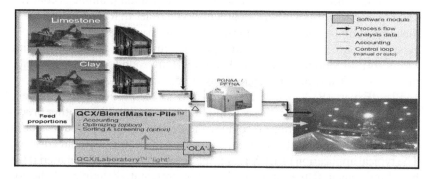

FIGURE 11.12   Schematic overview of BlendExpert – pile control.

"Optimizing" feature aims at fulfilling the requested stockpile target chemistry to the currently active stockpile, the "Sorting & Screening" option aims at sorting the feed to multiple accessible piles or to screen away unwanted material from reaching a stockpile. Classic examples are sorting of limestone to high- and low-grade stockpiles and screening against too high MgO content on a stockpile.

11.5.5.2 BlendExpert – Mill

It controls the blending of raw materials and corrective additives fed to raw mills. The advanced model based predictive programming system calculates the ideal feeder set points taking a whole range of factors into consideration including chemistry, tonnages, process limitations, and material costs. When adjusting the feeder set points to accommodate the correct chemistry, it can estimate and automatically compensate for any feeder off-set. This advanced logic is vital to come as close to target chemistry as possible in a process with varying limestone qualities and fluctuating additive quality. The advanced software can handle filter dust and other additives which are dosed to the process stream after the raw mill.

Further advantage of the BlendExpert package is that, in addition to improvements to the quality of clinker, it results in less wear on machinery and greater fuel economy, adding up to real cost savings and less downtime. This type of control over your chemistry can open new possibilities for the use of alternative fuels (Figure 11.13).

The overall benefits can be summarized as follows:

- Lower kiln specific heat consumption due to stable chemical quality
- Higher strength potential in total clinker production due to less off-speck clinker produced
- Higher kiln Alternative Fuel substitution rate as tight control enables handling of ashes and chemistry alteration in burning process

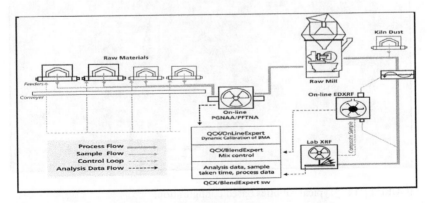

FIGURE 11.13 Schematic overview of BlendExpert – mill control.

- Longer lifetime of mechanical parts due to less thermal stress resulting from stable burning
- Reduced need for blending silos as the homogeneity of kiln feed quality is higher at the raw mill outlet
- Faster trouble shooting in of material feeders through direct monitoring of feeder operation
- Material blending can enhance optimal control during upset conditions, where use of less advanced solutions would necessitate switchover to manual control

11.6 AUTOMATED CEMENT LABORATORIES

An automated laboratory solution enables the cement plant laboratory to run more samples and implement more advanced sample preparation and analysis techniques than manual methods. The automated systems can easily be scaled to meet specific needs and budget. In today's cement industry, the increased use of alternative fuels and additives requires stricter controls and special analysis. Accurate results have never been more important. By automating sample preparation, the RoboLab system gives you reliable results. Sample preparation is more consistent. And the system also avoids sample contamination with in-built cleaning mechanisms (Figure 11.4).

RoboLab is an advanced sample preparation and analysis automation system designed to ensure that the process laboratory delivers fast, accurate, and safe analysis – with as few operators as possible. Consisting of integrated software and laboratory equipment, the system is designed to keep it up and

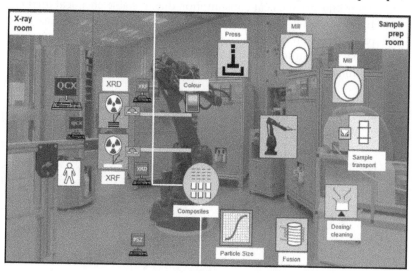

FIGURE 11.14   Typical robotics based configuration.

running all the time. It has a reliable performance, in-built equipment monitoring, and the FLSmidth global support network at call. By automating the processes, RoboLab reduces human error and accidental sample contamination while eliminating the labor-intensive manual processes that can lead to operator injuries.

It receives and coordinates samples from the AutoSampling system. Then, depending on your requirements, it can handle some or all the sample preparation and analysis tasks. Capabilities include dosing samples into sub-samples, particle characterization, sample preparation (e.g., pressed pellets or fused beads), and analysis methods such as carbon, sulphur and Blaine determination, and XRD or XRF analysis. RoboLab is a tireless worker in the laboratory, facilitating fool-proof sample analysis, and enhanced productivity. The key benefits are summarized as follows:

- Error-free sampling, sample preparation, and analysis
- Complete documentation and reporting
- Fast and accurate results leading to greater QC
- Labor saving system that promotes safe working environment
- Scalable in accordance with the specific user needs

11.7 PLANT/ENTERPRISE MANAGEMENT SYSTEMS

In addition to production, a cement plant also needs to manage the maintenance, inventory, safety, procurement, finance, and human resource (HR) associated with the plant operation. The systems used commonly across multiple plants either operated at enterprise level or sometimes individually at the plant level are ERP, CMMS, EAM, Management Information System (MIS), etc.

11.7.1 ENTERPRISE ASSET MANAGEMENT SYSTEM

The focus of the EAM system is generally on preserving assets for their full lifecycle and maximizing their potential for enterprise use. Generally, the EAM solution includes functionality for the following:

- Equipment and component definition
- Managing, planning, and scheduling work
- Work order creation
- Warranty claims and tracking
- Maintenance history tracking

- MRO (maintain, repair or overhaul) inventory management
- Equipment, component, and asset tracking for assemblies of equipment

In some plants, the EAM functionality is extended by the addition of basic financial management modules such as accounts payable, cost recording in ledgers, and HR functions such as maintenance skills databases to support more integrated approach toward resource planning, budget versus actual reporting, simplified analysis/reporting. The salient system capability for implementing the EAM solutions includes the following:

A. **Asset induction and record**
   - Detailed asset registry with specification of the asset, combined part details, and support descriptions with asset and component structures
   - Catalog of replacement parts linked to assets and components with inventory records of spare parts
   - Item number tracking and tracing for replacement parts and install history
   - Warranty tracking to component levels and support for manufacturers' records requirements for equipment under warranty
   - Standard reports and analytics for oversight of operations including consumables, wear parts, standard parts, etc.
   - Detailed cost analysis reports
B. Preventive and predictive maintenance
   - Record of all the unplanned shutdown or failures or issues
   - Registry for the recurring maintenance jobs based on a schedule for the assets
   - Overview of complete maintenance plans, project, and work schedules, including assigning tasks to resources for job completion over a period
   - Resource planning capabilities to match skills, training, and availability with work requirements
C. Shutdown planning
   - Integration to asset health/condition-based triggers and alerts creating work orders for maintenance team
   - Integration to the augmented solutions with ability for the maintenance service engineer to be able to use the system when on field
   - Safety (Lock out/tag out) and permit managements
   - Inspection and calibration records for trending and analytics
   - Maintenance budgets and estimation planning
D. Parts management
   - Support for complex inventory relationships for indirect maintenance, repair, and overhaul (MRO)

- Demand planning toward maintenance and repair
- Material requirements planning (MRP)-based inventory
- Flexible, supported integration with ERP and financial packages
- Usage and planning of consumables

The difference between EAM and CMMS comes down to their functionality and most common uses. The CMMS are small-scale, single-site applications with more focus on the maintenance management and less functionality around parts management and resource scheduling. EAM caters to larger business needs rather than toward specific department or site. Many of the EAM in the market are also hosted in Cloud to support easy integration to various supplier and partner solutions.

11.7.2 ENTERPRISE RESOURCE PLANNING AND ITS COMPONENTS

Generally speaking, implementation of ERP in cement plants is primarily a corporate priority, enforcing stricter controls for reporting and compliance. The EAM integration to ERP with finance, procurement, and HCM systems is critical to get an overview of the total performance of the plant, operation, and people.

The finance system includes fixed assets and accounting solutions to offer an overview of the fixed assets including asset plan, capitalization, insurance, inventory, depreciation, general accounting, planning, budgeting, report preparation, etc. The HCM system encompasses workforce planning, skill mapping for people, talent management, employee development, recruitment, payroll, planning, and other related activities.

11.7.3 MANAGEMENT INFORMATION SYSTEMS

The increasing amount of data at the plant level and the growing demand for status information in the management level requires adequate MIS. These MIS systems make critical plant data available to the people who need it most and play an important role of being the information gateway between the plant control system network and the management office network in a safe, reliable, and user-friendly manner.

These systems support decision-making processes at operational, management, and executive levels. It operates from a single central database to gather data from process, QC, energy and emission monitoring systems. This allows plant engineers, managers, and other personnel to visualize the data they need in the format they require, and trend long-term historical data. Data can be automatically or manually collected and stored in the comprehensive system for about 5–10 years for access by any user who is connected to the plant network.

The use of plant MIS systems leads to the following:

- Avoids human errors and wasting time through automatic, immediate, and accurate data collection mechanisms
- Enables asset management through centralized and safe data storage for historical analysis and documentation of crucial information
- Moves data efficiently from the plant floor to the control room for better data analysis and reporting
- Automatic generation of online Key Performance Indicators (KPIs)
- Provides plant managers, operators, and engineers with the specific information they need, regardless of their location
- Promotes the communication and collaboration between the plant personnel by presenting and sharing information through the same dashboards and portal
- Provides insight into specific process problems through trending and analysis of operational data
- Improves plant productivity (energy, production, quality) by making available real-time and historical data
- Safe storage of crucial data for historic analysis

MIS is one of the areas more impacted with new digital technologies. It is no more limited to data reporting, as it provides insights obtained with the help of digital tools, based on mobile/Cloud-based solutions.

11.8   FOUNDATION OF IOT PLATFORM

In today's advanced digital environment, access to real-time process data is critical with sensors and transmitters installed on every machine. The ability to process data has been simplified. Plants are continuously producing and recording a huge amount of process data. Data establishes equipment's performance in the field and can also show process issues while assisting in possible plant efficiencies and much more.

In order to use digitalization to improve processes and performance of people and machines, we need to be able to collect, store, and analyze data. Today, at field we have various data sources in different types and format. Aggregation makes the data collection, storage, and analysis more challenging.

So, how do we handle this challenge? We need to have an IoT Platform (Internet-of-Things solution – IoT) as a foundation or an enabler, and it is about securely collecting data from the sites, processing, and storing them in Cloud with high security (Figure 11.15).

The following are the key components of the IoT platform which is the foundation for processing and storage of data and then retrieve from here to provide value.

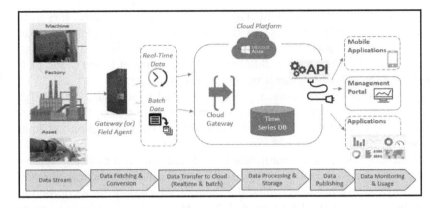

FIGURE 11.15 Typical components of an IoT platform and their interconnection.

11.8.1 Edge Gateway or Field Agent

We need to have an edge device or a gateway which is the entry point to the Cloud-based services. The device should be sturdy enough, easy to deploy, and remotely controllable for any updates, etc. The edge devices gather and transmit data to the Cloud from sensors, machines, and other intelligent components in the plant. It can gather data from one or several machines and sends the data to the Cloud system, where it is automatically stored in the correct directory. There could be two types of data collection, i.e., real time and batch data where the real-time data collection to Cloud platform is recommended to be done only in case of real business requirement as it would not be cheaper while the batch data could be collected once in certain time duration, for example, once in an hour or once in a day depending upon the business needs.

11.8.2 Cloud Platform

The Cloud is a huge, virtual storage space, where data is structured so that it is easy to retrieve and prepare for use. There are various Cloud service providers and one of them could be used to store and manage data. By storing data in the Cloud, the suppliers or service providers and the plants can use various applications to create value. The equipment suppliers can run analytics on an extremely large pool of data and build digital solutions and applications to help improve productivity and reduce cost. At the same time, the data also can be used to continuously improve the performance of the equipment supplied to the plants using their domain know-how and equipment design details.

Further, the way the data is stored in the Cloud is very important, and one must follow very clear definition, strategy, and process, considering various attributes about each data point. The data sets and their attributes vary from plant to plant and in some cases, within the plant itself. This approach makes the retrieval and usage of data smoother and efficient. Security is another aspect when the Operation Technology (OT) environment to the Cloud platform and the following general aspects are taken care of.

All communication between plant system and Cloud must be through Edge gateway. There should not be any direct connection from the plant control system or control network to the internet. All communication to Cloud should be encrypted.

Gateway devices are appropriately hardened and automatically receive security updates so that they do not rely on users to apply security patches. Gateway must be installed behind the Firewall, which only allows outbound communication to the IP addresses/DNS addresses required by the service provider.

11.9   CONNECTED ASSET INSIGHTS

The way we access data is changing; from home appliances to cars, almost everything, is now connected to the internet and accessed via a smartphone or tablets or through a computer. Today important performance data in a cement plant are scattered across numerous systems and people. Knowledge flow is not linear and requires lot of in-between communication. What if plant managers and personnel in a cement plant can look for this information from their mobile or tablet or computer? Would it make them more productive? Understanding plant performance is a time-consuming task. Data that can tell the story is often raw and needs to be manipulated. In a typical plant, a person spends about 4–7 days per month on the average for vertical reporting. Would manager and engineers be able to make the time they spend on data manipulation more productive if reporting is automated? Every manager in the plant has different priorities. Tracking the standard set of KPIs that make the most sense for the corporate establishment is no longer enough for the managers in the cement plant. Would a personalized dashboard empower managers in knowledge flow with less effort? Connected Insights is all about unlocking new learnings from the information by gathering, prioritizing, and visualizing the data and to empower managers with real-time operational KPIs. This requires that the data is collected and centralized so that right set of information is provided to the right persona when they need it through the right channel (Figure 11.16).

It may be relevant to mention here about a facility such as the FLSmidth SiteConnect Mobile app, which delivers on-demand and remote access to asset performance and health insights. It empowers users to monitor performance of plant and equipment on-the-go. It provides the opportunity of tracking performance trends and optimizing operations. It allows faster response to critical events with more accurate information. Typical screen shots of the app are displayed in Figure 11.17.

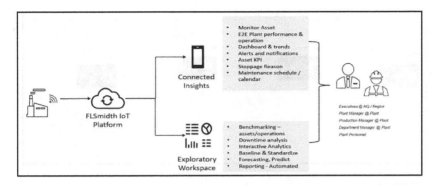

FIGURE 11.16 Illustration of FLSmidth Connected Asset Insights Architecture.

FIGURE 11.17 Typical screen shots from insights.

11.9.1 ASSETS PERFORMANCE INSIGHTS

When you do not measure, it makes it hard to improve. That is why it is important to use KPI to measure performance of assets and to further improve asset. It indicates how well the asset is doing at attaining goals, design intent both with performance and quality standards. Also, it is important to track the performance of these KPIs within the week, month, or year to evaluate the impact of changing conditions. Some of the KPI to measure across assets could include Overall Equipment Efficiency (OEE), machine availability, Run Factor, and performance factor. In addition, there are some specific KPI for individual assets which are more important to monitor (Figure 11.18).

11.9.2 EXPLORATORY WORKSPACE

Benchmarking assets across similar assets within the same plant or across plants provides an overview on the how an asset is performing as compared to the rest. The "FLSmidth secured asset benchmarking" solution is an example of how the data can be made completely anonymous to eliminate the identity

FIGURE 11.18   Typical KPI screen shots.

of the asset and focusing on the KPI across the best and worst performing similar assets. The embedded simulator also allows simulating the performance across the assets in varied plant conditions.

Unplanned equipment downtime is the single largest source of lost production time in most cement plants operation. Every time when there is a downtime, the equipment is underutilized, and maintenance demands increase. Production outputs and productivity are significantly diminished. When stoppages happen, it is vital to understand the cause and make sure they do not recur. Because of such negative impacts of stoppage and downtime, there is a strong need for a centralized powerful automated analysis software tool that gathers, presents and analyzes plant stoppage data. This must be a singular source of truth across the plant operation and should not be distributed with separate logbooks to CCR, maintenance, and product managers. This will help the plant staff with a detailed understanding of the main causes of the equipment failure. The information would help improve all the processes that are prone to stoppages, as in many cases, actions are required across various departments in a plant. Actions include what to monitor, where to increase maintenance focus, or when to interact with production planning. An illustration of the benchmarking solution is given in Figure 11.19.

11.9.3   Connected Asset Health

On an average, the cement plants lose anywhere from 5% to 20% of their productivity due to downtime. In addition to financial losses, unexpected failures and unscheduled stoppages result in loss of time, and consequently, the customer's demands may not be met. Most of the time there might be substantial pressure on the production team due to "fire-fighting".

When a stoppage occurs, your priority is to get your operations back online. The immediate next step should be to find out what went wrong and recording your response. It is important that this is recorded in a systematic way integrated to the operations systems including control system,

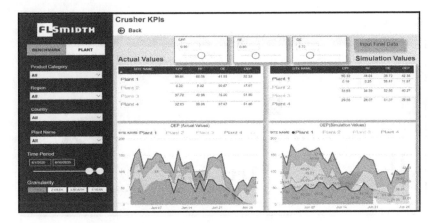

FIGURE 11.19 An example of benchmarking exercise.

maintenance management system, etc., to avoid any ambiguity with both event and reasons. This intelligence will provide insights for your future decisions and business improvements.

The plant downtime dashboard showing the category, type and equipment of failure and the enterprise downtime dashboard showing the total stoppages and the plant downtime are displayed in Figure 11.20.

In typical plant operation, very large volume of data is generated by systems and operators across the plant, but the data mostly remains in different pockets within the organization without being used to full potential. The "Connected Insights" is all about getting right information to right people at the right time through right channel in a systematic way including the process of recording events, gathering data, and transforming those into insights.

In general, a plant follows the following four maintenance approaches, depending on its maturity:

- **Reactive or Breakdown Maintenance**: Run to failure. "Fix-it-when-it-breaks" or corrective maintenance.
- **Preventive (Scheduled):** Planned, scheduled, coordinated activities, at fixed intervals of time. "Maintain-it-so-it-doesn't-break" (e.g., planned shutdown, planned inspections, maintenance planning). This reduces costs and extends equipment life.
- **Predictive:** "Fix-it-before-it-breaks" (with more accuracy than preventive approach). This predicts potential faults and optimizes maintenance activities => higher uptime and lower cost than corrective (Breakdown) and preventive maintenance.
- **Reliability-Centered Maintenance (RCM)** (and other types of proactive maintenance): It determines what must be done to ensure that the asset continues what the customer wants them to do. This is a

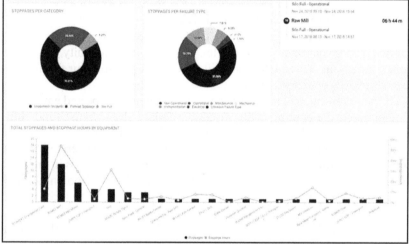

FIGURE 11.20   Downtime dashboard analysis with FLSmidth UptimeGO solution.

customized maintenance strategy for each equipment, combining input from online condition monitoring with asset criticality, failure mode and effects analysis, plant's focus, and maintenance approach (how critical the equipment is, how easy/fast it is to repair it, and to purchase parts, etc.). This ensures the equipment is maintained well and in a cost-effective way.

Based on field studies, it has been estimated that a properly functioning predictive maintenance program using new digital capabilities can provide a saving of 8% to 12% over a program utilizing preventive maintenance alone, as indicated in the report on "Operations & Maintenance Best Practices" of the US Department of Energy in 2010. The surveys indicate that the following orders of industrial benefits may flow from the practice of predictive maintenance program:

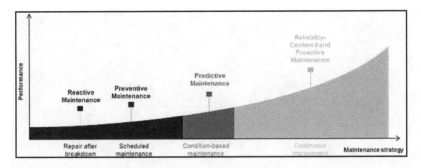

FIGURE 11.21 Effects of maintenance strategies on plant performance.

- Return on investment: 10 times
- Reduction in maintenance costs: 25% to 30%
- Elimination of breakdowns: 70% to 75%
- Reduction in downtime: 35% to 45%
- Increase in production: 20% to 25%.

The performance impacts of different maintenance strategies are shown in Figure 11.21.

Early insights from condition monitoring help detect failures and provide longer time for preventive action, thereby avoiding actual failure. Online condition monitoring helps predict failures in a way that is not possible with only preventive onsite services – either because the potential failures and root causes cannot be spotted without data, or because they would be spotted when it is too late.

But data on its own is not that useful. It requires the right expert to continually interpret the data, using their plant and technology experiences and to add context to those numbers. Actionable insights from the combined application of data, technology, and expertise reduce the chance of failure, optimize performance, and reduce operating costs (Figure 11.22). Further,

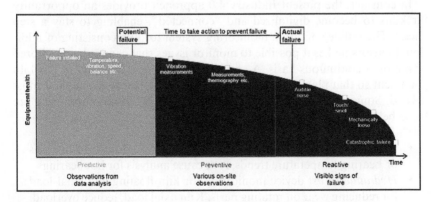

FIGURE 11.22 Prediction of potential failure from actual failure.

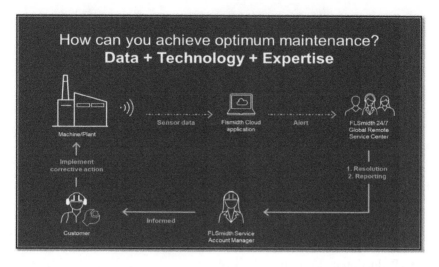

FIGURE 11.23   Actionable insights from data, technology, and expertise.

based on such learning, a model can be built which would predict similar failure in advance (Figure 11.23).

A few applications of online condition monitoring solutions in the cement plant are discussed below.

11.9.3.1   Smart kiln – Online Condition Monitoring

Kiln is one of the most critical and expensive assets in a cement plant. In most of the cement plants today kilns are monitored by just a few isolated sensors. A complete diagnosis is typically carried out like "spot check" every three years during a kiln inspection. When the kiln is not operating efficiently, the rest of the plant operations suffer. The entire production line shuts down. Breakdown in kiln will cause up to 4 days downtime purely in cooling down and heating up the kiln.

In contrast, the present Industry 4.0 approach provides an opportunity for kilns to become digitalized and "connected", enabling to stay a step ahead. The online condition monitoring system for kiln consisting of additional sensors makes it possible to monitor issues that cannot be monitored onsite on a continuous basis. A specific mention may be made of hot kiln alignment so that plant maintenance team could attend to it when it is most needed. Some of the other key devices and their condition monitoring are listed below:

- *Bearings* – identify overloads, identify improper lubrication, analysis of bearing temperature trends, root cause analysis for hot bearings
- *Hydraulic thrust* device monitoring for kiln floating and axial load for reducing wear on rotating parts, Kiln axial load, reduce overload,

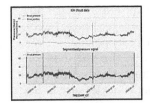 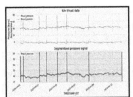 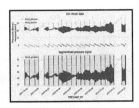

FIGURE 11.24.1 Kiln thrust data along with segment wise pressure signal data for three different plants.

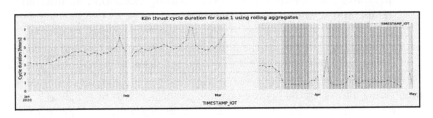

FIGURE 11.24.2 Kiln thrust cycle (change of duration from 24 h to 1 h indicates possible leak).

improve component life, health of hydraulic system, identify leaks and operation issues

- *Kiln drive monitoring* – identify cyclic loads from thermal crank or coating build-up
- *CemScanner integration*: early detection of hot spots, detection of refractory failures, coating formation, and analysis based on process insights
- *Kiln Crank:* detection of overload on rotating parts, prolong lifetime of rotating parts and avoid failures, avoid overload of kiln shell, tire, and supporting rollers
- *Kiln Shell Ovality:* monitoring based on the live ring migration, prolong lining life and reduce risk of kiln shell cracks, control kiln heating up process
- *Kiln drive vibration:* girth gear misalignment, detection of reduced tooth root clearance, detection of teeth failure
- *Axial Balance:* stabilize kiln floating, lower bearing temperatures, avoid overload of thrust roller (Figures 11.24.1 and 11.24.2).

11.9.3.2 Condition Monitoring Services for Vertical Roller Mills

Performance of a vertical roller mill is impacted by multiple variables that contribute to the mill's capability to produce quality product, reliably and consistently, at the expected capacity and with the best energy efficiency. To achieve and sustain fully optimized performance, it is important to address three specific areas:

- Operating parameters such as pressure and temperature of gas flow, hydraulic grinding pressure, and separator rotor speed must be consistently maintained and adjusted to the correct set points.
- The mechanical settings are regularly adjusted for the dam ring height, roller mechanical stop position, and nozzle ring open area as the mill internal components wear or the other variables such as feed moisture change.
- Critical components are maintained in good condition and working order by regular inspection of details such as lubricating fluid temperatures and contamination levels, bearing vibration and internal temperatures, and oil seal leakage.

More frequent monitoring of the critical operating parameters and mechanical conditions of mill internal components significantly improves the overall performance of the mill in daily operation and long-term reliability. Continuous monitoring allows for the highest possible performance to be achieved.

Focused analytics add detailed focus on specific critical components or subsystems in the mill circuit that negatively impact the mill operation over time. Advanced analytical methods expand the view of mill performance by incorporating more variables and data points than what can be analyzed by human capability alone. Some examples of the results of focused analytics are as follows:

- Mill air circuit and fan efficiency monitor for changes in the system airflow such as internal wear in the fan or leaking mill body air seals.
- Grinding hydraulic system monitoring identifies when the condition of cylinder oil seals begins to deteriorate so that they can be replaced before production is negatively impacted.
- Wear liner life uses historical data and the known interactions of various operating variables to indicate when mechanical adjustments are needed or when liner maintenance or replacement is needed.
- Separator top seal similarly uses a range of operating data points to indicate if the separator air seal needs to be serviced or replaced.

The above types of analytics either need more developments or require a "run in period" for site specific conditions (Figure 11.25).

11.9.3.3 Mill Case Studies

A few case studies of improving the mill performance through condition monitoring are illustrated below. The first one pertains to enhancing the uptime of a mill through monitoring of the grinding hydraulic system (Figure 11.26). Using data science matched with product expertise, the early onset of instability in grinding pressure caused by cylinder seal leaks was automatically detected before the more severe symptoms of lost production or product quality were seen. In this case, the piston seal was replaced on a

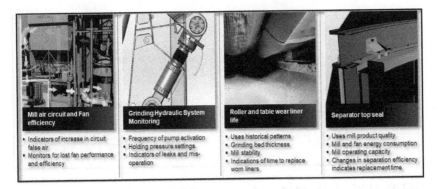

FIGURE 11.25 Illustration of VRM condition monitoring cases achieved through focused analytics.

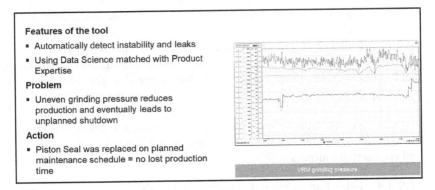

FIGURE 11.26 Illustration of improving the uptime of a VRM by monitoring the grinding hydraulic system.

planned maintenance schedule so that regular production schedule could be maintained.

The second case study pertains to the early detection of bearing defect in a gear box (Figure 11.27).

The online condition monitoring identified the early development of a bearing defect (developed when a replacement was not available) in a gear box. The pink trend shows the output from CMS, the dark trend line is the vibration sensor RMS feedback. Note the RMS signal never gave indication of the developing problem. Because of early detection development of the defect was controlled for over a year so the bearing could be replaced at an optimal time to avoid lost production. After a year, after all preparations were made, the faulty gearbox was replaced with spare one. The execution was planned and the downtime of mill was minimized. The gearbox with defect was overhauled and bevel pinion bearings were replaced. The bearing inner race was damaged.

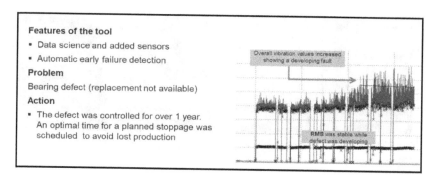

FIGURE 11.27   Early detection of bearing defect in a gear box.

Another case study of the connected online monitoring service for mill systems relates to a cement mill in which the KPIs were close to 60% of the target. The plant was contacted and analysis was done using 24/7 online access. It was learned by the service team that operators were disengaging the expert controller during every recipe change leaving operating parameters in manual control for extended periods of time. The plant was assisted to implement updated expert control logic for product recipe changes, raising the low performance factors to expected levels and improving the average of all performance factors from 66% to 95%. This is illustrated in Figure 11.28.

The above case studies demonstrate how the process know-how is applied through the computer maintenance system, and it is not just about the data and connectivity alone.

11.9.3.4   Filter Bag – Online Condition Monitoring

The filter bag is an essential wear part of a baghouse. By using the correct, high-quality bag, the plant can increase the filtration efficiency and reduce your dust emissions. In addition, the revenue expenditure can be substantially reduced.

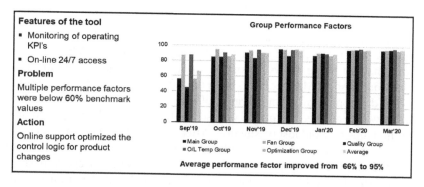

FIGURE 11.28   Improving the mill running factor by optimizing the mill settings.

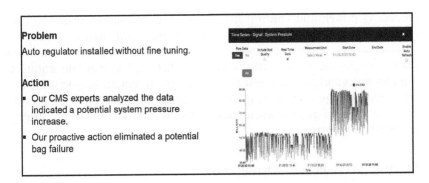

Problem

Auto regulator installed without fine tuning.

Action

- Our CMS experts analyzed the data indicated a potential system pressure increase.
- Our proactive action eliminated a potential bag failure

FIGURE 11.29   Monitoring of the system pressure for increased bag life.

With the online bag monitoring solution, several information from the filter could be monitored and analyzed to improve the life of the bags. The data monitored include the lifecycle counter, total bag life shots versus actual, pulse counters in various cleaning modes, filter running hours in various modes, filter inlet temperature trends, differential pressure trends, system pressure trends, burst bag detector trends, bag wear tracking using BBD sensors, vale failure percentage monitoring, etc. An illustration of pressure monitoring is shown in Figure 11.29.

11.10   CONNECTED OPERATION – PYROPROCESSING SECTION

For decades, cement plants are being digitized with field instruments, measurements, PLCs, SCADA systems, communications, and, in some cases, advanced process controls. Though it has improved plant operation, monitoring, controls, and visualizations for operators, most process systems have not kept up with the latest advances in analytics and in decision-support solutions that apply the Data Analytics (DA), ML, Artificial Intelligence (AI), etc.

Operators still rely on their experience, intuition, and judgment. For example, today's downsized teams of control room operators are expected to manually monitor a variety of parameters on several HMI screens and adjust process settings as needed. Meanwhile, they must verify and validate a few of the process control and optimization tasks that strain the limits of their individual capacity. As a result, many operators take shortcuts and prioritize urgent activities that do not necessarily add value. This inevitable and unavoidable dependency on experience makes it difficult to replace a highly skilled operator, even when it is unavoidable.

With ability to preserve, improve, and standardize knowledge, ML and AI can make complex operational set point decisions on their own and can outperform conventional decision-support technologies (a few times better than the talented operator). Also, thanks to new, high-performance software tools, processing power, and cheap memory, ML and AI enables plants to cost-effectively create and maintain their own algorithms and intellectual property in-house, which is cheaper, more versatile, and more adaptive to constantly changing equipment and market conditions. ML and AI can fully automate complex tasks and provide consistent and precise optimum set points in auto-pilot mode. It requires less manpower to maintain, and – equally important – it can be adjusted quickly when management revises manufacturing strategy and production plans.

11.10.1 AI IN PROCESS OPTIMIZATION

When Neural Networks (NN) and Genetic Algorithms (GA) are combined with Predictive Modeling, a NN will externally have an appearance as depicted in Figure 11.30. The figure is based on the data of a cement process and is generated by a powerful in-house simulator and not from prior publications.

11.10.2 PREDICTING THE KILN RED SPOT BY USING TEMPERATURE ANOMALY

Basically, the solution is an outcome of solving the problem of multiple refractory failures in the kiln. The approach is based on continuous monitoring of the kiln shell temperature which is a key parameter indicating the health of the kiln shell. The primary objective is to alert the system directly or the client for corrective action before the failure occurs. A pictorial display of hot spot in a kiln is shown in Figure 11.31.

Broadly speaking, the impacts of refractory failures can be illustrated as follows:

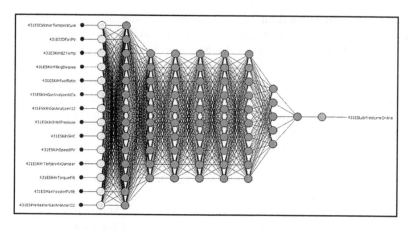

FIGURE 11.30   Neural Networks created by a simulator for the cement process.

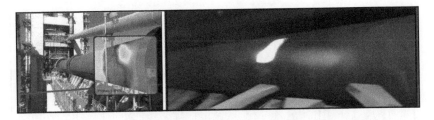

FIGURE 11.31   Hot spot in a kiln.

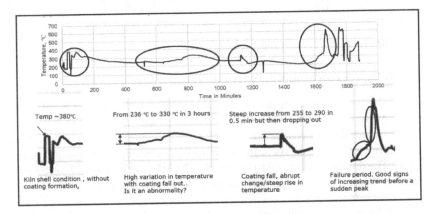

FIGURE 11.32   Typical temperature profile of a kiln shell and its variations.

- Unexpected breakdown: 5–15 days resulting in production volumes of 30,000t to 90,000t for a 6000tpd plant
- Repair of kiln refractory/shell: replacement, service and efforts requiring immediate attention
- Refractory thermal shock: affecting the lifetime of parts

The solution of the above problems is offered by the following approach. When the kiln shell temperature rises, an alert is issued in advance to take quick actions before the next failure happens. The challenge is that the kiln shell temperature changes cannot be a sole indication that the refractory is damaged. This is because the coating pieces in the kiln will fall and get developed again based on the process condition (Figure 11.32).

Empirical relations were generated from a model developed with the help of the data of the kiln shell temperature across the kiln, and the data included both good and failure data (Figure 11.33). It was observed that when the rate of change of difference between the maximum and minimum temperatures at a specific zone of the kiln was greater than 50 degrees in 10 min, then it could lead to hot spot (failure temperature) in the next 1 hour and that was used to set an alarm.

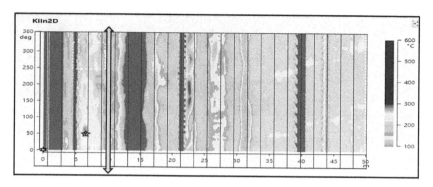

FIGURE 11.33   Kiln shell temperature map.

11.10.3 MONITORING KILN COATING STABILITY INDEX

Since the shell scanner system database continuously records the kiln temperature data along the full length and across the circumference of the kiln, it creates a two-dimensional (2D) array of time series data. If time is considered as the third dimension, the data set could be considered three dimensional (3D). The database was used to create 2D and 3D visualizations. For the standard solution, various other calculations such as minimum, maximum, and average temperatures are carried out. The kiln coating thickness was estimated, based on the brick type and thickness data. The Kiln Coating Stability Index is calculated as below:

- Every 10 min record of an array of temperature data: for example, at each 1 m of axial length take a reading each 10 degrees of kiln rotation. For an 80 m length kiln there would be 800 readings each 10 min.
- Calculate the variance of all points over a time period, say 1 h and 24 h.
- The variance is the Kiln Coating Stability Index.

The kiln shell scanner system has a comprehensive user interface that provides many useful ways to view and analyze the database for the kiln shell temperature, brick lining condition, coating stability, etc.

11.11   AI-ENABLED INTELLIGENT PRODUCTION MANAGEMENT SYSTEM

Achieving the production target at optimal power use and cost is often a challenging task to plant managers due to varying demand, fuel consumption, cost, unplanned maintenances, and Manual Process, etc. Though advanced process controls and MIS are quite popular in process optimization and

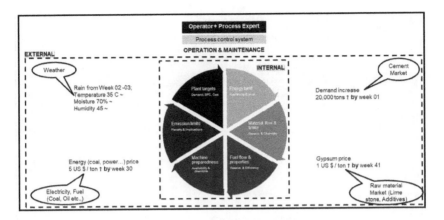

FIGURE 11.34   Challenges in production forecasting and planning.

planning, there are significant internal and external factors that still influence the power use and cost extensively which are out of vicinity of aforementioned solutions. Most of the external factors (Figure 11.34) such as demand, fuel, power, and weather data are dynamic data and reside on different web sources/Cloud. Therefore, it is important to consider the key influencers of operation and management (O&M) in dynamic prediction, planning, and optimizing their impacts on power use and cost by fixing timely targets.

Use, cost, significance, data collection, analytics, etc., of the parameters influencing the O&M scenario further elaborated in Fig. 11.35. It is important to manage such scenarios at each phase by suitably dealing with the data source, feed, tools, analytics, visualization, and Interface to external system.

An intelligent plant-management system, enabled with AI features, is capable of self-learning and correlating patterns from various data sources (weather, energy, market, prices etc.). It will also carry out dynamic planning for plant maintenance, based on history and real-time data; optimization of production schedule, based on demand and deadline; fuel and power use timelines.

Primarily, the system delivers optimized production schedule (at various time horizons with actual and forecast) to plant operator for a given set of prevailing demand constraints and conditions (such as energy, raw material, machine, market, emission and labor cost). It dynamically produces an output which will enable the operator to choose certain production lines to achieve the desired production levels while optimizing the energy consumption required for the desired production. Further, it improves forecast accuracy of plant production management.

In addition, the system delivers key "Optimized" planners/schedules which help in advanced management of the following inputs covering utilization, planning, procurement, and storage:

Key affecting scenarios at plant O&M	Impacting plant Energy		*Why it matters?	Data Collection		Analytics		Reporting
	Use	Cost						
Load schedule – Forecasting & Optimization	YES	YES	1. To achieve economic power consumption and cost 2. To achieve balance between power source and loads 3. To ensure uninterrupted power supply 4. To decide which stream of production line or department to start / stop	EMS/Production reports / ERP	Man	Spread sheets	Spread sheets / Man	Spread sheets
Power purchase planning & decision Electricity bidding		YES	1. To achieve economic power contracts and costs 2. To make accurate and timely "power estimate" for day/week/month bidding	EMS/Production reports / ERP	Man	Spread sheets	Spread sheets / Man	Spread sheets
Stock levels and flow (Raw materials, additives, coal)	YES	YES	1. To balance the livestock levels of hopper, yard and silo (min, max and buffer) w.r.t power tariff and surplus power availability. 2. To decide which stream of production lines (up or down) to operate till the stock level drops to minimum stock level w.r.t power tariff and surplus power availability.	CS / ERP	Man	Spread sheets	Spread sheets / Man	Spread sheets
Machine – efficiency, Availability times	YES	YES	1. To maximize and prioritize the use of energy efficient machines / production lines as much as possible (when the plant has multiple production lines (multiple crushers, mills & kiln, fans and drive) w.r.to their availability / maintenance schedules	CS / MMS/ERP	Man	Spread sheets	Spread sheets / Man	Spread sheets
Under production due to Emission limits and penalties	YES	YES	1. To manage the effects of underproduction into production target and power consumption as "Under production" results in reduced machine efficiency	CS	Man	Spread sheets	Spread sheets / Man	Spread sheets
Prioritize production – OPC/PPC/??		YES	1. To prioritize which product type could be produced (consumes less power) without compromising the production targets, When the power cost is fluctuating.	CS/ERP	Man	Spread sheets	Spread sheets / Man	Spread sheets
Machine idling power due to feed cuts & mill vibration	YES	YES	1. To optimize the idling power of Mill & connected circuits when unexpected feed cuts and mill vibration occurs 2. To manage the effects of "idling time" into production target and power consumption	CS	Man	Spread sheets	Spread sheets / Man	Spread sheets
Market demand		YES	1. Planning, Pricing & Distribution management	ERP	Man	Spread sheets	Spread sheets / Man	Spread sheets

FIGURE 11.35 Current state of operation and maintenance scenarios affecting energy use and production planning.

- Energy (Power and Fuel)
- Resource (Raw material and additives)
- Machine performance (efficiency) and downtime (when plant has multiple lines)
- Labor and Logistics

11.12 CONNECTED PEOPLE

People are the key players in the industry 4.0. It is not about the software, big data, or analytics that will take away people's power to make decisions or their responsibility. Industry 4.0 will enhance people's ability to make decision by providing relevant information in real time, thus enabling continuous improvement of processes. Digital and analog assistant systems will support people better than ever, taking over dangerous or difficult work. Human–machine collaboration will increase in a safe and intuitive way – but machines will continue to play a subordinate role. People's health and well-being will be safeguarded and enhanced through adaptive workplace ergonomics, digital assistance functions, and ability amplifiers.

One of the key enablers from the digital technologies is ability to work from anywhere with more insights about the plant conditions and tools. This enables access to the experts around the world and with the ability to gain knowledge of various similar conditions and issues (Figure 11.36).

11.12.1 REMOTE OPERATIONS-CONTROL ROOM

In the recent past, the concept of operating plants from a central control room has given way to dispersed control of unit processes near their sites. The new direction, however, is to adopt remote operation. Remote Operations-Control Room (RO-CR) is a secure remote connectivity solution, which will

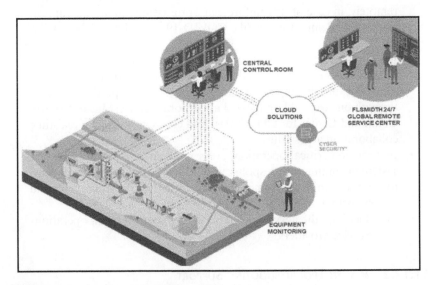

FIGURE 11.36 Connecting machines, plants, plant service team to global remote service centers.

enable the plant operator to operate securely without being physically present in the site control room.

This solution will ultimately mean that the experts can monitor, operate, and optimize the plant operation from anywhere in the world. The RO-CR option enables users to have a central command center for multiple plants or production facilities and to handle complex interdependent operations. A central team can streamline process workflow and operations across all the production facilities. The remote operation control room offers greater flexibility and functions as a valuable backup to the existing onsite control room and can mitigate situations where physical presence of specialist is not possible. Broadly speaking, the RO-CR option can have extended benefits such as the following:

- Corporate expert team can monitor/operate/optimize plant operation to better handle process upsets/emergencies/adapt to varying production, quality and emission targets and thereby improve operational efficiency and help meet various corporate goals.
- RO-CR can be situated in cities/towns, while the plant site could be at a remote location, and skilled resources are not necessarily required to be moved to site locations. One can avoid situations, where there could be problems of availability of experienced personnel who often may not be willing to relocate to remote site locations due to poor infrastructure or long commute or hard-working and living conditions near sites.
- Uninterrupted production can be achieved by operating plants remotely, in case of any unforeseen or disastrous situations (unrest, curfew, pandemic diseases like COVID-19, lock-down, labor strike, etc.).
- Backup to existing control room setup, to mitigate risk of potential hazard/accidents that could disallow local control room operation.
- Real-time monitoring of numerous locations of production sites through one central hub located at corporate office (Master Control Command Center for all facilities) and crew can engage in real-time collaboration across sites.
- In a mission critical operation, every moment is important. RO-CR system facilitates swift response and decreased downtime and operators at the plant know someone is watching their control actions and thereby help in improving their productivity.
- May facilitate the adoption of third-party control room operations from service providers.

11.12.2 Remote Troubleshooting Support

Given that the plants are mostly in remote locations, it is not always possible to have all the experts available in the plant when there is an emergency. Though in many cases the plants have strong experience base within their

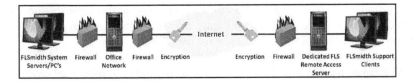

FIGURE 11.37 Simple representation of remote support architecture.

command, it may still be necessary to obtain expert support from outside or OEMs. The question, therefore, is how fast the expert advice or services can be obtained from OEMs or other sources. The solution lies in the new digital tools, which not only have the ability to connect with machines and systems but also to monitor, maintain, and upgrade them directly in a highly secured environment.

There are many remote support tools available in the market, but it is important for the plant operators to select highly secured solutions, possibly maintained by technology experts with experience in remote technology and tools. The remote control architecture is pictorially shown in Figure 11.37.

In other words, the remote support services can be deemed to belong to the family of "Augmented Field Engineer (AFE)" solutions having "Remote Assistance Tools" for rendering services (Figure 11.38). It is a collaboration tool for personnel at field to remotely connect specialists and augmented contents to troubleshoot, assess, and rapidly resolve issues in the field. The field personnel wear a helmet-mounted AR hardware with inbuilt camera, microphone, and headset. It can be voice-controlled, waterproof, and dust tight and has unmatched noise cancellation so that the field personnel always have their hands free. The tool also supports other hardware like smart phones, tablets, and computers.

The collaboration software is installed in the aforesaid hardware and works on "See What I See" technology, which allows the field personnel and a specialist to collaborate remotely with interactive audio and video, live

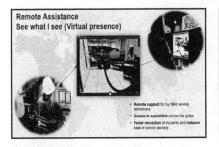
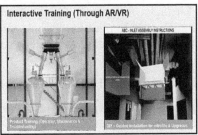

FIGURE 11.38 Augmented field engineer – remote assistance, training and do it yourself.

annotation, and sharing of media content. The interaction between them is stored in a knowledge database for future reference, and they can get data on the product from Cloud-based asset monitoring system.

The hardware of the Remote Assistance Tool (Head Mounted Tool – HMT) can be used in wet, dusty, hot, dangerous, and loud industrial environments. A fully rugged head-mounted device, it optionally snaps into safety helmets or attaches to bump caps and can be used with safety glasses or corrective eyewear.

Designed to collaborate on "things" in various field environments, teams can rapidly diagnose, inspect, and resolve issues in ultra-low bandwidth situations – even in a basement over cellular or on an offshore oil rig over satellite. It is used globally and can scale with any specific operation. Controlled and secured by a robust back-end management solution and infrastructure, it meets the needs of the most stringent IT and security environments. The wide range of capability of the technology is summarized below:

- Talk, stream video, share high-quality snapshots, content snips, add text, telestrate onscreen, and optimize A/V quality based on internet bandwidth variations.
- Remotely control the field camera adjusting zoom or lighting and taking pictures and record sessions for future record keeping or training.
- Bring in multiple participants to the call. Intuitive design with built-in tool tips and display options.
- Integrate with existing workflows using features such as Call Continuity and APIs.
- Deliver enterprise grade security with end-to-end encryption, authentication and SSO.

In addition to the above features, the remote assistance services enable the field person with inspection/maintenance checklist required for initial investigation before he could reach out to the specialist. It could be very well used to have interactive training using AR/VR technologies to improve the knowhow of the field personnel and during training.

Further development for the remote assistance solution could possibly include some application-specific capability to measure vibrations, color detection, sound meter, AR Ruler to measure the distances quickly, etc.

11.13 APPLICATION OF DRONE TECHNOLOGY

With the advent of drone technologies and the suitability of commercial drones, industries have already started using drones which are slowly transforming the industries. Drones are being used in industry applications, where there is a need to gather large amounts of qualitative data that cannot be captured by the human eye or access confined places. Recently there have been many discussions globally on the possibilities of using drones in cement industries.

Drones can be categorized as indoor flying and outdoor flying types. Presently, the outdoor drones are widely used for photogrammetry as explained below:

- Photogrammetry is a technique in which photography is used for surveying and mapping.
- Photogrammetry could be used for surveying of quarries and subsequent mapping and calculation of extracted volumes, using drone photography.
- For stock management, photogrammetry can also be of great help to the cement plants, making it easier and faster to efficiently manage stocks.
- It can also help with geotechnical monitoring such as slope stability.

The main advantages of drone-based photogrammetry are as follows:

- Faster and easier topographical surveys of quarries
- Faster and easier volume calculation of stocks
- Minimization of exposure to hazardous environments as dusty piles, rocky piles, or steep slopes
- Increased accuracy as millions of points are captured as opposed to field survey

Indoor flying drones may find many applications in a cement plant, relating to inspection and maintenance activities. Some possible uses could be in the following areas:

- *Inspecting refractory condition for Tertiary Air Duct (TAD):* Entering the TAD for accessing the condition of the refractory can be highly unsafe. Drones can inspect and provide relevant information on the internals of TAD, thereby preventing any injury or loss of life to humans while entering and investigating the TAD refractory.
- *Inspecting refractory conditions in calciner:* It takes around 3 to 4 days to build the scaffolding inside the calciner. Drones can provide details on the condition of the refractory, coating in the walls, etc. If the information obtained by the drone is enough for the investigation, it means that the drones have the potential to eliminate the construction of scaffolding for investigation and thereby saving at least 3 days of production for the plant.
- *Cooler ESP:* Condition of the different ESP plates from the passage between two chambers: the drone inspection may reveal the condition

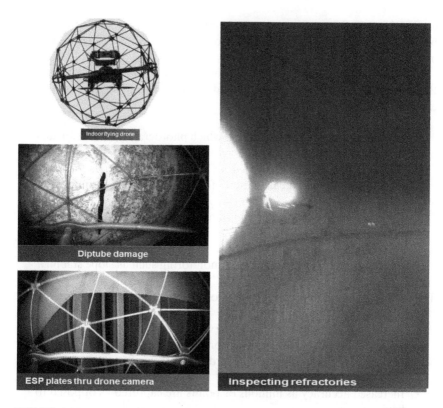

FIGURE 11.39 Inspection drone, inspecting refractories inside cyclone, dip tube damages, ESP plate condition monitoring.

of the equipment in the areas which are almost inaccessible to men, such the top of ESP plates.

- *Inspecting the condition of the refractory and the dip tube in cyclone preheater:* Dip tube at lowermost and the second lowermost stage is one of the critical equipment inside the preheater cyclone which needs to be monitored continuously. Drones can do the inspection without the necessity of scaffolding to reach the dip tubes from the inspection door of the cyclone. Entering a cyclone without assessing the coatings and the roof condition is highly unsafe and time consuming (Figure 11.39).

- *Inspecting Riser duct refractory Condition:* Dispersion plate in the distribution box of the riser duct is something that is very hard to reach, which can now be accessed by drones. Inspecting for any leakage or holes in down comer duct with drones from inspection door in the preheater fan casing. Detecting a leakage in the down comer duct can be possible only where the leakage becomes big and significant in increasing the pressure drop across the fan flow. Human inspection of down comer internals is highly impossible and unsafe. Drones can inspect the complete down comer duct within hours to the required quality.

- Inspecting the condition of the walls and the roof beams in raw mill silo: Raw mill silo roof is one of the important structures in the plant to be monitored for any kinds of deterioration so that preventive action can be done well in advance to avoid the major disaster of silo collapse. Since it is almost not reachable by the quality inspectors/ maintenance managers, the drone technology could be of immense help.
- The potential applications of drone technology are summarized in Figure 11.40.

It is pertinent to note that drones are already used in many locations to augment site maintenance engineers. The capability of drones, when combined with the other advance digital technologies like computer visioning and ML will revolutionize the way of working in cement plants, showing the way how inspection and maintenance will be performed in the plant in future.

11.14 CONNECTED BUSINESS PROCESS

As mentioned earlier, the digital transformation is not just about technology. It also brings significant change in the business process including maintenance, operation, procurement, people, and management of the plants. It will bring some fundamental changes in the way the people and things interact and reduce more operational process. For example, both the reactive and scheduled maintenance in plant environment will reduce with more predictive and prescriptive maintenance processes. There will be more automation with

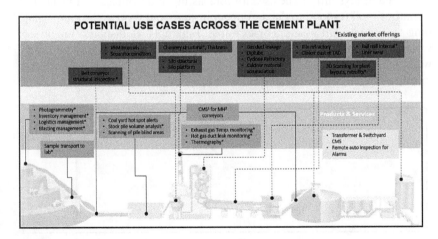

FIGURE 11.40 Potential applications of the drone technology in cement plants.

the operations of the plant where human force is more deployed in building models for what/if scenario instead of firefighting the daily operations. Given the fact that the technology will bring more transparency in operations, the procurement process would be oriented more toward the total cost of ownership and return on investments than the cost of goods. The management will focus more on partnering with suppliers in new engagement with "pay by use" or "risk and reward".

Unlike the earlier days, the end-to-end operation now not only extends from mine to packer, but much beyond. It includes distributors, end customers with personalized needs, manufacturing with digital twins, trading the electricity dynamically in exchange or stamping of the carbon footprint in cement. The latter helps support dynamic pricing of the product not only based on the delivered quality but also its carbon footprint in manufacturing.

Some of the current applications in plants include the following.

11.14.1 AUTOMATED DIAGNOSTIC SOLUTIONS

It is interesting to reflect on the fact when a plant operator last looked at the guide or the operation manual provided by the supplier to resolve an issue. The most likely answer to this question is "not that often", because these documents could not be found when and where they were needed. How do the knowledge base built over years be put to use by anyone who needs to resolve problems? This is where an automated diagnostic solution is relevant. This solution will have the built-in capability with the help of AI to index and search the knowledge base or various content across the organization and contextualize the search based on the issue requested. The system will easily capture important operational and maintenance information either through voice feed or text entry to be used for forecasting and troubleshooting later. A "Chatbot" will respond to the user query in steps and process the inputs mapping to the relevant assets. For example, Figure 11.41 displays how a user would capture the troubleshooting notes of a kiln with the help of a Chatbot walking through the user for step by step capturing and processing of relevant information and make it available for future analysis.

FIGURE 11.41 Automatic diagnostic solution snaps.

The system will continuously learn the right resolution for the issues and make it even more intelligent next time when requested in s the same plant or plants of other customers. Such will be the power of connectivity, data processing and technology!

11.14.2 Asset Performance Management

APM enhances the EAM foundation, using AI, advanced analytics, and real-time data from sensors and devices on plant floors to help propel the business forward. This enables moving the whole process from reactive and preventive maintenance to more predictive and prescriptive maintenance. Some of the key capabilities of the APM system are the following:

- Optimize asset health when you understand your equipment's needs
- Predict asset failure and avoid surprises
- Extend the useful life of assets
- Prioritize repairs and replacements
- Improve asset strategy and asset management processes

11.14.3 Remote Plant Operations

The remote plant operations will bring in a major transformation in plants with significant benefits including reduction of up to 15% in operating costs, enhanced optimization of work force, lower maintenance costs, and improvement in safety and working conditions. Most of the operation and control of the cement plant including process set values are controlled from the CCR. With advancements in digital tools, now most of the plant operation could be simulated and automated using dynamic models embracing the limestone quality, the cement process, and several other parameters including cost of energy, usage of alternative fuels, asset health conditions to proactively simulate the desired ideal condition for operation.

The fully enabled future digital cement plant could run the entire cement plant remotely including monitoring machines, predicting failures, scheduling maintenance requirements, operating the plant including end-to-end optimization of the operation. Also, the same set of infrastructures could also support running multiple plant operations from the central remote operation centers.

11.15 CONNECTED INNOVATION

The new digital capability brings unlimited possibilities with innovations. The advancement with AI, VR, and AR capability has been explored in different business areas. It is now possible to virtually simulate changes to understand various outcomes before making expensive changes with the real cement plant systems. For example, this could include decision on upgrades or process changes within the plant operations where the plant operator now can visualize the changes before implementation. Similarly, digital twin along with AR/VR technologies changes the way employees could get trained and work on the new system or plants.

11.15.1 Digital Twin

Digital twin is a significant area of interest, from manufacturing, design, service to operations. It is and will be disrupting the entire cement plant lifecycle starting from the marketing, product design, plant operation, and with managing customer demand. Digital twin is a dynamic concept growing in complexity along the product, process, and plant lifecycle of a cement plant lifecycle. Digital twin has a structure that consists of connected elements and meta-information as well as semantics. Digital twin has virtual representation of physical manufacturing elements like personnel, products, assets, and process definitions.

The digital transformation with industrial 4.0 is driving all manufacturing industries toward on-demand services with high reliability, scalability, and availability in a distributed environment. Digital twin with integration with various other digital technologies like IoT, Cloud, AR, and AI has potential to make replica of wanted characteristics, appearance, and functionality, along with processes and systems. It can represent large number of things like product, process, or the entire plant depending on the ability to handle the data sources, variation in parameters and processing capabilities with modeling.

The digital twin can bridge between physical and digital gap by providing the instant access to digital models that virtually represents physical product, process, or plant. It allows overlaying virtual objects in the real world to experience real-time tailored features and appearances. With AI, ML, and deep learning capability digital twin could be applied beyond monitoring and controlling toward predictive and personalization capabilities. It supports both predictive future states based on conditions and predicting the conditions to support future state depending on how you want to use the capability.

The digital twin could be applied from different business perspectives as indicated below:

- Simulation combined with analytics to make prediction of specific applications
- Virtual plant replication twins for design and simulation
- Evaluating production processes in virtual models
- Digitalizing assets and processes for forecasting performance (what–if analysis)
- Modeling to predict equipment failure, to optimize maintenance schedules, and to make better decisions
- Simulating the interactions between people, places, and devices.

Digital twins are being used to monitor and evaluate wear and tear to understand how this could support design and process. It involves aggregating real-time data from physical sensors and use analytics in cyberspace to manage the actuators in an incremental process through physical–digital–physical loop.

The application of the digital twin in a cement plant could be segregated into three areas: product, process, and plant level. Product digital twins focus on the design and operation for the product based on several operating conditions to increase effectiveness and predict potential failure in advance based on different scenarios. Process twins focus on process simulation solutions like cement clinker quality prediction, overall process flow, etc. Similarly plant twins focus on complete plant operation simulation.

We believe that half of the major industrial companies will be using digital twins in some form and increasing their effectiveness by about 10%. Like most of the technologies, digital twin will also have a prolonged life when practical implementations and value-added services are seen in the industry (Figure 11.42).

11.15.2 AUGMENTED REALITY/VIRTUAL REALITY

AR/VR is another breakthrough area in innovation which in combination with the IoT, AI, digital twins is reshaping the way people, process, and tools interact. As you could have seen in connected people AR is currently used in plants for augmented user manuals, guided work instructions, computer vision-led quality assurance (QA) strategy, etc. The standard operating procedures documented in manufacturing systems, the product design parameters, the IoT sensors that monitor send all the data in the form of a single digital

FIGURE 11.42   Digital twin illustration example.

thread with holistic view across products and processes, based on which the tasks of the service engineer are planned. The creation of the complete digital threads from the engineering systems to QA technicians will support augmenting the plant personnel capability to drive productivity dramatically in the coming period.

11.16 CONCLUDING OBSERVATIONS

The industry 4.0 exploits new digital technologies to connect between physical devices, people, and processes. Cement plants can also be benefited in multiple ways with these technologies. Some of the examples are as follows:

- Reduction in unplanned downtime through predictive maintenance and increased availability
- Savings in energy consumption through end-to-end demand planning
- Optimizing energy consumption by optimizing the mill and kiln control loops
- Production savings by automated cement process
- Uniform and predictable clinker and cement production
- Savings in spares parts consumption and inventory
- Reduction in overall plant maintenance costs
- Savings in plant commissioning, training and operations cost
- Reduction in engineering process
- Higher safety and security of work environment using digital tools and remote management
- Reduce manual "gut" feel approach in quality prediction by moving towards end-to-end quality prediction and quicker operational adjustments
- Better transparency and control for plant operators

Though some industries like automotive and aeronautics are advanced with their implementation for some time, cement and mining industries are relatively in the early part of the transformation. Many cement players are fully aware the benefits of 4.0, but only few have structural approach toward it to derive full advantage of the value. Like other industries, the first movers in the cement sector will not only gain substantial business value but will also build up the transformation culture which will be critical for the future success of the digital journey.

Management buy-in, cultural aversion to change within the organization, data protection, data clean-up or structuring, access to skills, lack of budget, access to operational data, fast changing technology landscape, vendor selection, traditional commercial approach are some of the key challenges encountered in adapting the digital transformation in cement plants.

In the fast changing technology landscape where everyone is going through the same learnings and pitfalls, the traditional "do-it-myself" approach will slow down the implementation without success. Hence, the key to successful implementation of digital transformation will lie in the ability of an enterprise to partner with supplier, competitor, and customers. Unlike a product or software, digital transformation is not something you can buy, install, and learn software or technology to use it; this is a lifecycle engagement with your partner with continuous learnings and improvements. So, make sure you select the right technology partner to support you in producing sustainable cement in the most cost-effective means with your product, process, and people.

ACKNOWLEDGMENTS

The authors primarily acknowledge the use of several articles and references from FLSmidth digital solutions (http://flsmidth.com) and also the following publications in preparing the chapter:

- Olfa Kanoun and H.-R. Trankler, *Sensor Technology Advances and Future Trends* –(https://www.researchgate.net/publication/3090944_Sensor_Technology_Advances_and_Future_Trends)
- Sumit Gupta, Suresh Subudhi, and Ileana Nicorici *Why Cement Producers Need to Embrace Industry 4.0* (https://www.bcg.com/publications/2018/why-cement-producers-need-embrace-industry-4)
- Digital transformation (https://www.i-scoop.eu/digital-transformation/digitization-digitalization-digital-transformation-disruption/)

12 Technological Forecasting for Commercializing Novel Low-Carbon Cement and Concrete Formulations

Sadananda Sahu
Solidia Technologies, Piscataway, NJ, USA

CONTENTS

12.1   Introduction 406
    12.1.1   Portland Cement and CO_2 Emissions 408
12.2   Supplementary Cementitious Materials...409
    12.2.1   Traditional Supplementary Cementitious Materials..........410
    12.2.2   Non-Traditional or Alternative Supplementary
               Cementitious Materials 413
12.3   Alternative Non-Clinker-Based Binders 418
    12.3.1   Alkali-Activated Binders (Geopolymers)...........................418
    12.3.2   Magnesium Carbonate-Based Cements.............................419
    12.3.3   C-S-H-Based Binder (Celitement®) 422
               12.3.3.1   Performance and Durability of Celitement........424
12.4   Alternative Clinker-Based Binders..425
    12.4.1   Belite-Rich Cement 425
    12.4.2   Sulfoaluminate Belite Cement 428
12.5   Calcium Silicate Cement and Concrete: Solidia Technologies.........430
    12.5.1   Energy Savings 430
    12.5.2   Carbon Dioxide Emissions Reductions 431
    12.5.3   Industrial Production 432
    12.5.4   Curing Process of Solidia Concrete..................................435
    12.5.5   Performance and Durability of Solidia Concrete 437
    12.5.6   Application and Performance in Pavers............................438
12.6   Direct Utilization of CO_2 in Concrete 439
    12.6.1   CarbonCure Technologies...439

DOI: 10.1201/9781003106791-12

12.6.2 Calera Cement/Blue Planet Aggregate 442
12.7 Comparative SWOT Analysis...445
12.8 Concluding Remarks...447
References...449

12.1 INTRODUCTION

There is a general consensus among scientists that global warming is directly related to increased greenhouse gases (GHGs) in the atmosphere. It is also well established that global warming is directly related to increased CO_2 concentration in the atmosphere, one of the major GHGs [1]. The ice core data for the past 800,000 years shows that there is a direct correlation between global temperature change and atmospheric CO_2 concentration [2, 3]. In the early 18th century, at the beginning of the Industrial Revolution, CO_2 concentrations in the earth's atmosphere were at approximately 280 ppm. Since then, the CO_2 concentration has steadily increased in the atmosphere, reaching up to 415 ppm in 2019. To keep the global temperature rise within 2°C of the preindustrial era (per The Paris Agreement, COP-21), significant action is needed to reduce CO_2 emissions. The industrial production sector is one of the major CO_2 emitters after transportation and electric power generation. Portland cement production is a significant contributor of CO_2 emissions in this sector.

The production of ordinary Portland cement (OPC) is a very energy-intensive process and a major contributor to GHGs emissions. The cement sector is the third-largest industrial energy consumer and the second-largest CO_2 emitter of total industrial CO_2 emissions. World cement production reached 4.1 Gt in 2019 [4] and is estimated to contribute about 8% of total anthropogenic CO_2 emission [5, 6]. During the 21st Conference of the Parties (COP-21) within the United Nations Framework Convention on Climate Change (UNFCC) on 6th December 2015, in Paris, the member nations adopted by consensus to keep the global temperature rise within 2°C of the preindustrial era by the end of the 21st century by reducing the GHG emissions (The Paris Agreement) [7]. In order to meet this 2 Degree Celsius Scenario (2DS) in accordance with guidelines set forth by the International Energy Agency (IEA), the World Business Council for Sustainable Development's (WBCSD) Cement Sustainability Initiative (CSI) group developed a Global Technology Roadmap called "Low-Carbon Transition in Cement Industry" (Figure 12.1) [8]. This roadmap has set a target to reduce the cement industry's CO_2 emissions from 2.2 Gt emitted in 2014 to 1.7 Gt by 2050, a 24% reduction from the current level. This must be accomplished despite a predicted 12%–23% growth in worldwide cement production is expected during this period. An increase in CO_2 emissions by 4% during this period is predicted in the Reference Technology Scenario (RTS).

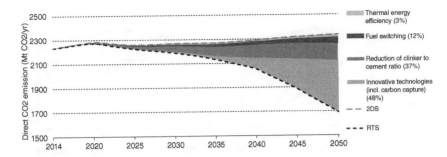

According to this roadmap, four levers have been identified to achieve the targeted goal of meeting 2DS by 2050:

1. Improving Thermal Energy Efficiency: Only a modest reduction in CO$_2$ emission of 3% can be achieved by adapting best available energy-efficient technologies, as the improvements are expected to be very slow and incremental without any major breakthroughs.
2. Switching Fuel Type: An ambitious 12% reduction in CO$_2$ emissions is targeted by switching the type of fuel. This target could be constrained by the local availability of suitable low-CO$_2$ emitting fuel.
3. Reduction in Clinker-to-Cement Ratio: A 37% total reduction in CO$_2$ emissions by 2050 is anticipated through reductions in cement clinker factor (CF). The cement and concrete industries are already practicing this approach by blending Supplementary Cementitious Materials (SCMs), such as fly ash, slag, calcined clay, silica fume, or fillers like ground limestone, with cement or by direct addition to concrete. Fly ash and slag are the most common SCMs currently in use, but their supplies are diminishing in some markets. There is a clear need for alternative sources of SCMs to augment the local needs. As most of the SCMs are industrial byproducts, their quality and supply are inconsistent or questionable. As a result, it is likely that synthetic SCMs will be better suited to fulfill this gap.
4. Innovative Technologies (including carbon capture): To meet the 2DS by 2050, reductions in CO$_2$ emissions of up to 48% must come from innovative technologies. Carbon capture is one option, but there is a need to find the proper utilization of the captured carbon.

However, in a recent publication, Energy Technology Perspectives 2020, IEA has revised the above roadmap and projected the technology pathways towards net-zero emission from 2019 till 2070 (Figure 12.2).

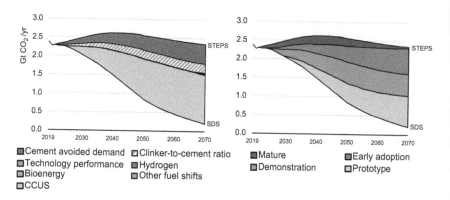

FIGURE 12.2   Global CO_2 emissions reductions in the cement sector by mitigation strategy (left) and current technology maturity category (right), 2019–70.

Notes: STEPS = Stated Policies Scenario; SDS = Sustainable Development Scenario. CCUS = carbon capture, utilization and storage. Cement avoided demand and clinker-to-cement ratio both fall within the broader category of material efficiency. The thermal energy used by chemical absorption CCUS deployment is subtracted from the CCUS contribution, and thus does not impact the efficiency contribution. See Box 2.6 in Chapter 2, Energy Technology Prospective 2020, for the definition of the maturity categories: large prototype, demonstration, early adoption and mature. (Figure 4.17, Energy Technology Prospective 2020, All rights reserved by IEA, Reproduced with permission of IEA).

Material efficiency and CCUS together play the leading role in reducing cement sector emissions in the Sustainable Development Scenario. The material efficiency category includes reduction in cement demand strategy, reduction in CF in cement. Over 60% of cumulative reductions come from technologies that are not yet commercially available.

In this chapter, current practices of reduction in Clinker-to-Cement ratio and new opportunities will be discussed. In addition, all emerging, innovative technologies for CO_2 reduction will be described.

12.1.1   PORTLAND CEMENT AND CO_2 EMISSIONS

Portland cement clinker is produced by burning a mixture of calcareous and siliceous raw materials in a rotary kiln at very high temperatures. Calcareous material, such as limestone, and siliceous materials, such as sand, clay, or similarly composed materials, are ground and processed at a sintering temperature of ~1450°C in a rotary kiln. The clinker nodules produced by the sintering process are then co-ground with ~5% gypsum to produce finished cement. Portland cement production emits CO_2 by two direct mechanisms:

- The primary source of CO_2 emission is the decomposition of $CaCO_3$ to produce $CaO_{(s)}$ and $CO_{2(g)}$ (raw material CO_2, RM-CO_2); and

- The heat necessary to achieve the high sintering temperature in the kiln is supplied through the burning of fossil fuels (Fuel derived CO_2, FD-CO_2).

CO_2-emissions are also indirectly created by the use of electricity. Mining, crushing, and grinding of raw materials to create the raw meal feed to the kiln and final grinding of cement clinker also contribute to this category. Emissions from electricity used during cement production can vary widely depending on the source of the electricity, but they are relatively small – about 10% of the total CO_2 emissions. Due to this, the CO_2 emissions from the use of electricity are omitted in the calculation here.

Portland cement clinker typically contains up to 70% CaO by weight. The calcination of limestone used to achieve this proportion of CaO releases ~540 kg of CO_2 gas per tonne of clinker [9]. The CO_2 emitted from combustion within the kiln can vary depending on the type and efficiency of the kiln as well as the fuel source. A high-efficiency kiln with a five-stage preheater and precalciner has an efficiency of 58%, which results in a CO_2 emission of ~270 kg CO_2 per tonne of clinker. An older wet-process kiln with an efficiency of 26% can have a CO_2 emission from combustion as high as 600 kg CO_2 per tonne of clinker [9]. The contribution from the calcination and combustion process of Portland clinker production yields an associated specific CO_2 emission of 810 kg to 1146 kg per tonne of clinker produced. The innovative technologies are trying to reduce the CO_2 emissions from both the fronts of RM-CO_2 and FD-CO_2.

12.2 SUPPLEMENTARY CEMENTITIOUS MATERIALS

Current estimated clinker-to-cement ratio (i.e., clinker factor) is 70%. The goal is to reduce this ratio to 60% by 2050 to meet the 2DS goals. Predominantly, SCMs are used to reduce the CF. SCMs are primarily amorphous silicate, aluminosilicate or calcium aluminosilicate powders used as partial replacements of clinker in cements or as partial replacements of OPC in concrete formulations. Most SCMs are pozzolanic and therefore react with calcium hydroxide produced during the hydration of OPC to provide supplementary strengthening of the concrete through formation of C-S-H and C-A-H or C-A-S-H phases (Eq. (12.1 to 12.3). Formation of C2ASH8 (strätlingite) is also reported in some instances.

$$3CaO.SiO_2 \left(2CaO.SiO_2\right) + 5.3\left(4.3\right)H_2O \rightarrow$$
$$CaO - SiO_2 - 4H_2O + 1.3\left(0.3\right)Ca\left(OH\right)_2 \qquad (12.1)$$

$$SiO_2 + Ca\left(OH\right)_2 + H_2O \rightarrow CaO - SiO_2 - H_2O\left(C - S - H\,gel\right) \qquad (12.2)$$

$$\left(Al_2O_3\right)_x \left(SiO_2\right)_y + Ca\left(OH\right)_2 + H_2O \rightarrow CaO - SiO_2$$
$$- H_2O\left(C - S - H\,gel\right) + CaO - Al_2O_3 - H_2O\left(C - A - H\right) \quad (12.3)$$

12.2.1 TRADITIONAL SUPPLEMENTARY CEMENTITIOUS MATERIALS

The majority of SCMs currently in use are byproducts of other industries, such as fly ash from pulverized coal-fired power plants, ground granulated blast furnace slag (GGBFS) from the iron industry, and silica fume from the silicon or ferrosilicon industry. Besides these, small quantities of natural pozzolans are also used, such as calcined clays (metakaolin), volcanic ash, pumice, etc. The bulk chemistry of these SCMs is shown in the CaO-SiO_2-Al_2O_3 ternary diagram (Figure 12.3).

In general, SCMs have a very low CO_2 footprint compared to OPC, as the CO_2 footprint associated with industrial byproducts are allocated to the primary industry. Natural pozzolans typically have low CO_2 footprints, as there are no decarbonation processes involved; most natural pozzolans only require hydroxylation during the calcination process, which is a relatively low-energy expenditure. For example, the calcination of kaolin to make metakaolin requires a thermal energy input of 0.35 GJ/t of clay compared to 3.5 GJ/t of OPC clinker [8]. Considering the low CO_2 footprint and participation in pozzolanic reactions, SCMs are very attractive materials for clinker substitution.

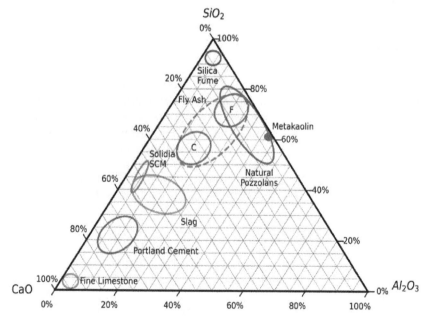

FIGURE 12.3   Composition of various SCMs in the system CaO-SiO_2-Al_2O_3.

SCMs have been used by the cement and concrete industry for decades, but recently the SCM supply chain is facing some challenges. However, according to WBCSD-CSI "Technological Roadmap for CO_2 Reduction," SCMs have been identified as one of the main strategies for reduction of the CF. The status of the important traditional SCMs is briefly reviewed below.

i. **Blast Furnace Slag**

Blast furnace slag is a byproduct of the ironmaking process from iron ore, limestone, and coke. The molten slag, which is collected on the top of molten iron, is tapped from the blast furnace and quenched with water to produce a glassy phase (i.e., granulation process). The granulated slag is further ground to use in cement or concrete. GGBFS has latent hydraulic properties and therefore can be substituted for OPC at higher levels. Substitution levels up to 70% are common; however, the supply of blast furnace slag globally is about 330 Mt/yr [10, 11]. Most of this granulated slag (>90%) is being used by the cement and concrete industry, as slag replacement in concrete provides unique durability performance and some concrete is specified for slag addition. In some markets, slag is in short supply and imported from long distances. Over the long term, the supply of slag is expected to decrease as the number of blast furnaces decreases and production processes become more modernized. In some markets, the cost of slag is more expensive than OPC. Therefore, there is very little room for slag to contribute further to CO_2 reduction within the cement industry.

ii. **Fly Ash**

Fly ash is a byproduct of pulverized coal-fired power plants. The quality of fly ash is quite variable depending on the type of coal used, the burning condition, and the environmental control measures taken to limit emissions of SO_x and NO_x. According to ASTM C 618: Standard Specification for Coal Fly Ash and Raw or Calcined Natural Pozzolan for Use in Concrete, the fly ash used in cement and concrete is divided into two categories depending on the oxide composition: Class F and Class C. When the total SiO_2 + Al_2O_3 + Fe_2O_3 ≥ 70%, the fly ash is classified as Class F; when 50% ≤ SiO_2 + Al_2O_3 + Fe_2O_3 ≤ 70%, it is classified as Class C. There are about 900 Mt/yr of fly ash available globally, but only about one third of this is used in cement and concrete. Most of the fly ash used in concrete is Class F. Class C fly ash has self-cementing properties, as it contains reactive alumina and sulfate, which can form ettringite during hydration and complicate management of setting and hardening behavior. As a result, Class C fly ash is less popular for use in concrete applications. There is a potential for increasing the use of Class C fly ash through better characterization and quality control. However, there is a concern about the long-term supply of fly ash, as burning coal to produce electricity is by far the largest source of anthropogenic CO_2 emissions, and, in some countries, there is resistance to installation of new coal-fired power plants and a push to phase out old plants. In recent years, several coal-fired power plants have been closed in favor of new gas-fired plants in the United States as the supply of shale gas is plentiful. This has led to shortages and increases in the price of fly ash in some markets. This is not unique to the United States; many other countries face similar challenges. Currently, the most popular fly ash usage is in the range of 20%–30% replacement of OPC, but there is a potential to increase this replacement level up to 50% by modifying the mix design (e.g., high-volume fly ash (HVFA) concrete) [12]. However, this push has been disincentivized considering the supply situation in several markets.

iii. Calcined Clays

Calcined clay, or burnt clay, was used by Romans as a substitute to natural pozzolans long before the invention of OPC. Recently, there is renewed interest in the utilization of this material as a cement replacement material in concrete to improve performance as well as to reduce CO_2 emissions. Many clay minerals can be thermally activated in air at 600°C to 900°C, leading to dehydroxylation, or breakdown of the crystal lattice structure, and formation of a transition phase (amorphous) with high reactivity. A typical example is the production of metakaolin ($Al_2O_3.2SiO_2$, AS_2) by calcining kaolinite. Although clay has been used extensively as a pozzolanic material, its utilization and supply as an SCM is limited, as metakaolin is the preferred material and its supply for the cement and concrete industry is limited and expensive. To expand the use of calcined clay as a CF-reducing material for OPC, more research is needed in utilizing alternative calcined clays that are less reactive than metakaolin.

iv. Limestone and Calcined Clays Cement (LC3)

A new binder system with a ternary blend of OPC-calcined clay limestone was patented by Alborg Portland Cement [13]. In this system there are synergies of combined addition. The advantage of this system is that it does not require pure metakaolin, and impure calcined clay combinations exhibit good strength. A detailed performance review of this material is provided in a 2018 publication [14]. A wide variety of products can be made by varying the proportion of the ternary components. The most optimized combination is called LC^3-50, which contains 50% clinker, 30% calcined clay, 15% limestone, and 5% gypsum. It is important to note that the hydration products of LC^3 are different from traditional SCMs and primarily contain calcium hemicarboaluminate ($C_3A.C\overline{C}_{0.5}.12H$), calcium monocarboaluminate ($C_3A.C\overline{C}.11H$), and calcium aluminosilicate hydrate (C-A-S-H) [15]. Long-term durability of this material has not yet been reported. Implementation of this technology is expected to reduce the CO_2 emissions by up to 30% compared to OPC [14].

v. **Silica Fume**

Silica fume (microsilica) is a byproduct of the manufacturing process of silicon metal and ferrosilicon alloys in an electric arc furnace at temperatures up to 2000°C. Silicon metal is manufactured by heating raw materials such as quartz and carbon in a smelting furnace. Smoke released from this operation is trapped in the bag house filters. Silica fume is then extracted and sold after densification. Silica fumes are amorphous in nature and are primarily composed of SiO_2 with minor amounts of impurities, such as Fe_2O_3, Al_2O_3, MgO, etc. Silica fume particles are spherical in shape with an average size of about 150 nm and a very high surface area (20 m^2/g). Silica fume is an excellent pozzolanic material due to its unique characteristics, and it improves the durability of concrete. Only a small amount of silica fume is available for use in concrete, and its supply is mostly controlled by a mere handful of companies globally. Silica fume is an expensive pozzolanic material. It is also used in small replacement levels (5%) and would not contribute significantly to CO_2 emission reduction goals.

vi. **Natural Pozzolans**

There are a variety of natural pozzolans available with reactive amorphous silica, poorly crystalline silica, or aluminosilicates, such as diatomaceous earth, volcanic ash, etc. The supply of these materials is very localized, and their quality is quite variable in composition and crystallinity. As a result, utilization of natural pozzolans is very limited.

12.2.2 Non-Traditional or Alternative Supplementary Cementitious Materials

There are a variety of industrial byproducts and agricultural wastes with some degree of pozzolanic activity, which are not utilized in concrete as they are not specified within the existing ASTM standards. In a recent RILEM State-of-the-Art Report, potential utilization of (ASCM) such as steel slag, bottom ash, sugarcane bagasse ash, "off-spec" products such as low-SiO_2 silica fume, and less-investigated byproducts such as waste glass is described [16]. To encourage utilization of ASCMs in concrete, ASTM has approved a new standard, ASTM C 1709: Standard Guide for Evaluation of Alternative Supplementary Cementitious Materials (ASCM) for Use in Concrete.

Some of the innovative synthetic SCMs developed recently will be described here.

i. **Carbonated Construction and Demolition Waste**
Recycling of construction and demolition (C&D) waste in concrete faces several challenges such as increased water demand, and subsequently, lower compressive strength. However, C&D waste is rich in calcium hydroxide and calcium silicate hydrate, which are potential calcium sources for CO_2 sequestration and form thermodynamically stable calcium carbonate [17]. It has been shown that the addition of carbonated C&D waste to new concrete yields improved mechanical performance compared to the addition of uncarbonated waste [18]. C&D waste can act as an effective carbon sink and can potentially store significant amounts of CO_2. Recently, a European consortium, FastCarb, was formed to promote carbonation of C&D waste using flue gases from cement plants [19]. The reported CO_2 storing capacity of these materials is quite variable ranging from 11 kg CO_2/t [1] to 30 kg CO_2/t [3]. A recent study shows that carbonated concrete fines have pozzolanic properties [20].

ii. **Carbonated Steel Slag**
Granulated blast-furnace slag produced during the ironmaking process is used as pozzolanic material in concrete. However, slags produced during the steelmaking process in basic-oxygen-furnace (BOF), electric-arc-furnace (EAF), and ladle-furnace-refining (LFR) are not directly used in concrete as these slags contain free lime and periclase, as well as high levels of sulfur, which can result in later-age volumetric instability in the hardened concrete. However, the presence of high levels of CaO in these slags enables the storage of CO_2 during a carbonation process [21]. These carbonated steel slags find application as aggregate in concrete or road base [22, 23]. Recently, a Canadian Company, Carbicrete, is making concrete blocks using steel slag through a carbonation process [24].

iii. **Synthetic Slag**
Recently, granulated synthetic slags were prepared in laboratory using a variety of waste materials such as municipal solid waste incinerator (MSWI) fly ash [25, 26], MSWI fly ash and LED sludge [27], MSWI fly ash/scrubber ash, glass frit [28], MSWI fly ash and scrubber ash [29], MSWI fly ash and semiconductor waste sludge [30], and C&D waste and/or shells from shellfish [31]. One of the main advantages of this approach is the reuse of waste materials which otherwise have no other applications. In general, these materials have lower emission from the decarbonation process and thus lower RM-CO_2. However, to make vitrified slag, the raw materials

need to be melted at high temperatures ($1400°C–1600°C$). During the granulation process, the molten liquid material is quenched with water, which significantly limits the ability to recover heat. A recent study shows that heat can be recovered by quenching using cold air in a packed bed of metal balls [32]. This study shows that maximum heat recovery is about 15% of the total energy consumption in the synthetic slag manufacturing process. The increase in FD-CO_2 from the increased process temperature and decreased heat recovery means that the overall reduction in CO_2 footprint is minimal.

By proper characterization and proportioning of the waste materials, these synthetic slags can be prepared with desired chemistry. The quality of these products can be maintained to the expected level by adjusting the oxide composition. The performance of these slags is comparable to commercially available GGBFS.

iv. **Solidia Supplementary Cementitious Material (Solidia SCM™)**
As a direct extension of its CO_2-cured precast concrete application, U.S.-based Solidia Technologies® developed a new approach creating a synthetic, low-CO_2 SCM that can serve as a substitute for traditional SCMs. Solidia SCM™ is created by directly reacting Solidia Cement™ with CO_2 in a wet (slurry) or semi-wet condition. This carbonation can be done at the cement plant utilizing the flue gas, creating an opportunity for direct utilization of CO_2 and permanent sequestration of carbon.

In an initial study, industrial grade CO_2 was used in experiments to carbonate Solidia Cement for the production of Solida SCM in a slurry form. As described below in Eq. 12.4 and 12.5, the main carbonation products are calcite and amorphous silica during the carbonation process of Solidia Cement. An energy dispersive spectroscopic map (EDS) of scanning electron micrograph (SEM) in the backscattered electron (BSE) mode is shown in Figure 12.4. The images show precipitation of calcite crystals on the surface of the cement particles and the formation of a silica rim around the core cement particle. In some instances, the smaller cement particles are completely carbonated, producing core silica surrounded by calcium carbonate as shown in Figure 12.4 (right image).

$$CaSiO_3 + CO_2 \xrightarrow{H_2O} CaCO_3 + SiO_2 \qquad (12.4)$$

$$Ca_3Si_2O_7 + 3CO_2 \xrightarrow{H_2O} 3CaCO_3 + 2SiO_2 \qquad (12.5)$$

The amorphous silica created during this carbonation process is pozzolanic in nature, that is, it reacts with calcium hydroxide produced during the hydration of OPC, creating additional C-S-H gel in the system and a minor amount of monocarbonate phase (Eq. 12.6 and 12.7).

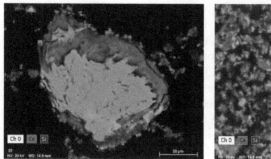

FIGURE 12.4　EDS map of SEM-BSE image showing precipitation of calcite on the surface as well as away from cement particles (left), complete carbonation of smaller particles with silica core (right).

$$SiO_2 + Ca(OH)_2 + H_2O \rightarrow CaO - SiO_2 - H_2O(C - S - H\,gel) \quad (12.6)$$

$$3CaO.Al_2O_3 + CaCO_3 + H_2O \rightarrow 3CaO.Al_2O_3.CaCO_3.11H_2O \quad (12.7)$$

The formation of additional C-S-H gel reduces the overall amount of calcium hydroxide in the cementitious matrix as well as in the interfacial transition zone (ITZ), thus forming a microstructure with a low amount of calcium hydroxide. This process helps to improve the strength and durability of the concrete over time. The formation of small amounts of monocarbonate takes place quite early in the hydration process and does not negatively affect the performance of concrete.

To evaluate the performance of Solidia SCM, the strength activity index (SAI) following ASTM C618 at 20% replacement was carried out. The SAI results are provided in Figure 12.5. SAI is an indirect measure of pozzolanic activity of SCMs. The minimum compliance requirement of SAI is 75% at 7 days or 28 days. Solidia SCM meets this SAI requirement at both ages.

Some targeted durability tests have been conducted using this material. Sometimes, to mitigate alkali-silica reaction (ASR) of aggregates in concrete, SCMs are used at various replacement levels. Class F fly ash is a popular SCM used for ASR mitigation. A study following ASTM C1567: Standard Test Method for Determining the Potential Alkali-Silica Reactivity of Combinations of Cementitious Materials and Aggregate (Accelerated Mortar-Bar Method), using Solidia SCM at 35% replacement shows that this is a very effective material to mitigate deleterious ASR-related expansion in mortar samples, as shows in Figure 12.6.

There is a marked potential for CO_2 emissions reduction with the use of Solidia SCM. The carbonation of one tonne of Solidia Cement is expected to

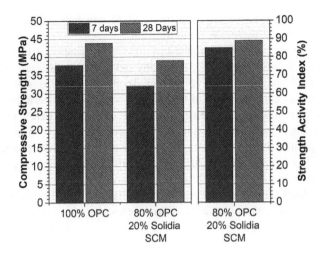

FIGURE 12.5   Compressive strength and SAI of Solidia SCM at 20% replacement.

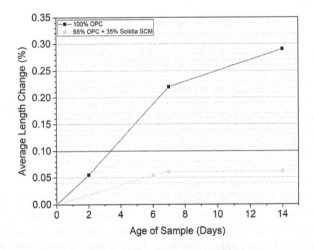

FIGURE 12.6   Length change measurements of mortar samples following ASTM C1567 of Solidia SCM and OPC.

utilize and store 240 kg of waste CO_2 gas creating 1.24 tonnes of a solid, carbonated product, or Solidia SCM. Thus, the CO_2 footprint associated with one tonne of Solidia SCM is approximately 260 kg (this includes 30% CO_2 reduction during the production of Solidia Cement plus 240 kg CO_2 utilization to produce 1.24 tonne of Solidia SCM), representing a reduction of over 550 kg of CO_2 for each tonne of OPC replaced in hydraulic concrete. This is equivalent to a 68% reduction in the CO_2 footprint compared to OPC-based concrete. The introduction of a synthetic SCM can help mitigate ongoing supply interruptions and price fluctuations.

Solidia Technologies was launched in 2008 to commercialize low-energy, low-emissions Solidia Cement™, originally developed at Rutgers, the State University of New Jersey, and to further develop applications for carbon-utilizing Solidia Concrete™. Solidia Cement is based on an alternative cement chemistry that reacts with gaseous CO_2 instead of H_2O (see below for further details). Solidia Cement was originally produced for market in partnership with Lafarge (now LafargeHolcim). The first Solidia Concrete precast products were sold commercially in 2017. Along with Solidia SCM, the company is developing other low-CO_2 solutions for the ready-mix market, with some currently in tests for infrastructure applications. The commercial launch of Solidia SCM is anticipated for 2021, following a series of commercial application tests completed in 2020.

12.3 ALTERNATIVE NON-CLINKER-BASED BINDERS

In recent years, several approaches have been tried to create cementitious binders without a high-temperature sintered clinker like OPC.

12.3.1 ALKALI-ACTIVATED BINDERS (GEOPOLYMERS)

Alkali-activated binders have been studied extensively in recent decades due to their relatively low carbon footprint and unique properties. The first application of alkali-activated binder based on slag was reported in Ukraine during the 1970s by V. D. Glukhovsky, an early pioneer in this field [33]. Subsequently, it was demonstrated that these binders, also known as geopolymers, can be made from aluminosilicates, such as fly ash and metakaolin, by using sodium silicate activators. The concept of geopolymers was first described by Davidovits [34]. This family of mineral binders is closely related to artificial zeolites consisting of a Si-O-Al framework and can be classified into two broad groups depending on the calcium content of the starting materials. The high-calcium content raw materials are mainly granulated blast furnace slags that can be activated at room temperature to produce calcium-aluminosilicate-hydrate (C-A-S-H), and, to a lesser degree, sodium-aluminosilicate hydrate (N-A-S-H). The low-calcium-containing starting materials, such as fly ash and metakaolin, are typically activated at elevated temperatures and produce sodium or potassium-aluminosilicate hydrate (N-A-S-H/K-A-S-H) by hydrolysis and a condensation process. Common alkaline activators used for geopolymers include sodium or potassium silicate, NaOH, KOH, Na_2CO_3, K_2CO_3, Na_2SO_4, and K_2SO_4. Of these activators, sodium silicate is the preferred activating agent.

As most alkali-activated binders are produced from industrial byproducts such as fly ash and slag, the overall carbon footprint of this class of materials is low. However, the carbon footprint of the most commonly used activator, sodium silicate, is quite high. And while the total carbon footprint of this material may be somewhat lower than OPC, there are numerous technical challenges for mass application of this technology within the construction industry, such as early setting time, a high degree of drying shrinkage, etc. Additionally, since the binders used for producing concrete are industrial byproducts, there is little control over composition, PSD, surface area, etc. As a result of material inconsistency, it can be difficult to produce concrete mixes designed to achieve targeted properties and performance specifications. To complicate use of this technology further, there are supply chain challenges as fly ash and slag are consumed as supplementary cementitious material in OPC concrete. Over the years, the availability of slag and fly ash has declined. Supply of the main activator, sodium silicate, is also limited. Due to these technical and logistical challenges, commercial application of alkali-activated binders is very limited on global scale.

Zeobond Pty. Ltd., based in Australia, is now marketing a concrete based on the fly ash activation technology with limited success.

12.3.2 Magnesium Carbonate-Based Cements

The concept of magnesium oxide-based cements is not new. For example, Sorel cements, invented in 1867 [35], are based on the hydration of magnesia (MgO) in highly concentrated solutions of either magnesium chloride or magnesium sulfate to precipitate hydrated magnesium oxychloride ($3MgO.MgCl_2.11H_2O$) or magnesium oxysulfate ($3MgO.MgSO_4.11H_2O$), respectively [36]. There even exists a British Standard (BS 776:1963) defining the properties of these materials. While magnesium oxide-based cements harden and gain strength fairly quickly, the formed compounds are susceptible to attack by water due to their appreciable water solubilities. These compounds have a strong tendency to lose $MgCl_2$ or $MgSO_4$ to the solution by leaching, which leaves brucite ($Mg(OH)_2$) as the main insoluble component, because it is relatively insoluble under neutral or basic conditions. As a result, these cements have been limited to use in flooring and other specialty applications.

Another important class of MgO cements is based on magnesium phosphates [37]. Magnesium phosphate binders are created by reacting dead-burned magnesium oxide with potassium phosphate or ammonium phosphate. The main reaction product is struvite ($MgNH_4PO_4.6H_2O$) or its potassium analog. Magnesium phosphate cement gains strength rapidly as it undergoes an exothermic acid-base reaction. Magnesium cement has a better water resistance than Sorel cements as struvite is fairly insoluble in water. However, this cement relies on MgO (periclase) obtained from the calcination of magnesite ($MgCO_3$) as the primary ingredient. Due to the high cost and limited supply of MgO, this cement has been limited to repair applications. While the CO_2 emissions from calcination of magnesite are similar to the calcination of limestone for OPC, the lower processing temperature of MgO-based cements allows the potential for a net reduction in FD-CO_2 emissions. While considering the CO_2 footprint associated with making phosphoric acid and its salts, there may not be any overall reduction in CO_2 emissions. A notable innovative development in this category is discussed below.

Novacem™: Researchers at Imperial College, London, developed a newer concept for extracting MgO from magnesium silicate minerals, such as olivine or serpentine, to produce new types of MgO-based hydraulic binders. These binders form magnesium oxycarbonate hydrates, such as, hydromagnesite ($MgO.4MgCO_3.5H_2O$), as important constituents [38, 39]. Olivine and serpentine are available in excess of 10,000 Gt and are potentially extractable from surface mining. The main advantage of using these minerals as raw materials is that they do not contain chemically bound CO_2 and, therefore, do not contribute to CO_2 emissions from calcination. A startup company, Novacem Ltd., was founded in London in 2008 to commercialize this technology. Novacem aimed to develop cementitious binders with performance and cost similar to that of Portland cement but with a reduced carbon footprint.

The Novacem™ production process, yielding magnesium hydroxy-carbonate cement, involves three distinct stages. During the first stage, magnesium silicates are carbonated under elevated levels of temperature and pressure (i.e., 170°C/15 MPa) to produce magnesium carbonate and SiO_2. In the second stage, magnesium carbonate is heated at low temperatures ($\sim$700°C) to produce MgO, with the CO_2 generated being recycled for use in the first stage. During the final stage, some of the MgO formed is used to produce hydrated magnesium carbonates using either the CO_2 contained in the flue gases or external CO_2. The schematic showing the production process is shown in Figure 12.7 [40].

The high-pressure, water-catalyzed reaction between supercritical CO_2 and finely-ground magnesium silicate rocks primarily yields magnesite and amorphous silica products (Eq. 12.8a and 12.8b) [41].

$$\left(\text{olivine}\right)Mg_2SiO_4 + 2CO_2 \rightarrow 2MgCO_3 + SiO_2 \qquad (12.8a)$$

$$\left(\text{serpentine}\right)Mg_3Si_2O_5\left(OH\right)_4 + 3CO_2 \rightarrow 3MgCO_3 + 2SiO_2 + 2H_2O \quad (12.8b)$$

Much of the prior work on such high-pressure processes was conducteds in the context of development for processes which would enable permanent sequestration of CO_2 through carbonation of basic magnesium silicate rocks [42]. In these processes, separation of the magnesite and silica was not necessary. However, the initial Novacem approach required separation of the two products as the second step relied on low-temperature ($\approx$700°C) calcination of magnesite to yield a reactive periclase (MgO) (Eq. 12.9).

$$MgCO_3 \rightarrow MgO + CO_2 \uparrow \qquad (12.9)$$

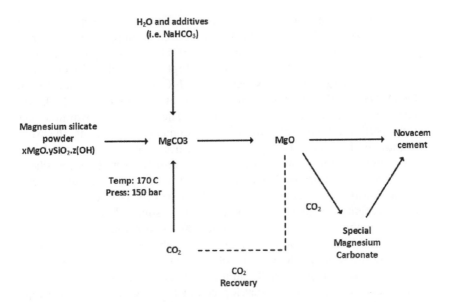

FIGURE 12.7   Process flow of Novacem production.

Any residual silica remaining in the products may be problematic to the process as it could potentially recombine with the MgO during calcination to form a magnesium silicate, thus defeating the purpose of the reaction cycle. As a result, a reasonable degree of separation was thought to be needed between steps illustrated in Eqs. 12.8a and 12.8b, and subsequently in Eq. 12.9. This was to be achieved by running reaction 12.8 at sufficient dilution to permit all of the $MgCO_3$ to dissolve in the liquid phase as "magnesium bicarbonate" (Eq. 12.10).

$$MgCO_3 + CO_2 + H_2O \rightarrow Mg_{2+}(aq) + 2HCO_3^-(aq) \qquad (12.10)$$

Under these conditions, SiO_2 was to be separated by filtration or sedimentation, and solid $MgCO_3$ was to be recovered by reducing the CO_2 partial pressure, thus reversing the reaction (Eq. 12.10). However, a subsequent, published Novacem patent application suggests the separation of silica prior to calcining the magnesite was not required [43]. The presence of some reactive silica in the cement may, in fact, be desirable. Regarding the composition and hydration mechanism of the cement itself, there is still relatively little information available in the public domain. This can be attributed in part to the fact that Novacem itself went out of business in 2012. Novacem's approach for extracting MgO closely resembles to the work done at the U.S. National Energy Technology Laboratory (NETL) [44].

The use of magnesium silicates as a raw material combined with a low-temperature production process to produce hydrated magnesium carbonates could aid in the reduction of CO_2 emissions. There are several potentially suitable hydrated carbonates that can be produced for application in hydraulic binders, including artinite ($MgCO_3 \cdot Mg(OH)_2 \cdot 3H_2O$), hydro-magnesite ($4MgCO_3 \cdot Mg(OH)_2 \cdot 4H_2O$), dypingite ($4MgCO_3 \cdot Mg(OH)_2 \cdot 5H_2O$), barringtonite ($MgCO_3 \cdot 2H_2O$), nesquehonite ($MgCO_3 \cdot 3H_2O$), and lansfordite ($MgCO_3 \cdot 5H_2O$) [45]. The final Novacem cement is a mixture of MgO, pozzolans, and hydrated magnesium carbonates. The addition of hydrated magnesium carbonates controls strength development by changing the MgO hydration mechanism and the physical properties of the hydration products.

Conceptually, the manufacturing of Novacem from magnesium silicate minerals appears to be novel. The claim of net negative CO_2 emissions needs to be demonstrated with detailed LCA. All three of the Novacem production stages have finite, positive carbon footprints. It is estimated that the CO_2 footprint of making this cement is about 300 kg per tonne of Novacem. Unfortunately, there are no performance or durability results available in the open literature to judge the viability of this material as a suitable binder for the construction industry. Finally, adoption of the Novacem chemistry would require cement companies to develop completely new raw material supply chains and manufacturing equipment.

Due to lack of funds, Novacem Company went into liquidation in 2012 and the company's technology and Intellectual Property have been sold by the Liquidator to Calix Limited.

12.3.3   C-S-H-Based Binder (Celitement®)

Another novel approach to cementitious binder production was developed at Karlsruhe Institute of Technology (KIT) and is based on hydraulic calcium hydrosilicates (hCHS). This binder is commercially known as Celitement® and is promoted by Celitement GmbH & Co. KG, based in Germany. While examining the hydration mechanisms of alite (C_3S), the main phase of Portland cement, KIT researchers found that the final hydration product C-S-H (calcium-silicate-hydrate), which is responsible for the strength development in Portland cement hydration, is formed via an "intermediate" or "precursor" phase. This phase is comparable in structure to the final hydration product of OPC, but with a slightly different chemical composition. It was designated as hydraulic calcium hydro-silicate (hCHS) to distinguish it from the typical C-S-H phase, which is the primary binder phase in OPC-based concrete.

Celitement patents are based on the concept of synthesizing and stabilizing this metastable, intermediate of C-S-H phase for use as a hydraulic binder [46]. When this metastable, intermediate phase further hydrates and transforms to the C-S-H phase typically found in OPC hydration, it imparts bonding and strength to the concrete. The crystal structure of hCHS contains both structural water molecules and hydroxyl groups bound to Ca and Si atoms

[47]. Production of this material in a rotary kiln is not feasible, since hCHS will dehydrate at high temperatures. Rather than a rotary kiln process, an autoclave process working at 200°C and saturated steam pressure (12 bar) is optimized to prepare large amounts of pure CSH (e.g., α-C2SH) phases of appropriate composition. In a second step, the material coming from the autoclave (α-C2SH, CSH), which is non-hydraulic, is transformed into the final hydraulic product hCHS by a special grinding operation, either in the presence of inert silicates such as quartz (core-shell type hCHS) or without silicates (pure hCHS). During this activation grinding process, stabilizing hydrogen bonds of α-C2SH or comparable CSH-phases originally formed in the autoclave process are destroyed ("tribochemical" surface reaction).

The product formed in this process is amorphous, reactive, and already contains structural water. When additional mixing water is added to the concrete mixture, the hydraulic reaction is induced, transforming the hCHS phase into C-S-H [48–50]. During this process a dense microstructure is created. An example of such transformation is shown in the SEM images of Figure 12.8. The Ca/Si ratio of the final C-S-H phase is lower than C-S-H formed during OPC hydration. The quality of Celitement can be tailored by adjusting the CaO to SiO_2 ratio of the raw mix during the autoclave process as well as the degree of long- and short-range structural rearrangement during the activation grinding process. Other important physical properties such as the particle size distributions (PSD) and specific surface area (SSA) can also be adjusted for better performance. By controlling the raw mix chemistry (mainly the ratio and reactivity of $Ca(OH)_2$ and SiO_2) and the activation grinding step, nearly single-phase hydraulic binders with tailor-made properties can be obtained. The product made in the Celitement pilot plant contains about 85%–90% of the hydraulic active phase (hCHS), small amounts of non-reactive phases from the raw materials (e.g., feldspars, unreacted limestone, and spurrites), and smaller levels of reactive phases (e.g., residual portlandite,

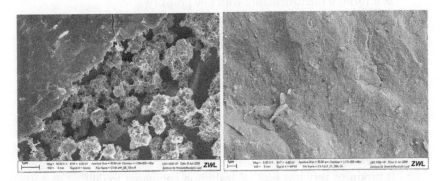

FIGURE 12.8 SEM image showing Celitement after 8h hydration (left) and after 7 days of hydration (right) creating dense microstructure.

belite, etc. <5%). According to the studies on the CO_2 balance of the Celitement process, up to 50% reduction in emissions can be achieved largely due to low molar ratio of CaO to SiO_2 (0.5 to 2.0) compared to OPC.

Celitement GmbH was founded in 2009, and the binder concept was developed from laboratory to an industrial scale together by KIT and Schwenk Zement KG as the industrial partner. In order to test and fine-tune the complete production process, Schwenk Zement KG financed the installation and operation of a small pilot plant with a production capacity of around 100 kg per day on Campus North of the KIT in Karlsruhe. The pilot plant is not only used for process and product development but also provides material samples for testing by third parties. The final decision to invest in a first industrial reference plant is expected by the end of 2021. This 50,000 t/yr installation is expected to be operational by year-end 2024. In this first industrial plant, Celitement will be produced using two starting raw materials: hydrated lime ($Ca(OH)_2$) and a quartz (SiO_2) sand slurry, which are typical raw materials used in the production of autoclaved aerated concrete (AAC). For one tonne of Celitement, approximately 600 kg of high-quality $Ca(OH)_2$ and 400 kg of pure sand are needed [51].

12.3.3.1 Performance and Durability of Celitement

The performance and durability of Celitement are quite encouraging as a low-emissions solution.

- Upon hydration, Celitement produces comparable strength to Portland cement of classes EN 42.5R and 52.5R.
- It has better sulfate resistance as no calcium-aluminate phase is present.
- It contains very small amounts of soluble alkalis, resulting in better resistance to ASR.
- The concrete made from Celitement has good freeze–thaw resistance as it forms a microstructure with low porosity.
- The corrosion resistance of this material on steel reinforcement in concrete is not reported but may have some issues as the pH of this cement is expected to be lower than OPC and there is no calcium hydroxide present or if present in small amounts in the system for stabilization of the passivation layer on steel.
- The starting materials for this cement are relatively pure and the produced cement is relatively white (L>90%).
- However, the relatively pure raw materials will certainly increase the cost of Celitement, as will the need for pressure vessels to create the processing environment.
- It produces very low heat of hydration compared to OPC. 120–150 [J/g] the average values determined by isothermal calorimetry are even well below the threshold of 220 [J/g] for Very Low Heat (VLH) cements such as CEM/III-B.

12.4 ALTERNATIVE CLINKER-BASED BINDERS

In recent years, several clinker-based binders have been developed containing lower lime at lower clinkerization temperatures compared to OPC clinker.

12.4.1 BELITE-RICH CEMENT

OPC is primarily composed of calcium silicate phases. A typical Type I OPC contains the following phases: alite (C_3S) 60%–65%, belite (C_2S) 15%–20%, calcium aluminate (C_3A) 8%–12%, calcium aluminoferrite (C_4AF) 8%–12%, periclase (MgO) 1%–3% and gypsum 4%–6%. The early-age strength of OPC is primarily derived from the hydration of alite (C_3S). Belite hydrates slowly and contributes to the later-age strength development. The hydration products of both alite and belite are calcium silicate hydrate (C-S-H) and calcium hydroxide (CH) and represented by the following Eqs. 12.11 and 12.12.

$$C_3S + 5.3H \rightarrow C_{1.7} - S - H_4 + 1.3CH \qquad (12.11)$$

$$C_2S + 4.3H \rightarrow C_{1.7} - S - H_4 + 0.3CH \qquad (12.12)$$

As the main hydration product (C-S-H) of belite is similar to that of alite, attempts have been made to produce high-belite-containing cement. Since belite forms at lower burning temperatures and requires less limestone as raw material, the potential for CO_2 savings could be significant.

Reactive belite-rich Portland cement (RBPC) is an alternative cement chemistry that has been introduced and accepted by the industry in several geographies around the world [52]. RBPC is sometimes referred to as high-belite cement (HBC). Unlike in OPC clinker, the predominant phase in RBPC is C_2S and is typically present in excess of 40% by mass. The chemistry and minerology of RBPC reduces CO_2 emissions during clinker production due to lower lime content and also offers unique material properties that can be beneficial in certain applications.

In order to produce RBPC clinker, the CaO content of the raw mix is reduced. Therefore, the raw mix will have a lower lime saturation factor (LSF). The LSF is the ratio between CaO and the other primary constituents of the cement, and is shown in Eq. 12.13.

$$LSF = \frac{C}{2.8S + 1.2A + 0.65F} \qquad (12.13)$$

LSF is related to the ratio of alite and belite in the clinker [53]. In OPC, the LSF is typically between 0.92 and 0.98, while the LSF in a RBPC may range from 0.78 to 0.83. RBPC can be prepared with an LSF of 0.75 without any alite. RBPC has been produced in the laboratory with an LSF of 0.67 [54]. The reduction in CaO content in the clinker (hence, the raw mix) leads to a

TABLE 12.1

Enthalpy of Formation for OPC Phases, Calculated as $CaCO_3$ + SiO_2 for C_3S and C_2S, $CaCO_3$ + Al_2O_3 for C_3A, $CaCO_3$ + Al_2O_3 + Fe_2O_3 for C_4AF, and $CaCO_3$ for Free Lime

	Proportion (OPC, %)	Proportion (RBPC, %)	ΔH (GJ/t)	RM-CO_2 (kg/t)	FD-CO_2 (kg/t)
C_3S	65.3	28.2	1.85	578	282
C_2S	15.6	51.8	1.34	511	204
C_3A	12	10.8	1.95	489	298
C_4AF	6.1	9.2	1.36	362	208
Free lime	1	1	3.18	785	486
OPC	100		1.76	546	271
RBPC		100	1.55	507	237

RM-CO_2 is the CO_2 Emission from Calcination Calculated from the Proportion of CaO in Each Phase. Fuel Calorific Value Is Taken at 27.9 MJ/kg, an Emission Factor for 2.47 kg CO_2/kg Coal and a Kiln Thermal Efficiency of 58% as Reported in [9]

lower level of emissions from calcination. The standard enthalpy of formation of the primary OPC phases can be calculated from the reactants, including the calcination of limestone [55, 9]. The formation of C_3S in the cement kiln requires significantly more energy input than C_2S. Reducing the quantity of C_3S in the clinker reduces the energy required for clinkering and further reduces the CO_2 footprint of RBPC. The heat of formation for each phase as well as for a theoretical OPC clinker and RBPC clinker are shown in Table 12.1.

The CO_2 footprint associated with RBPC clinker decreases with LSF. The further the LSF of the raw mix can be reduced, the greater is the CO_2 footprint reduction. This relationship is shown in Figure 12.9.

Five polymorphs of C_2S commonly occur in cement clinker. Descriptions of these polymorphs are shown in Table 12.2. In order to attain the desired performance in a RBPC cement, it is sometimes necessary to introduce methods to stabilize the high-reactivity β and α'_L phases. C_2S phases have a high capacity to form solid solutions with other species [56], which affects their stability and hydraulic reactivity. Rapid cooling and control of certain additives can improve the stabilization and yield of β and α'_L phases. It has also been found that that control of K_2O, Na_2O and SO_3 in β-C_2S can improve its hydraulic reactivity [54].

The high proportion of C_2S phases and the absence of appreciable C_3S in RBPC results in slower initial rates of hydration and strength development. However, the 28-day strength of RBPC is able to meet or exceed the performance of OPC concretes due to the ongoing C_2S hydration process. When required, early strength in RBPC concretes can be improved by finer grinding and reduced W/C ratio through the use of superplasticizers [57].

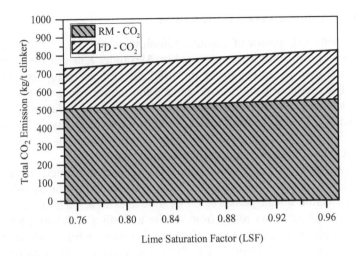

FIGURE 12.9 Total CO_2 emissions of RBPC clinker as a function of LSF. Composition calculated according to the Bogue formula with an Al_2O_3 content of 5.8% by mass and a Fe_2O_3 content of 2.0% by mass.

TABLE 12.2
Various Polymorphs C_2S and Their Characteristics

Phase	Family	Temperature (°C)	Density (g/cc)
A	Hexagonal	1425	2.94
α'_H	Orthorhombic	1160	3.11
α'_L	Orthorhombic	630–680	3.14
B	Monoclinic	<500	3.20
Γ	Orthorhombic	780–860	2.94

The heat of hydration associated with C_2S is nearly half that of C_3S, and, as a result, the evolution of heat in RBPC concretes is much lower than in OPC concretes [55]. In large concrete pours where thermal cracking is a concern, the low heat of hydration in RBPC is an advantage. Additionally, it has been found that the rate of strength gain in RBPC concretes is enhanced at elevated temperatures. Therefore, the heat that is generated in a large pour will contribute to better rates of strength gain. Finally, the durability of RBPC concretes is expected to be similar to OPC-based concretes, but evidence of improved resistance to sulfate attack and reduced drying shrinkage behavior has been observed [58]. The overall CO_2 savings in the production of RBPC is limited to within 10%. However, the slower initial rates of hydration and strength development limit the utilization of RBPC concrete, particularly in speedy construction applications.

12.4.2 SULFOALUMINATE BELITE CEMENT

The cementing behavior of calcium sulfoaluminate (CSA, $C_4A_3\bar{S}$, ye'elimite) has been known since it was patented by Alexander Klien in 1963 [59]. CSA cement is used as an expansive or shrinkage compensating component in OPC. The production and use of cements containing primarily the ye'elimite phase has been reported in China since 1970s [60] and in Japan. The actual volume of CSA cement produced is not known. However, due to its rapid setting, high early strength, and shrinkage compensation characteristics, this cement is used primarily in specialized applications.

As far as CO_2 emission is concerned, CSA cement is attractive as the CO_2-footprint of ye'elimite phase is much lower compared to the high-lime containing calcium silicate phases of OPC. Alite releases 0.578 g CO_2 per g of the cementing phase when made from calcite and silica; calcium CSA clinker releases only 0.216 g CO_2 per g of cementing phase when made from limestone, alumina and anhydrite. The firing temperature used to produce CSA clinker is typically 1250°C, about 200°C lower than that used for Portland cement clinker. Thus, there is a significant CO_2 savings in producing this cement from both the RM-CO_2 as well as the FD-CO_2 perspectives. Depending on the composition of the CSA cement, the overall CO_2 emissions are about 25 to 30% less compared to OPC [61]. One of the key drawbacks of CSA cement is the high raw material cost. This cement requires a significant amount of expensive bauxite (30%–40% of the raw material mass) to achieve the target alumina concentration in the raw mix. Additionally, a good source of sulfate is needed to make CSA clinker, which may also increase the cost.

As described in the previous section, belite cement is an attractive alternative to OPC for reduction of CO_2 emissions. However, belite cements suffer from low strength gain in early age, limiting their use in many common applications. CSA cements address this problem by combining the slow hydrating belite phase with a highly reactive calcium sulfoaluminate, yielding a variety of cements that can be prepared with more acceptable performances. Examples of various CSA cements providing high early strength, normal hardening, and slow hardening are shown in the ternary diagram illustrated in Figure 12.10 [62].

To tailor new cements in this ternary system, with a high amount of belite instead of ye'elimite, recent research and developments focus on sulfoaluminate belite cements (SAB), with belite being the main phase, and some amount ye'elimite [61, 63, 64]. In this process, the Al_2O_3 content of the cement can be lowered to values between 14% and 17% with some excess amount of Fe_2O_3. To overcome the need for bauxite as the source of Al_2O_3, attempts have been made to use industrial byproducts with significant Al_2O_3 as raw materials in SAB cements [65, 66]. The SAB clinkers are inter-ground with different levels of calcium sulfate in order to obtain rapid-hardening, high strength, expansive, or self-stressing cements. In recent years, LafargeHolcim has tried to produce SAB clinker by stabilizing high-temperature belite phases, such as

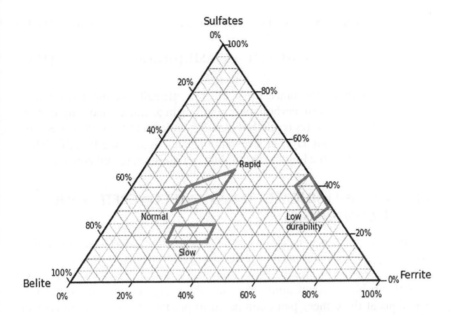

FIGURE 12.10   Ternary plot of belite (C_2S) – ferrite (C_4AF) – sulfates ($C_4A_3\bar{S} + C\bar{S}$) showing various types of cement depending on the phase composition [62].

α-C_2S and α'-C_2S, for better reactivity. For stabilization purposes, a B_2O_5 dopant is particularly suitable. But other dopants, such as Na_2O and P_2O_5, can also be used [64]. Technically, this concept was implemented as part of a large-scale experiment in which 5500 t of SAB clinker (Lafarge AETHER™) with the composition 55% α'-C_2S, 25% $C_4A_3\bar{S}$ and 15% C_4AF were prepared [61].

Another approach is the technology of belite-calciumsulfoaluminate-ternesite (BCT), in which belite, CSA, and ternesite ($C_5S_2\bar{S}$) is a main phase [67, 68]. Ternesite, a sulfate spurrite, has long been considered to not be a hydraulic phase. However, recent studies show that in the presence of the reactive CSA phase, ternesite is highly reactive and produces hydraulic characteristics.

Although these cements have been around for quite some time, there are not many reports describing the long-term durability of the concrete. The hydration products of these cements are ettringite ($3C_3A\cdot3C\bar{S}\cdot32H$), strätling-ite ($C_2ASH_8$), and C-S-H, and the proportion of the hydration products will vary depending on the phase composition of the base cement, as shown in Eqs. 12.14 to 12.17.

$$C_4A_3\bar{S} + 18H \rightarrow C_3A\cdot C\bar{S}\cdot 12H + 2AH_3\,(\text{monosulfate}) \qquad (12.14)$$

$$C_4A_3\bar{S} + 2C\bar{S}H_2 + 34H \rightarrow C_3A\cdot 3C\bar{S}\cdot 32H + 2AH_3\,(\text{ettringite}) \qquad (12.15)$$

$$C_4A_3\bar{S} + C\bar{S}H_2 + 6CH + 74H \rightarrow 3C_3A \cdot 3C\bar{S} \cdot 32H \text{ (ettringite)} \quad (12.16)$$

$$C_2S + AH_3 + 5H \rightarrow C_2ASH_8 \text{ (strätlingite)} \quad (12.17)$$

For long-term performance, the stability of ettringite is critical. It is known that ettringite is prone to carbonation, resulting in degradation of strength. As the pH of these concretes is relatively lower compared to the OPC system, the stability of passivation layer of steel reinforcement is also questionable. These factors may reduce the long-term durability of steel-reinforced concrete.

12.5 CALCIUM SILICATE CEMENT AND CONCRETE: SOLIDIA TECHNOLOGIES

Solidia Cement, a new calcium silicate cement (CSC) developed by Solidia Technologies, is a low-lime, non-hydraulic cement, primarily containing wollastonite/pseudowollastonite ($CaO \cdot SiO_2$ (CS)) and rankinite ($3CaO \cdot 2SiO_2$ (C_3S_2)) capable of significantly reducing the energy consumption and CO_2 emissions at the cement plant compared to production of Portland cement. This new cement is produced in existing cement plants without any modification to the equipment or production process. The raw mix is designed in a way to contain 30% less lime (< RM-CO_2), and the clinkerization temperature is reduced by 200°C (< FD-CO_2). It offers cement manufacturers considerable savings in energy consumption and lower CO_2 emissions. Additionally, Solidia Cement cures via a reaction with gaseous CO_2 producing $CaCO_3$, thus offering the ability to utilize CO_2, permanently and safely sequestering carbon. Combined, the CO_2 emissions reduction at the cement plant and the CO_2 utilized during curing at the concrete plant can reduce the CO_2 footprint of cement by up to 70% compared to OPC.

Both OPC and Solidia Cement manufacturing require significant amounts of energy and emit significant quantities of CO_2. Heat energy is needed to dry the raw meal, calcine the limestone, make the oxide components react, and form the cement clinker. The electrical energy needed to crush and grind the raw materials, to operate the clinkering process, to comminute the clinker, and to transport materials throughout the process are not considered here as the electrical energy consumption for Solidia Cement and OPC are roughly the same. The CO_2 emitted from electricity is dependent upon the source, such as renewable, non-renewable or mix. To illustrate the benefits associated with the production of Solidia Cement, the differences in energy consumption and CO_2 emissions are discussed below.

12.5.1 ENERGY SAVINGS

In modern cement plants, on average, the production of one tonne of OPC clinker requires heat energy totaling 3.2 GJ [69]. From a theoretical perspective, the thermal energy consumed in producing one tonne of OPC clinker is about 1.8 GJ [55]. The difference between the actual and theoretical heat requirements is due to heat retained in clinker, heat loss from kiln dust and exit gases, and heat loss from radiation. The pyro-processing step that consumes the most heat energy is the endothermic decomposition of calcium carbonate (calcination).

The total lime content of Solidia Cement clinker is in the range of 45–50 wt.%, representing approximately a 30% reduction from that required for OPC. This reduction in lime content translates directly to a 30% reduction in the major component of the theoretical enthalpy, i.e., the calcination step. Solidia Cement as well as OPC raw meals require roughly equivalent amounts of enthalpy to decompose the clay component and the exothermic reaction associated with the formation of the cement phases. Dominated by the large difference in the calcination step, the total theoretical enthalpy of formation of Solidia Cement clinker is expected to be about 1.1 GJ/t, almost 40% lower than that of OPC clinker [70]. From a practical perspective, Solidia Cement clinker is burned at temperatures approximately 200°C lower than OPC clinker, leading to significantly reduced system-wide heat losses. This is expected to translate into a reduction in fossil fuel consumption by as much as 30% [71].

12.5.2 Carbon Dioxide Emissions Reductions

The low-lime content of Solidia Cement clinker enables two separate opportunities to reduce the CO_2 emissions associated with cement production: (1) Reduction in the lime content of the cement from approximately 70% (for OPC clinker) to approximately 50% (for Solidia clinker) enables a proportionate reduction in CO_2 emission (540 kg/t for OPC clinker vs. 375 kg/t for Solidia clinker); and (2) Reduction of 200°C in the clinkerization temperature of 1450°C vs.1250°C enables CO_2 emissions reduction coming from fossil fuels (270 kg/t for OPC clinker vs. 190 kg/t for Solidia linker). The details of CO_2 savings have been summarized in Table 12.3 [71, 72]. It also translates into a 10%–20% reduction in total manufacturing cost when compared to OPC production. This cost reduction is primarily due to reduced energy costs (lower kiln temperature), reduced dependence on high-priced raw materials (such as high-grade limestone, bauxite, iron ore, and gypsum), reduced raw material

TABLE 12.3
CO_2 Emissions during the Production of OPC and Solidia Clinker

CO_2 Emissions	Per Tonne of OPC Clinker	Per Tonne of Solidia Clinker
Limestone decomposition (RM-CO_2)	540 kg	375 kg
Fossil fuel combustion (FD-CO_2)	270 kg	190 kg
Total CO_2 emissions	810 kg	565 kg

Note: The CO_2 associated with the electrical energy usage in the cement making process is not considered.

consumption (reduced loss-on-ignition) when compared to OPC production, and increased throughput.

12.5.3 INDUSTRIAL PRODUCTION

Industrial-scale production of Solidia Cement was carried out on two separate occasions at a North American and one time in a European plant of the LafargeHolcim group [73, 74]. These campaigns sought to prove the production feasibility in a modern industrial kiln and provide cement for market introduction and sale. Approximately 5000 tonnes of Solidia clinker were produced in each trial. The raw mix was adapted to meet the chemical specifications, in addition to the target wollastonite (CS) and rankinite (C_3S_2) clinker phases of Solidia Cement. During the production process, CO_2 emissions and energy consumption (specific heat consumption) were tracked in order to assess the relevance of the theoretical numbers indicated above. In order to adequately compare the production of OPC and Solidia clinker, stable production periods were considered for each clinker type, not only in the same plant, but also in the same kiln. The measurements of the first of the three campaigns are highlighted in Table 12.4 confirm the predicted energy and CO_2 savings.

In terms of energy, a 20% savings was measured for the specific heat consumption (SHC). This SHC savings is slightly lower than expected because the production rate of Solidia Cement clinker in the kiln was not fully optimized. Room for considerable improvement in Solidia clinker production remains. Measurements at the stack of the plant confirmed that conversion from OPC production to Solidia Cement production resulted in CO_2 emissions savings of over 30%. In conclusion, measured reductions in the SHC and CO_2 emissions during the industrial Solidia clinker production campaigns matched with predictions. Further improvements of these parameters are expected as clinker production is further optimized. A picture of Solidia Cement clinker is shown in Figure 12.11. The clinker nodules are similar to OPC clinker, but somewhat lighter in color.

The average elemental composition of the clinker as measured by XRF and expressed as oxides, is shown in Table 12.5. Note the CaO/SiO_2 of the cement is about 1.

TABLE 12.4

Industrial Solidia Cement Clinker Production Measurements and Comparison to OPC Clinker during Steady-State Production Period

Heat Consumption and CO_2 Emissions	OPC Clinker	Solidia Clinker
Specific heat consumption (SHC) (GJ/t clinker)	3.89	3.16
Stack CO_2 (%)	24.4	14.2
CO_2 emissions (Nm³/t clinker)	474	334

FIGURE 12.11   Solidia Cement clinker nodules from the first industrial production.

TABLE 12.5
The Average Composition of CSC

Oxides	CaO	SiO$_2$	Al$_2$O$_3$	Fe$_2$O$_3$	MgO	Na$_2$O	K$_2$O	SO$_3$	LOI	Trace
wt.%	42.76	43.20	6.00	2.47	2.03	0.14	1.00	1.06	0.27	0.23

The average phase composition of the clinker as determined by quantitative X-ray diffraction with Reitveld refinement is shown in Table 12.6.

SEM examination in backscattered mode to determine the spatial distribution of various phases in the clinker reveals that pseudowollastonite and rankinite cohabits; it was very difficult to separate them from the backscattered

TABLE 12.6
The Average Phase Composition of Solidia Cement as Measured by X-Ray Diffraction

Component	Formula	Concentration, Wt.%
Pseudowollastonite	CaSiO$_3$	22.3
Wollastonite	CaSiO$_3$	0.2
Rankinite	Ca$_3$Si$_2$O$_7$	18.1
Belite	Ca$_2$SiO$_4$	1.3
Amorphous		22.2
Melilite	(Ca,Na,K)$_2$(Al,Mg,Fe^{2+})[(Al,Si)SiO$_7$]	30.5
Bredigite	Ca$_7$Mg(SiO$_4$)$_4$	0.3
Silica	SiO$_2$	5.0
Lime	CaO	0.1

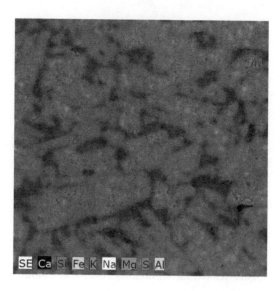

FIGURE 12.12   SEM – EDS map on a polished clinker showing distribution of elements as indicated in the insets, calcium silicate phases (light gray) and melilite phase (dark gray).

intensity of the image. SEM-EDS map shows subtle differences in calcium content between these phases (Figure 12.12). The melilite phase occurs as interstitial phase in between the calcium silicate phases.

A ternary plot of CaO-SiO_2-Al_2O_3 showing the relative composition of Solidia Cements in two different campaigns of North America with respect to the OPC compositions is shown in Figure 12.13. In the second trial, the liquid and CaO content of the mix was adjusted slightly, which is why the chemistry falls slightly outside the area of the first trial.

The clinker produced in these industrial trials was ground using a conventional ball mill with separator as well as a vertical roller mill to a Blaine fineness of 5000 cm^2/g (PSD, d_{50} = 12μm). Gypsum was not added during the grinding process. Gypsum addition for controlling the setting behavior of this cement is not needed as this clinker does not contain reactive aluminate phase (C_3A). Therefore, cooling of cement at the mill outlet is not necessary to prevent dehydration of gypsum.

Solidia Cement is generally lighter in color compared to OPC with L-value ranging 78 to 89 (for reference, the L-value of white cement is greater than 85). The naturally lighter color of Solidia Cement helps to improve the pigment efficiency in architectural concrete. Cost savings from eliminating the need for a white cement and/or a lower pigment dosage requirement have been realized in some applications.

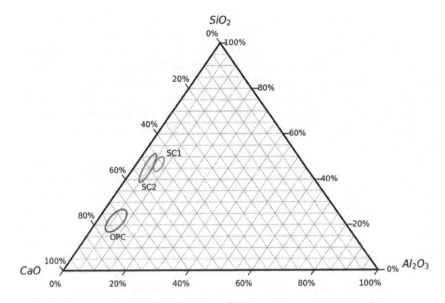

FIGURE 12.13 Ternary diagram of CaO-SiO_2-Al_2O_3 showing composition of Solidia Cements with respect to OPC.

12.5.4 Curing Process of Solidia Concrete

The low-lime, CS and C_3S_2 phases of Solidia Cement do not hydrate when exposed to water during the concrete mixing and forming processes. Cast Solidia Cement-based concrete elements will not cure unless they are simultaneously exposed to water and gaseous CO_2. Solidia Cement-based concrete curing is a mildly exothermic reaction in which the low-lime calcium silicates in the Solidia Cement react with CO_2 in the presence of water to produce calcite ($CaCO_3$) and silica (SiO_2) as follows (Eqs. 12.18 and 12.19) [75]:

$$CaSiO_3 + CO_2 \xrightarrow{H_2O} CaCO_3 + SiO_2 \qquad (12.18)$$

$$Ca_3Si_2O_7 + 3CO_2 \xrightarrow{H_2O} 3CaCO_3 + 2SiO_2 \qquad (12.19)$$

The above reaction process requires a CO_2-rich atmosphere. However, the process can be conducted at ambient gas pressures and at moderate temperatures (~60°C). These parameters are well within the capabilities of most precast concrete manufacturers. Unlike the hydration reaction in OPC-based concrete, the carbonation reaction in Solidia Cement-based concrete is a relatively rapid process. For example, Solidia Cement-based pavers can be cured to full strength within 24 hours compared to the up to 28 days required for OPC-based concrete to reach full strength. Unlike OPC-based systems, concrete products hardened with CO_2-cured Solidia Cement do not consume water. In fact, up to 90% of the water used in the Solidia Cement-based concrete formulation can be recovered during the CO_2-curing process [76]. The remaining water is retained in the pores of the cured concrete. The recovered water can be recycled in future concrete mixes. This provides concrete manufacturers the opportunity to recover and reuse concrete mix water and a potential for savings.

Each tonne of Solidia Cement used in the precast concrete formulation will utilize and store up to 240 kg of CO_2. Taken together, the CO_2 emissions reduction at the cement plant coupled with the CO_2 utilized during curing at the concrete plant can reduce the CO_2 footprint associated with Solidia Cement production and use in precast concrete by up to 70% compared to OPC-based concrete, depending on the product and its formulation. The total CO_2 emissions reduction compared to OPC is depicted in Figure 12.14. Note that in this depiction, the electricity-related CO_2 emissions for both Solidia Cement and OPC are considered to be same 90 kg/tonne of cement.

A microstructural evaluation of Solidia Cement-based concrete shows the reaction products calcite ($CaCO_3$) and amorphous silica (SiO_2) as well as uncarbonated cement particles (Figure 12.15). The calcite fills the pore space within the concrete, creating a dense microstructure. As the silica is relatively insoluble in the prevailing conditions of the carbonation process, it forms as a layer at the outer surface of the reacting cement particle. In some instances, the smaller cement particles are completely carbonated leaving the silica, mimicking the original cement particle.

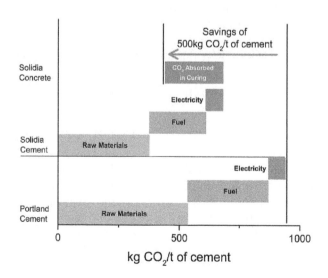

FIGURE 12.14 CO_2 reduction potential of Solidia Concrete compared to OPC.

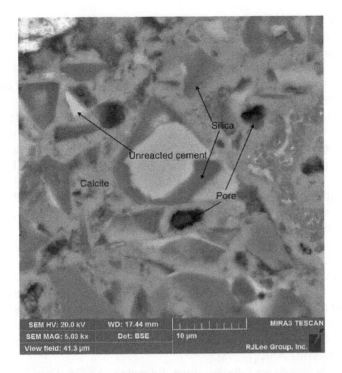

FIGURE 12.15 Microstructure of CO_2-cured Solidia Concrete (light gray area is calcite ($CaCO_3$), dark gray area is amorphous silica (SiO_2), and bright area is unreacted cement particle ($CaO \cdot SiO_2$)).

The incorporation of Artificial Intelligence (AI) through the use of electronic sensors in the curing chambers enables precise measurements of completion of curing as well as CO_2 consumed in the curing process of Solidia Concrete.

Solidia's AI-controlled, CO_2-curing process allows unprecedented monitoring inside the curing chamber and a means of rapid-fire testing that can expedite production upgrades and efficiencies, new recipes, and improved performance in concrete. The AI CO_2-curing process is remotely monitored by Solidia and allows data collection. Employing data science, machine learning and AI, the process allows manufacturers to improve production and products by trialing myriad concrete formulations quickly during pilot and commercial phases.

12.5.5 PERFORMANCE AND DURABILITY OF SOLIDIA CONCRETE

The performance and durability of Solidia Concrete in the presence of reactive aggregates (ASR) as well as various exposure conditions were independently studied by Purdue University, Turner-Fairbank Highway Research Center of the U.S. Federal Highway Administration (TFHRC- FHWA), New

York State Department of Transportation (NYSDOT), National Concrete Masonry Association (NCMA), and University of South Florida (USF). The findings are summarized below.

- For a paving mix, OPC was replaced with Solidia Cement and cured under a CO_2 atmosphere for three days provided similar or better mechanical properties compared to 28-day cured OPC concrete [77]. The compressive strength was 9,145 psi (63.05 MPa), split tensile strength was 931 psi (6.42 MPa), flexural strength was 783 psi (5.40 MPa), modulus of elasticity was 7,192,400 psi (49,589.85 MPa), and Poisson's ratio was 0.17. Pavers produced in commercial scale exhibit compressive strength consistently >10,000 psi (69 MPa) for the ASTM C 936 requirement of >8,000 psi (55 MPa).
- Air-entrained Solidia Concrete performed similar or better compared to OPC concrete of a similar mix design for freeze–thaw resistance tests following ASTM C 666, Procedure A (Relative dynamic modulus of elasticity >90% after 540 freeze–thaw cycles) [77, 78].
- Alakli-silica reaction (ASR) studies done using fused silica sand as well as natural reactive sand (Jobe sand) shows excellent resistance with minimum expansion. As there is no calcium hydroxide present in the system, and the pH is about 9.5, the environment is optimal for suppressing ASR [77, 78].
- Sulfate resistance studies done following ASTM C1012 shows Solidia Cement has excellent resistance to sulfate attack as there are no reactive aluminate (C_3A) and calcium hydroxide in the system to form expansive phases like gypsum ($CaSO_4.2H_2O$) or ettringite ($Ca_6Al_2(SO_4)_3(OH)_{12}.26H_2O$) [79].
- After full curing, the pH of Solidia Concrete is about 9.5. This pH is not expected to provide the required passivation of steel reinforcement for protection from corrosion in a chloride environment. However, Solidia Concrete has unique characteristics with low ionic strength of the pore solution and high electrical resistance. These characteristics prevent the macrocell corrosion process. In addition, Solidia has embarked upon a new approach by improving the pH of the carbonated concrete as well as introduction of corrosion prevention admixtures to the concrete mix, thereby improving the overall corrosion resistance of steel reinforcement. Corrosion rate measurement and service life modelling studies have shown the predicted service life is on par with OPC concrete [80].

12.5.6 Application and Performance in Pavers

Applications of Solidia Cement have been demonstrated on an industrial scale at commercial OPC-paver manufacturing facilities in both North America and Europe. Solidia Cement was added to the existing production lines, replacing OPC but utilizing all of the existing standard equipment for storage, mixing, pressing, and material handling. The only change to the plant process equipment was the modification of the curing chamber to allow for introduction of CO_2 for Solidia Cement carbonation as well as conditioning of the gas – the process in which Solidia Cement is cured and CO_2 is consumed (typically a 24-hour process for Solidia Cement). The mix designs for OPC pavers were slightly modified by changing admixtures and included a one-to-one replacement of OPC with Solidia Cement.

Four different batches of 4×8-inch Solidia pavers produced in the United States were tested for conformity to the requirements of ASTM C 936: Standard Specification for Solid Concrete Interlocking Paving Units. These tests were carried out independently by a third-party laboratory. Comprising compressive strength, abrasion resistance, freeze–thaw resistance, and water absorption, the test results demonstrate that Solidia Concrete pavers exhibit performance characteristics that meet or exceed the requirement of ASTM C936. CO_2-cured Solidia pavers are available for residential and commercial installations in the United States.

Applications of CO_2-cured, Solidia Concrete have been demonstrated in other products, such as blocks, roof tiles, railroad ties, hollow core slabs, architectural slabs, aerated concrete, and more.

12.6 DIRECT UTILIZATION OF CO_2 IN CONCRETE

While Solidia Concrete formulation is based on using CO_2 as a curing agent for Solidia Cement, which is a non-hydraulic cement, there are a few new technologies trying to utilize CO_2 in concrete either in fresh state, or as a mineral additive after CO_2 sequestration, or as a curing agent for carbon capture, utilization and storage (CCUS).

12.6.1 CARBONCURE TECHNOLOGIES

Another novel technology being developed as a potential component in a portfolio of strategies for emissions reduction involves the use of CO_2 as an admixture during the production of concrete. At the forefront of this technology is a Canadian company, CarbonCure Technologies (CCT), which is pursuing adoption in both the ready-mix and precast concrete markets. CCT has demonstrated that the addition of CO_2 into fresh concrete mixes can lead to permanent binding of CO_2 in the form of calcium carbonate nano-particles. CCT also reports improvements in early strength resulting from the use of CO_2 as an admixture, which provides an opportunity for concrete producers to optimize their mix designs with reduced cement content.

The CCT process involves the utilization of liquid CO_2 through installation of a dispensing system retrofitted to integrate with the existing process equipment of a concrete plant. During batching and mixing, liquid CO_2 is metered for injection into the mix where, subsequently, it converts to a mixture of carbon dioxide gas and solid carbon dioxide snow. Eventually, this snow converts into CO_2 gas. Upon incorporation of CO_2 into the fresh concrete mix, a carbonation reaction occurs in the presence of water, carbonate ions generated from CO_2 react with calcium ions supplied by the calcium silicate cement phases. This mechanism of carbonation in cementitious systems has been studied extensively for decades and is well supported in the literature [81–83]. The formation of calcium carbonate and calcium silicate hydrate gel can be described by Eqs. 12.20 and 12.21 below:

$$3CaO.SiO_2 + (3-x)CO_2 + yH_2O \rightarrow xCaO.SiO_2.yH_2O + (3-x)CaCO_3 \quad (12.20)$$

$$2CaO.SiO_2 + (2-x)CO_2 + yH_2O \rightarrow xCaO.SiO_2.yH_2O + (2-x)CaCO_3 \quad (12.21)$$

In addition, any calcium hydroxide available in the system may also undergo carbonation to form calcium carbonate as shown in Eq. 12.22.

$$Ca(OH)_2 + CO_2 \rightarrow CaCO_3 + H_2O \quad (12.22)$$

These carbonation reactions which are reported to occur during mixing and during the early stages of hydration result in the formation of nano-calcite particles [84, 85]. These particles may in turn act as seed crystals to accelerate the conventional hydration reactions which occur in Portland cements such as the formation of C-S-H gel from dicalcium and tricalcium silicate phases. Additionally, these nano-calcite particles may also promote the formation of hydration phases in pore spaces away from cement grains, thereby reducing passivating effects at cement-water interfaces and shortening induction periods. This mechanism is evidenced by measured increases in the heat of hydration for mixes wherein CO_2 is injected during the batching process [84]. The result is an accelerating effect which enhances the early strength as well as the final strength of CCT concrete mixes relative to control mixes. It is worth noting that these accelerating effects are similar to those observed when incorporating fine seed crystals into concrete mixes such as nano-$CaCO_3$ [86]. Operationally, this impact on strength development provides concrete producers with the option to either leverage improved mix properties (i.e., reduce times required for demolding or reduce dosages of accelerating admixtures) or to utilize leaner mixes for cost savings.

There are several potential benefits associated with the use of CCT in the context of CO_2 emissions reduction strategies. Firstly, liquid CO_2 injected into a concrete mix, presumably sourced as a recycled byproduct, reacts with the cementitious constituents to permanently sequester a fraction of the CO_2 in the form of calcium carbonate. While this quantity of sequestered CO_2 is relatively small, and quite difficult to directly quantify, at least 50% of the injected CO_2, which is in solid form, is incorporated into the concrete with an estimated efficiency of approximately 90%. The remaining fraction of injected CO_2 is in gaseous form and has a lower estimated efficiency of approximately 30%. In all, the reported efficiency of conversion from CO_2 to permanently sequestered calcium carbonate is about 60%, or about 0.289 kg of CO_2 sequestered per m³ of concrete. This is a very insignificant amount of CO_2 utilization in the concrete.

The second potential benefit in regard to emissions reductions results from the claimed improvements in compressive strength with use of CO_2 injection. This improved strength allows for the design of mixes with reduced binder, most often the highest contributor to the carbon footprint for a given volume of concrete. Field trials have demonstrated that at least a 5% reduction in cement content is feasible for ready-mixed concrete using CCT, which equates to ~17.6 kg of CO_2 emissions mitigated per m³ of concrete. Together with the CO_2 absorbed during injection, the reported net CO_2 reduction equates to ~17.9 kg of CO_2 per m³ of concrete. A facility with annual production levels of 50,000 m³ could potentially achieve net emissions reductions of 897 tonnes of CO_2 per year.

One other claimed benefit resulting from the improved early-strength development in CCT mixes is the potential for greater utilization of slow-reacting SCMs having lower carbon footprints in relation to conventional Portland cements. Some SCMs also have the added benefit of enhancing concrete durability through pozzolanic reactions.

Field testing of CCT technology has demonstrated that this approach is not without its own set of unique challenges. For example, one report suggests an optimum dosage, above which CO_2 injection may begin to negatively impact strength development. The authors proposed that at increasing dosages, there could be increased risk of passivation or inhibition of hydration reactions due to excessive calcium carbonate formation at cement interfaces [84]. However, at optimized dosages of CO_2 injection, concrete mix designs produced at industrial scale using CCT technology have passed important durability tests such as linear shrinkage OPS LS-435: similar to ASTM C157 with 28 days drying at 50% RH after 7 days of moist curing, air void spacing factor (CSA A23.1), rapid chloride permeability (ASTM 1202), and freeze–thaw resistance (ASTM C666) [84].

While use of liquid CO_2 as a concrete admixture shows promise as part of a multicomponent strategy for carbon emissions reduction, there are clear challenges to implementation, such as relatively narrow optimal dosage ranges and competing functionality with other well-established admixture technologies, such as dispersants and accelerators. Additionally, it is likely that more field data from broader utilization is still needed to demonstrate the feasibility and robustness of CCT technology before widespread adoption within the industry can occur.

12.6.2 CALERA CEMENT/BLUE PLANET AGGREGATE

In 2009, Calera Corporation developed a process for taking the calcium and magnesium in seawater or brines and captured carbon dioxide from flue gases to form carbonates. The idea is that CO_2-rich gases are bubbled through sea water. The calcium and magnesium from the sea water will react with CO_2 to produce calcium and magnesium carbonate, which has cementing value. Many other laboratories and scientific institutes are looking into the same idea; however, Calera seems to be at the most advanced stages of development. The Calera process, developed by B. Constantz, is described in Figure 12.16 [87, 88]. In simple, the Calera process can be represented by the chemical reaction in Eq. 12.23.

$$CO_2 + 2NaOH + CaCl_2 \rightarrow CaCO_3 + NaCl + H_2O \qquad (12.23)$$

Calera's calcium carbonate can be used in a variety of applications. The material is produced as a fine, free-flowing white powder. It can function as an SCM in traditional concrete mixes, where a portion of Portland cement can be replaced, helping reduce the overall carbon footprint of traditional concrete. The Calera calcium carbonate can also be used as the sole cement or binder system in concrete products. Calera has tested concrete made from a blended cement of OPC and Calera cement. Test results show the Calera

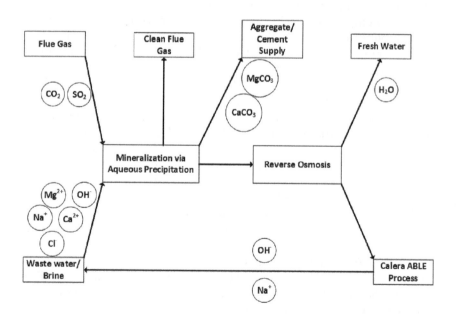

FIGURE 12.16　The Calera process of CO_2 capture from flue gases by sea water brine to precipitate $CaCO_3$ and $Mg CO_3$.

cement as it has a low strength, but when used at 20 wt.% replacement in concrete, the SAI at 7 days is 60% [88]. There was an expressed interest from the California Department of Transport and Moss Land Power plant to utilize this cement-making process, which could never be realized. To promote the carbonation process described in Eq. (12.1, it needs a base (NaOH), which is obtained by an electrolysis process. Besides requiring high electrical energy input, this method is problematic because the calcium and magnesium ions in the resulting brines occur almost entirely as chloride, so it produces an equivalent amount of a chloride-rich waste stream, which in turn presents a significant disposal problem as there are insufficient large-scale uses for this byproduct. It is proposed that this product is carbon negative. However, considering the electricity needed for electrolysis and large-volume water handling, the process may not able to produce a carbon-negative product unless the entirety of electricity used comes from renewable sources.

While interest in using Calera cement has waned, the basic chemistry described in Figure 12.15, with ammoniated water to capture CO_2, has been repurposed to synthesize aggregate for use in concrete. The challenge associated with making discrete, carbonate-based aggregate particles is far simpler than those for fabricating net-shaped concrete parts or structures. This technology is now being practiced by a new company, Blue Planet, at a pilot plant located in California [89]. While not a cement, Blue Planet aggregate offers a promising route to utilize waste CO_2 in aggregate production, thereby reducing the overall carbon footprint of concrete. For the reasons described above, coupled with Blue Planet's early-stage capabilities, it is still difficult to make accurate CO_2 calculations.

To meet the 2DS goals for the cement sector, the WBCSD-CSI has determined that, in addition to new innovative technologies, the majority of CO_2 emissions reduction must be achieved through carbon capture and storage (CCS). CCS technologies in the cement industry are in nascent stages and will require significant investments in research and equipment. Currently, commercial deployment of CCS is very limited; however, a number of innovative efforts are underway.

- Chemical absorption and separation: The flue gases are passed through a column of solvent absorber (monoethanolamine, MEA), which reacts with CO_2 and removes from the gas stream. The solvent saturated with CO_2 is then passed through a separation or regeneration process. The pure CO_2 recovered from this process is ready for storage, and the regenerated solvent is recirculated. Instead of solvent, a solid sorbent (KOH) can also be used to absorb CO_2 (RTI process). Feasibility of these processes has been demonstrated in cement plants in Texas as well as in Norway [90–94]. Some demonstrations have shown that NH_3 and water mixture is a potential sorbent.

- Oxyfuel combustion: Combustion in a high-purity oxygen atmosphere without any air produces a flue gas rich in CO_2 and water vapor. CO_2 is separated via condensation after compression and cooling. In a joint research project, four European cement producers began collaborating in late 2019 with plans to build a semi-industrial oxyfuel test facility in Germany [95].
- Ca-Looping: In this process, CaO particles react with CO_2 to from $CaCO_3$ at high temperatures in a reactor (carbonator). The $CaCO_3$ is decomposed in a second chamber (calciner) to capture pure CO_2 and CaO is recycled back to the carbonator. The CLEANKER project is developing a pre-commercial demonstration of a calcium-looping carbon capture process at a cement plant in Italy. It is expected to begin operation in 2020 [96].
- Membrane Separator: A polymeric membrane is used to increase exhaust CO_2 concentration by passing the flue gas through the membrane. CO_2 is separated through condensation after compression and cooling. The feasibility of this process has been studied through Canada's CO_2ment project [97].
- Direct Separation: This technology captures process CO_2 (RM-CO_2) emissions by applying indirect heating in the calciner. It was successfully demonstrated in pilot-scale by the LEILAC project at a cement plant in Belgium in 2019 [98].
- Electrification of Cement Production: Rather than burning fossil fuel to generate the heat needed for clinkerization, low-carbon emission electricity could potentially be used instead. Combining this technology with CO_2 capture from the decomposition of limestone (RM-CO_2) could result in substantial emissions reductions. The CemZero project in Sweden is looking into the feasibility of such a process [99].

The above-described CCS processes are all expensive, the technologies are not matured, and there is no assured permanent solution for the storage of CO_2 in underground geological formations. Therefore, there is a need for an alternative strategy for reducing CO_2 emissions by applying the principles of CCUS, which take CCS one step further and include the utilization or consumption of CO_2. One alternative strategy is direct mineralization of CO_2 at the point source. Ideally, the mineralized product should have some useful application and commercial value. There are some companies already working on this strategy.

- Solidia Technologies is working on the production of concrete products using its low-carbon-emission Solidia Cement that utilizes CO_2 during the curing process, producing $CaCO_3$, thereby permanently storing the CO_2 molecules. Solidia is also working on the production of Solidia SCM by carbonation of Solidia Cement utilizing flue gases.

- Calera Corporation/Blue Planet Ltd.'s technologies are based on direct mineralization of sea water/brine with flue gases to produce cement/aggregate.
- Carbon8 Systems and O.C.O Technologies in the UK are trying to carbonate waste materials containing calcium by flue gases to produce products with some valuable applications such as aggregates.
- Skyonic Corporation had a demonstration project in Texas, converting CO_2 from flue gas from a cement plant into commercial products such as sodium carbonate, sodium bicarbonate, hydrochloric acid, and bleach.

12.7   COMPARATIVE SWOT ANALYSIS

An analysis of strength, weakness, opportunity, and threat (SWOT) is a strategic tool used to identify an organization's strengths and weaknesses, as well as broader external opportunities and threats. The strengths and weakness are internal factors of the organization, while opportunities and threats are external factors. This tool is frequently used to address complex business decisions in order to streamline information and support the decision-making process.

It is difficult to forecast the technological feasibility and success of a particular technology, as the details of these technologies are not available in the public domain and many are still in the developmental stage. Using the SWOT approach, the feasibility of several emerging low-CO_2 binder technologies was analyzed in 2017 [100]. A similar approach has been utilized to analyze all the technologies described in this chapter. As it is challenging to quantify all relevant parameters, a feasibility factor of High (H), Medium (M), or Low (L) is used. Comparative SWOT analysis parameters are provided in Table 12.7.

In summary, this analysis of the emerging technologies demonstrates that Solidia Cement and belite-rich cement (BRC) have the highest potential for success in terms of measurable level of carbon impact, commercial adaptability (e.g., cost, type of raw materials, no need for new equipment), current commercial traction, publicly available test results, adherence to or surpassing the requirements of established industry standards, especially compared to some of the more nascent technologies. The concept of belite-rich cement is well established, and industrial production trials have taken place for decades, but BRC has not gained widespread application in construction. However, the overall reduction in CO_2 emissions is low, at approximately 10%. Industrial-scale production of Solidia Cement has been demonstrated on three separate occasions and pavers made using this cement are commercially available in the United States. The second group of technologies with moderate-level of feasibility includes Celitement, sulfoaluminate cement, sulfoaluminate belite cement, CarbonCure product, and LC3. The main constraints for Celitement, sulfoaluminate cement, sulfolauninate belite cement, and LC3

TABLE 12.7

Comparative SWOT Analysis

Criteria	Novacem	Celitement	Belite-rich Cement	Sulfoaluminate Cement/Sulfoaluminate Belite Cement	Solidia Cement	CarbonCure	Calera/blue Planet	LC3
Raw materials required	Magnesium silicate	Same as OPC	Same as OPC	Limestone, clay/sand, bauxite, gypsum	Same as OPC	OPC	Sea brine, alkali hydroxide	OPC, calcined clay, limestone
CO_2 reduction	>100% (63%*)	Up to 50%	Approx. 10%	Approx. 30%	Up to 70%	Approx. 0.6%	Approx. 50%	Approx. 30%
Raw material availability	L	H	H	M	H	H	H	H
Scientific soundness of the concept	H	H	H	H	H	H	H	H
Simplicity of the process	M	M	H	H	H	H	M	M
Stage of development	Pilot plant	Pilot plant	Industrial	Industrial	Industrial	Industrial	Pilot plant	Industrial
Potential viability	L	M	H	H	H	L	L	H
Product versatility	M	H	H	H	H	H	L	H
Up-scaling potential	L	H	H	H	H	M	L	H
Preventive barriers	Many	Some	Some	Some	Some	Some	Many	Some
Time to industrial adoption	Long	Long	Short	Short	Short	Short	Long	Short
Overall feasibility	L	M	H	M	H	M	L	M

Note: Most of the CO_2 emissions reduction numbers are taken from the publication [101].

* Estimated reduction in CO_2 emission after considering emissions related to processing.

are availability of suitable raw materials and overall costs of the system. CarbonCure technology appears to have an easy pathway to implementation but has a relatively low impact on CO_2 emissions. The third group of technologies with low-level feasibility includes Novacem and Calera Cement. These technologies need new equipment as well as new materials supply chain.

12.8 CONCLUDING REMARKS

As discussed in this chapter, there are many existing and developing technologies which will help the cement industry in meeting the 2DS goals set forth by the WBCSD-CSI roadmap, "Low-Carbon Transition in Cement Industry" as well as IEA Energy Technology Prospective 2020. While some of these approaches will have limited impact due to technical, logistical and cost constraints, some potential technologies have been identified that are expected to have a significant impact on the industry's efforts to reduce CO_2 emissions. However, major innovations are still needed in several areas to achieve the targeted goal.

- Optimization of current, Portland cement-making processes to reduce CO_2 emissions has reached its climax, and there are very few opportunities for further improvement without major breakthroughs. One potential approach is clinkerization in electrical furnaces/kilns where the electrical energy can be supplied by renewable sources.
- Use of alternative fuels to reduce CO_2 emissions has some potential but is primarily influenced by the availability in local markets.
- Reduction in CF has the potential to significantly reduce CO_2 emissions. Recent data shows that, on average, SCMs are used in cement and concrete at replacement levels of approximately 20%. There exists a potential to increase this substitution level up to 50%. However, more research and optimization are needed to reach this high level of replacement. In the meantime, supply of good quality fly ash has suffered from the shutdown of pulverized coal-fired power plants in some localities. GGBFS is not available in some markets, requiring transport from long distances. If this trend continues, SCM supplies are expected to become increasingly limited. There is a need for synthetic low-carbon SCMs with consistent quality to fill the supply-demand gap in industrial byproducts used as SCMs.
- According to the "Low-Carbon Transition in Cement Industry" roadmap as well as Energy Technology Prospective 2020, the large majority of CO_2 emissions reductions needed to achieve the industry goals will rely on innovative technologies as well as CCS. Most CCS technologies are in their infancy for cement applications and, to date, have been largely cost-prohibitive. Even if the carbon capture part of

this puzzle is solved, there are not many easy storage and utilization solutions. Moreover, the long-term consequence of the stored CO_2 in geological formations is not known. This leads to an urgency and increasing need for accelerating the innovation of new technologies.

The cement industry has been working for several decades to reduce the large energy consumption associated with cement manufacturing. During this time, the industry saw many major breakthroughs in process technologies and has managed to reduce energy consumption from 6.3 GJ/t of clinker in a wet-process kiln down to <2.93 GJ/t using a dry-process kiln with 6-stage cyclone preheater and precalciner systems. However, in recent years, there have been very few advancements in energy efficiency.

Because the basic use and requirements of cement and concrete have changed very little over the years, the cement industry has largely been resistant to major raw mix modifications. However, there were attempts made as early as the 1970s to modify the raw mix in order to produce alternative sustainable cements, such as belite-rich cement, sulfoaluminate cement, and sulfaoluminate belite cements. In recent years, after the adoption of the Kyoto Protocol in 1997, which was implemented in 2005, several startup companies appeared with new disruptive technologies aimed at CO_2 emissions reductions within the cement industry. All of these new startups have some common challenges while competing with a well-established cement industry relying on a core production technology and raw materials developed almost 200 years ago.

- All of these startups are spinoffs from university research programs with venture capital investments. Some of these technologies had challenges for scaleup or did not have the required funding to maintain operations, and therefore, were short-lived, e.g., Calera Corporation and Novacem.
- As these alternative cements are based on new chemistries, they don't have the track record of performance and durability for immediate acceptance by consumers.
- The construction industry is relatively conservative, and any new material must meet all of the regulatory standards and codes. As these new materials may not qualify within the existing standards and codes, most will not be eligible for immediate use. The development of new standards for new materials is a major hurdle in construction and can take years or almost a decade to ratify a standard in the ASTM committees. The same is true with other international standards organizations.
- The startups do not have production and supply chain infrastructures required to stand alone and be successful. The development of any greenfield production facility would require massive funding

and many years to establish. To overcome these challenges and be successful, the startups must form partnerships with existing players within the OPC industry.

REFERENCES

1. https://www.ncdc.noaa.gov/global-warming/temperature-change/. Accessed on 11 August 2020.
2. D. Lüthi, et al., High-resolution carbon dioxide concentration record 650,000–800,000 years before present. *Nature*, Vol. 453, pp. 379–382, 15 May 2008.
3. J. Jouzel, et al., Orbital and millennial Antarctic climate variability over the past 800,000 years, *Science*, Vol. 317, No. 5839, pp.793–797, 10 August 2007.
4. https://www.iea.org/reports/cement/. Accessed on 11 August 2020.
5. https://www.carbonbrief.org/qa-why-cement-emissions-matter-for-climate-change/. Accessed on 11 August 2020.
6. https://edgar.jrc.ec.europa.eu/news_docs/jrc-2016-trends-in-global-co2-emissions-2016-report-103425.pdf/. Accessed on 11 August 2020.
7. https://unfccc.int/process-and-meetings/the-paris-agreement/the-paris-agreement/. Accessed on 11 August 2020.
8. https://www.wbcsdcement.org/index.php/key-issues/climate-protection/technology-roadmap/. Accessed on 11 August 2020.
9. L. Barcelo, J. Kline, G. Walenta, E. Gartner, Cement and carbon emissions, *Mater. Struct.*, Vol. 47, pp. 1055–1065, 2014.
10. https://www.worldsteel.org/en/dam/jcr:1b916a6d-06fd-4e84-b35d-c1d911 d18df4/Fact_By-products_2018.pdf/. Accessed on 11 September 2020.
11. K. Scrivner, V. M. John, E. M. Gartner, Eco-efficient cements: potential economically viable solutions for low-CO2 cement-based materials industry, *Cem. Concr. Res.*, Vol. 114, pages 2–26 (2018).
12. P. K. Mehta, High-performance, high-volume fly ash concrete for sustainable development, *Proceedings of the International Workshop on Sustainable Development and Concrete Technology*, Pages 3–14 (2004). http://publications.iowa.gov/2941/1/SustainableConcreteWorkshop.pdf#page=14.
13. D. Herfort, J. S. Damtoft, Portland limestone calcined cement, EP2253600A1, 2010.
14. K. Scrivener, F. Martirena, S. Bishnoi, S. Maity, Calcined clay limestone cements (LC³), *Cem. Concr. Res.*, 114 (2018) 49–56.
15. M. Antoni, J. Rossen, F. Martirena, K. Scrivener, Cement substitution by a combination of metakaolin and limestone, *Cem. Concr. Res.*, 42 (2012) 1579–1589.
16. K. Sobolev et al. Chapter 7, Alternative supplementary cementitious materials N. De Belie et al. (eds.), *Properties of fresh and hardened concrete containing supplementary cementitious materials*, RILEM State-of-the-Art Reports 25 (2018). https://doi.org/10.1007/978-3-319-70606-1_7
17. S. K. Kaliyavaradhan, T-C Ling, Potential of CO_2 sequestration through construction and demolition (C & D) waste—An overview, *J. CO_2 Utiliz.* 20 (2017) 234–242. http://dx.doi.org/10.1016/j.jcou.2017.05.014
18. D. Xuan, B. Zhan, C.S. Poon, Assessment of mechanical properties of concrete incorporating carbonated recycled concrete aggregates, *Cem. Concr. Compos.* 65 (2016) 67–74.
19. https://fastcarb.fr/en/home/. Accessed 12 September 2020.

20. M. Zajac, et al., Effect of carbonated cement paste on composite cement hydration and performance, *Cem. Concr. Res.* 134 (2020) 106090. https://doi.org/10.1016/j.cemconres.2020.106090

21. W.J.J. Huijgen, G.J. Witkamp, R.N.J. Comans, Mineral CO2 sequestration by steel slag carbonation, *Environ. Sci. Technol.* 39 (24) (2005) 9676–9682.

22. Z. Ghouleha, R. I. L. Guthrieb, Y. Shao, Production of carbonate aggregates using steel slag and carbon dioxide for carbon-negative concrete, *J. CO₂ Utiliz.* 18 (2017) 125–138. http://dx.doi.org/10.1016/j.jcou.2017.01.009

23. H. Yi, et al., An overview of utilization of steel slag, *Procedia Environ. Sci.* 16 (2012) 791 – 801.

24. http://carbicrete.com/. Accessed on 12 September 2020

25. K. L. Lin, K. S. Wang, B. Y. Tzeng, C.Y. Lin, The hydration characteristics and utilization of slag obtained by the vitrification of MSWI fly ash. *Waste Manag.* 2004, 24, 199–205.

26. K. L. Lin, K. S. Wang, C. Y. Lin, C. H. Lin, The hydration properties of pastes containing municipal solid waste incinerator fly ash slag. *J. Hazard. Mater.* 2004, B109, 173–181.

27. K. L. Lin, D. F. Lin, W. J. Wang, C. C. Chang, T. C. Lee, Pozzolanic reaction of a mortar made with cement and slag vitrified from MSWI ash-mix and LED sludge. *Constr. Build. Mater.* 2014, 64, 277–287.

28. T. C. Lee, W. J. Wang, P. Y. Shih, Slag-cement mortar made with cement and slag vitrified from MSWI fly-ash/scrubber-ash and glass frit. *Constr. Build. Mater.* 2008, 22, 1914–1921.

29. T. C. Lee, M. K. Rao, Recycling municipal incinerator fly—and scrubber—ash fused slag for the substantial replacement of cement in cement—mortars. *Waste Manag.* 2009, 29, 1952–1959.

30. T. C. Lee, Recycling of municipal incinerator fly-ash slag and semiconductor waste sludge as admixtures in cement mortar. *Constr. Build. Mater.* 2009, 23, 3305–3311.

31. M. Rodriguez-Galan, et al., Synthetic slag production method based on solid waste mix vetrification for manufacturing of slag-cement, *Materials* 2019, 12, 208. doi:10.3390/ma12020208.

32. F. M. Baena-Moreno, Novel study for energy recovery from the cooling-solidification stage of synthetic slag manufacturing: estimation of the potential energy recovery, *Processes* 2020, 8, 1590. doi:10.3390/pr8121590.

33. V.D. Glukhosky, G.S. Rostovskaja, G.V. Rumyna, 1980. High strength slag-alkaline cements. In: *Communications of the 7th International Congress on the Chemistry of Cement*, vol. 3, pp. 164–168.

34. J. Davidovits, Mineral polymers and methods of making them, US 4,349,386. 1982.

35. S. Sorel, Sur un nouveau ciment magnésien (On a new magnesium cement), *Comptes Rendus Hebdomadaires des Séances de l' Académie des Sciences* 65(1867) 102–104.

36. F. M. Lea, *The chemistry of cement and concrete*, 3rd Edition, Chemical Publishing Company, (1971) pp.17.

37. A.S. Wagh, *Chemically bonded phosphate ceramics*, Elsevier, Oxford UK, 2004 pp. 283.

38. N. Vlasopoulos and C. R. Cheeseman, World Patent Application WO2009156740, (2009).

39. N. Vlasopoulos, "Low Carbon Cements: recent research and developments on alternatives to Portland cements", *Novacem Carbon Negative Cement, Society of Chemistry and Industry Conference*, (London, UK), 2010.
40. E. Gartner, D.E. Macphee, A physico-chemical basis for novel cementitious binders, *Cem. Concr. Res.* 41 (2011) 736–748.
41. E. Gartner, H. Hirao, A review of alternative approaches to the reduction of CO_2 emissions associated with the manufacture of binder phase in concrete *Cem. Concr. Res.* 78 (2015) 126–142.
42. J. Sipilä, S. Teir, R. Zevenhoven, Carbon dioxide sequestration by mineral carbonation literature review update 2005–2007, Åbo Akademi University Faculty of Technology Heat Engineering Laboratory Report VT 2008-12008.
43. N. Vlasopoulos, J.P. Bernebeu, Production of magnesium carbonate, WO 2014/ 009802 A2, 2014.
44. E. Nduagu, J. Bergerson, R. Zevenhoven, Life cycle assessment of CO_2 sequestration in magnesium silicate rock – a comparative study, *Energy Convers. Manag.* 55 (2012) 116–126.
45. R. J. Flatt, N. Roussel, and C. R. Cheeseman, Concrete: an eco-material that needs to be improved, *J. Eur. Ceram. Soc.* 32 (2012) 2787–2798.
46. G. Beuchle, et al., Single-phase hydraulic binder, methods for the production thereof and building material produced therewith, DE102007035259A, 2007; PCT/EP 2008/005785.
47. K. Garbev, et al., Preparation of a novel cementitious material from hydrothermally synthesized C–S–H phases, *J. Am. Ceram. Soc.*, 97 7. 2298–2307 (2014).
48. P. Stemmermann, U. Schweike, K. Garbev, G. Beuchle, and H. Möller, "Celitement – a sustainable prospect for the cement industry," *Cem. Int.*, 8 [5] 52–67 (2010).
49. P. Stemmermann, U. Schweike, K. Garbev, and G. Beuchle, "Celitement"-a new sustainable hydraulic binder based on calcium hydrosilicates. *13th International Congress on the Chemistry of Cement (ICCC)*, Madrid, 2011.
50. K. Garbev, G. Beuchle, U. Schweike, and P. Stemmermann, Hydration behavior of celitement: kinetics, phase composition, microstructure and mechanical properties. *13th International Congress on the Chemistry of Cement (ICCC)*, Madrid, 2011.
51. H. Möller, *Celitement-a novel cement based on hydraulic calcium hydrosilicates (hCHS)*, Celitement Internal Document, June 2020.
52. E. Gartner, T. Sui, Alternative cement clinkers, *Cem. Concr. Res.* 114 (2018) 27–39.
53. C.D. Lawrence, The production of low-energy cements, in: P.C. Hewitt (Ed.), *Lea's chemistry of cement and concrete*, 4th Edition, Arnold, London, 1998, pp. 421–470.
54. L. Kacimi, A. Simon-Masseron, S. Salem, A. Ghomari, Z. Derriche. Synthesis of belite cement clinker of high hydraulic reactivity, *Cem. Concr. Res.* 39 (2009) 559–565.
55. H. F. W. Taylor, *Cement chemistry*, 2nd Edition, Thomas Telford, ISBN 978-0727725929 (1997) pp. 59
56. A.K. Chatterjee, High belite cements—present status and future, *Cem. Concr. Res.* 26 (8) (1996) 1213–1225.
57. T. Sui, L. Fan, Z. Wen, Properties of belite-rich Portland cement and concrete in China, *J. Civ. Eng. Arch.* 4 (2015) 384–392.

58. T. Sui, L. Fan, Z. Wen, J. Wang, Z. Zhang, Study on the properties of high strength concrete using high belite cement, *J. Adv. Concr. Technol.* 2 (2) (2004) 1–6.

59. A. Klein, Calciumaluminosulfate and expansive cements containing same, US Patent No. 3, 155, 526, 1963, 4 pp.

60. Y. Wang, M. Su, The third cement series in China, *World Cem.* 25 (1994) 6–10.

61. G. Walenta, C. Comparet, New cements and innovative binder technologies BCSAF cements – recent developments, *ECRA-Seminar Barcelona, Presentation*, http://www.aether-cement.eu/press-room/publications/aether-cement-ecra-barcelonapresentation-2011-05-05.html 2011.

62. P. K. Mehta, Investigations on energy-saving cements, *World Cem. Technol.* 11: 166–177 (1980).

63. M.A.G. Aranda, A.G. De la Torre, 18 – Sulfoaluminate cement, in: F. Pacheco-Torgal, S. Jalali, J. Labrincha, V.M. John (Eds.), *Eco-efficient concrete*, Woodhead Publishing 2013, pp. 488–522.

64. E. Gartner, G. Li, High belite-containing sulfoaluminous clinker, method for the production and the use thereof for preparing hydraulic binders, US785077B2 (Dec. 14, 2010).

65. S. Sahu, J. Majling, Preparation of sulphoaluminate belite cement from fly ash, *Cem. Concr. Res.*, 24(6), 1065–1072 (1994).

66. P. Arjunan, M. R. Silsbee, and D. M. Roy, Sulfoaluminate-belite cement from low-calcium fly ash and sulfur-rich and other industrial by-products. *Cem. Concr. Res.*, 29(8), 1305–1311 (1999).

67. W. Dienemann, D. Schmitt, F. Bullerjahn, B.M. Haha, Belite calciumsulfoaluminate ternesite (BCT)-a new low-carbon clinker technology, *Cem. Int.* 11 (2013).

68. F. Bullerjahn, D. Schmitt, M. Ben Haha, Effect of raw mix design and of clinkering process on the formation and mineralogical composition of (ternesite) belite calcium sulphoaluminate ferrite clinker, *Cem. Concr. Res.* 59 (2014) 87–95.

69. N. A. Madlool, et al., A critical review on energy use and savings in the cement industries, *Renew. Sustain. Energy Rev.* 15 (2011) 4, pp. 2042–2060, ISSN 1364-0321.

70. S. Sahu, N. DeCristofaro, (2013) *Solidia cement*. Solidia Technologies. https://solidiatech.com/white-papers/. Accessed on 7 August 2019.

71. N. DeCristofaro, S. Sahu, *CO_2 reducing cement, part one: solidia cement composition and synthesis*, World Cement, Jan. 2014, 571.

72. V. Atakan, et al., Why CO_2 matters—advances in a new class of cement, *ZKG Int.*, Vol. 67, No. 3, Jan. 2014, pp. 60–63.

73. N. DeCristofaro, et al., Environmental impact of carbonated calcium silicate cement-based concrete, *Proceedings of the 1st International Conference on Construction Materials for Sustainable Future*, A. Baricevic, ed., Zadar, Croatia, Apr. 19–21, 2017, pp. 65–70.

74. S. Sahu, et al., CO_2-reducing cement based on calcium silicate, *14th International Congress on the Chemistry of Cement (ICCC)*, Beijing, China (2015).

75. W. Ashraf, J. Olek, and S. Sahu, Phase evolution and strength development during carbonation of low-lime calcium silicate cement (CSC). *Constr. Build. Mater.* 210, 473–482 (2019). https://doi.org/10.1016/j.conbuildmat.2019.03.038.

76. J. Jain, O. Deo, S. Sahu and N. DeCristofaro (2014) *Solidia concrete™ part two of a series exploring the chemical properties and performance results of sustainable solidia cement™ and solidia concrete™*. Solidia Technologies, Piscataway, NJ, USA. http://solidiatech.com/wp-content/uploads/2014/02/

Solidia-Concrete-White-Paper-FINAL-2-19-14.pdf. Accessed on 3 October 2017.

77. S. Sahu, R. C. Meininger, Sustainability and durability of Solidia Cement Concrete based on an alternative CO_2-functional cement chemistry, *Concr. Int.* (August 2020) 29–34.

78. J. Jain, A. Seth, and N. DeCristofaro. Environmental impact and durability of carbonated calcium silicate concrete. *Proc. Inst. Civ. Eng. Constr. Mater.* 172(4), 179–191 (2019). https://doi.org/10.1680/jcoma.17.00004.

79. R. Tokpatayeva, J. Olek, S. Sahu and J. Jain (2016) Comparative study of sulfate attack resistance of carbonated calcium silicates and plain Portland cement mortars. In *7th Advances in Cement-Based Materials (Cements 2016)*, Northwestern University, Evanston, IL, USA.

80. C. P. Jiménez, et al., Corrosion related durability of steel reinforcement in a novel concrete formulation, *NACE Conference*, April 18–22, Salt Lake City, UT, 2021.

81. R. L. Berger, J. F. Young, and K Leung. 1972. "Acceleration of hydration of calcium silicates by carbon-dioxide treatment." *Nat.: Phys. Sci.* 240 (Dec 8): 16–18. doi:10.1038/physci240016a0.

82. C. J. Goodbrake, J. F. Young, and R. L. Berger. 1979. "Reaction of beta-dicalcium silicate and tricalcium silicate with carbon dioxide and water vapor." *J. Am. Ceram. Soc.* 62 (3–4): 168–171. doi:10.1111/j.1151-2916.1979.tb19046.x.

83. D. R. Moorehead, 1986. Cementation by the carbonation of hydrated lime." *Cem. Concr. Res.*, 16(5), 700–708. doi:10.1016/0008-8846(86)90044-X.

84. S. Monkman, M. MacDonald, R. D. Hooton, & P. Sandberg (2016). Properties and durability of concrete produced using CO2 as an accelerating admixture. *Cem. Concr. Compos.*, 74, 218–224.

85. S. Monkman, M. MacDonald, On Carbon dioxide utilization as a means to improve the sustainability of ready mixed concrete, *J. Clean. Prod.* 167 (2017) 365–375.

86. T. Sato, and F. Diallo. 2010. "Seeding effect of nano-CaCO3 on the hydration of tricalcium silicate." *Transp. Res. Rec.* 2141 (1): 61– 67. doi:10.3141/2141-11.

87. B. Constantz, Calera-using CO_2 to make useful materials, *Carbon Capture J.* (2010) 23–25.

88. B. Constantz, C. Ryan, L. Clodic, Hydraulic cements comprising carbonate compound compositions, US 7735274B2, 2010.

89. http://www.blueplanet-ltd.com/#technology. Accessed on 1 November 2020.

90. https://www.iea.org/reports/cement/. Accessed on 11 July 2020.

91. L. M. Bjerge and P. Brevik, CO_2 capture in the cement industry, Norcem CO_2 capture project (Norway) *Energy Procedia* 63 (2014) 6455–6463.

92. M. Naranjo et al., CO_2 capture and sequestration in the cement industry, *Energy Procedia* 4 (2011) 2716–27232717.

93. K. Jordal et al., CEMCAP – making CO_2 capture retrofittable to cement plants, *Energy Procedia* 114 (2017) 6175 – 6180.

94. D. Leeson et al., A Tecno-economic analysis and systematic review of carbon capture and storage (CCS) applied to iron, and still, cement, oil refining and pulp and paper industries, *Energy Procedia*, 114 (2017) 6297–6305.

95. https://www.worldcement.com/europe-cis/11122019/cement-producers-have-founded-an-oxyfuel-research-corporation/. Accessed on 11 July 2020.

96. http://www.cleanker.eu/the-project/project-contents/. Accessed on 11 July 2020.

97. https://www.lafarge.ca/en/project-CO2ment/   https://www.project-leilac.eu/carbon-capture/. Accessed on 11 July 2020.
98. https://www.project-leilac.eu/carbon-capture/. Accessed on 11 July 2020.
99. https://energyindustryreview.com/construction/cemzero-project-the-next-step-towards-a-climate-neutral-cement/. Accessed on 11 July 2020.
100. A. K. Chatterjee, Alternative cements and concretes in development with recycled plant-emitted carbon dioxide, *Cement, Wapno, Beton*, 2017 (2):120–137.
101. A. Nagi, J. G. Jang, Recent progress in green cement technology utilizing low-carbon emission fuels and raw materials: a review, *Sustainability* 2019, 11, 537. doi:10.3390/su11020537.

Epilogue

In regard to use and application, the Portland cements have surpassed all other man-made construction materials. In 2018, the year for which the data is readily available, the per capita consumption of cement was 521 kg as compared to 225 kg for crude steel, 28 kg for plastic and 11 kg for primary aluminum. The global production of cements increased from 1000 million tons in the year 1990 to 4000 million tons in 2013, and the annual world production continues to remain above 4000 million tons even in recent years, despite the economic slowdowns and other socio-political problems. China alone accounts for about 56% of world production followed by India in the second position with 7% production.

The manufacturing base for Portland cements is spread over more than 180 countries having more than 2500 production units of varying capacities. The individual daily capacities of these production units are tentatively estimated to range from 1500 tons to 7000 tons. The single kiln capacities generally range from 3000 tons to 9000 tons, though the highest daily output of an operating single rotary kiln has reached now 14,500 tons. Most of the production units are integrated from mines to market, though the producers often set up split grinding units to derive various business benefits. Even the capacities of grinding units are large, ranging approximately from 3000 tons to 6000 tons per day.

The production of cement is overwhelmingly dependent on natural raw materials, and more particularly limestone, and is highly resource-intensive. Even with the significant advancements made by the cement industry in producing different types of blended cements with lower consumption of clinker, the intermediate product in the manufacturing chain, the total consumption of materials is estimated to equal the volume of finished cement produced. In addition to resource intensity, the cement sector is known to be the third largest industrial energy consumer with 7% of the total industrial energy use of 10.7 Ej. Furthermore, due to the intrinsic nature of the thermo-chemical process of manufacture, the sector has the second largest share of direct industrial emissions of carbon dioxide, which was 2.2 $GtCO_2$/yr in 2014. While the present levels of resource use, energy consumption, and CO_2 emissions in the cement industry are already very high, the situation will be appalling with the predicted rise of cement demand in the next three decades. As of now, it is conservatively estimated that the cement demand in 2050 will be 4682 million tons per year, and therefore it is obvious that there will be significant upward swing in the use of resources, consumption of energy, and emissions of greenhouse gases. It is estimated that even under the RTS (Reference Technology Scenario) the CO_2 emissions are likely to rise by 4% by 2050 due to the predicted increase in cement demand and production.

Since such an increase in the CO_2 emissions will further contribute to the rise in the average global temperature, a technology roadmap was prepared jointly by IEA and WBCSD-CSI to define the pathways to be adopted by the cement sector in order to keep the global temperature rise below 2-Degree Celsius, compared to the pre-industrial era, by the end of the 21st century. This roadmap identifies four important levers to reduce the CO_2 emissions by 24% by 2050: a modest reduction of 3% from improvements in thermal efficiency, an ambitious reduction of 12% from switching over to alternative fuels, a significant reduction of 37% from lowering the consumption of clinker in cement, and the balance 48% reduction coming from the commercialization of innovative technologies including CCUS (carbon capture, use and storage). The roadmap has been further revised by IEA in 2020 to define the pathways for net-zero emissions of CO_2 from 2019 to 2070. This plan emphasizes more on materials efficiency, which additionally includes reduction in cement demand strategy, and extensive proliferation of CCUS technologies.

Based on the above roadmap and continuing practices, the cement industry has taken several technological initiatives to grow with more resource efficiency, high energy conservation, and lower CO_2 emissions. The initiatives include the setting of plant-level or sector-level energy efficiency improvement targets, deployment of circular economy, use of alternative fuels and raw materials, emission reduction, production of blended cements with lower clinker-to-cement ratio, product standardization and optimizing cement use in concrete construction. For implementing these steps, the industry had to upgrade itself in many ways. The effective elements of advancement may be summarized as follows:

- Scale of production as discussed above
- Design and installation of energy-saving grinding mill systems
- Low pressure-drop multistage preheating strings, efficient precalcining systems, short rotary kilns, and enormously improved clinker grate coolers
- Especially engineered facilities for storing, handling, feeding, and burning alternative fuels and raw materials available as waste from other sectors of economy
- Installation of waste heat recovery for power generation in addition to use of captive power plants and renewable energy sources, both captive and via grids, along with rationalization of multi-source power distribution system
- Use of energy-efficient motors, drives, and process fans
- Application of process control systems with DCS, SCADA, and PLC
- Introduction of 'Expert Systems' and neural network based 'Model Predictive Controllers'
- Extensive use of smart sensors for temperature, pressure, humidity, level, proximity, motion, vibration, angular velocity, noise, image, etc.

- High levels of process instrumentation and online bulk material composition with the help of gamma rays, X-rays, near infrared rays
- Use of robotics in laboratory functions
- Adoption of digitalization all along the process from mine to packer
- Closed-loop real-time process controls at sub-systems level
- Computerization of the maintenance management systems
- Off-line assets health checks
- Varying levels of adoption of Enterprise Resource Planning software for logistics, inventory management, stock handling, accounting, and integration of vendors and customers
- Production of blended cements with a view to reducing clinker factor in cement
- Introduction of products and systems standards
- Continuous monitoring of emissions of dust and polluting gases
- Compliance with regionally prevalent policies of environment, health, and safety.

The above list is long but certainly not exhaustive. But, on the whole, the comparative yardsticks today for the cement plants are the following:

- Specific heat consumption
- Specific electricity consumption
- Emissions of dust, NO_x, and SO_2
- Plant availability
- Customer satisfaction with quality and delivery assurance.

Notwithstanding the great strides made and being pursued on the above directions, the cement industry appears to be reaching a plateau in terms of its performance indices. The need of the hour for the industry is to acquire a low carbon footprint and achieve significant reduction in CO_2 cmissions. Further, recognizing the inevitability of continuing with the future use of Portland cements without any visible and viable alternatives and impending expansion of production volumes in the next three decades at least, the cement industry will have to look for new, innovative, and disruptive technologies for its growth path. The possibilities seem to be emerging from the realm of 'artificial intelligence'. The industry is now seized with the potential being unfolded by AI technologies, preparing the way toward 'Industry 4.0' revolution. In 2011, a German research group headed by Sigfried Dais of Robert Bosch GmbH coined the phrase 'Industrie 4.0' and introduced a working framework for its implementation. The framework highlighted the importance of an interconnected environment driven by smart equipment sharing data and a decentralized decision-making process.

The concept has matured over the last decade. The important elements of AI, such as machine learning, natural language processing, computer vision processing, problem-solving, etc. are now the best providers of the essential tools and techniques to construct a framework that can be applied to specific cases of asset performance management, process optimization, autonomous operation, and predictive maintenance with minimal human intervention. The step toward this direction by any industry, however, will require a basic infrastructure, expertise, and domain knowledge. It was, therefore, important to introspect to what extent the cement industry is prepared and what more is needed further to supplement the efforts. The book has attempted through all its chapters to look at the different facets of the cement industry from the perspective of its preparedness to move into the Industry 4.0 level. Since cement manufacturing is a continuous process, some overlaps in the sectional status reviews by the contributors could not be avoided, but it was interesting to find the convergence of thoughts and a consensus to move into the realm of artificial intelligence to reinforce intelligent and sustainable cement production.

It is understood that the basic requirements in advancing toward this goal are the databases, internet connectivity, Cloud computing and storage services, and the application of machine learning algorithms. Use of sensors for gathering data is widely prevalent in cement plants. Data archiving is also practiced and the past data is in most cases used for only trend visualization. The distributed control systems typically store short-term information required by the operators and plant engineers for different analyses. Most often this information is inadequate to train a model for automation and optimization. Hence, the cement plants will have to be prepared for standardized storage of data and easy access to it. The datasets need to be time-stamped and maintained in time-series in process data historians that continuously retrieve data from the DCS using connectivity protocols. In addition, there has to be an interface between the plant DCS and process data historian, on one side, and installation and configuration of the Cloud connector, on the other, that would enable communication with the AI platform. It may be relevant to mention here that Cloud engineering is a systematic approach to developing, operating, and maintaining Cloud computing systems and encompasses systems, software, web, IT engineering, security platform, risk dimensions, and so on, for which a level of expertise will be necessary in the cement industry. Further, along with interconnectivity with a centralized remote platform, it is possible to achieve decentralization as well via edge computing solutions, in which IoT, sensors, smart devices, and human–machine interfaces support the capture and analysis of factory data at the edge. It must be remembered that the internet connectivity is the foundation of internet-of-things. This is also not a new feature in cement plants. Communication between interacting machines with suitable protocols is being adopted at the plants. However, an inadequate internet bandwidth at the plant locations is often an obstacle.

It is known that some sections of the cement plants are already automated, and it is often claimed that control of these processes is autonomous with continuous monitoring in the central control room, operators intervening only during emergency situations. It is stated that the automation system in a cement plant with standard instrumentation accept and process 10,000–20,000 channels of analog and digital signals at a frequency of about 1 Hz. While this volume of data is often considered by some as large, there is a strong view that the amount of data is inadequate for automation because of the stochastic nature of the manufacturing process and variability of measured data. Further, the measured data in a cement plant is only a fraction of the total information of the process as many parameters are not directly measurable. In addition, many parameters are measured at single points, while in realty they vary widely in space and time. The sum and substance of all these reflections is that for autonomous operation of cement process much larger amount of data, measured at close intervals, will be necessary. At the same time, the simulation and modeling approaches have to be adopted for stochastic and non-linear processes, instead of normally used steady-state linear functions. The study of the dynamics of each of the sub-processes such as grinding, pyro-processing, cooling, etc. must be based on large volume of historical data.

This brings in the essentiality of delving into machine learning that results in predictive modeling. As already stated, machine learning is a subset of AI, while deep and reinforcement learning is a subset of machine learning. It uses a variety of algorithms that iteratively learn from data to improve and predict outcomes. For machine learning one requires the right set of data that can be applied to the learning process. Since deep learning uses hierarchical neural networks to learn data in an interactive manner, it is useful to learn patterns from unstructured data and is useful for image recognition and computer vision applications. Reinforcement learning is behavioral learning model that learns through trial and error. A sequence of successful decisions results toward solving the problem at hand. Although some software introduced in the cement sector claims the incorporation of machine learning, its wide-scale application is still awaited.

While on the subject of digitalization, it is worthwhile to note that the cement industry must keep a track of the new technology of 'digital twin' to create an accurate virtual replica of the process with back and forth data transfer between the digital twin and the physical factory. It will provide a virtual environment for evaluating the production process, gaining insight into the process complexity, and taking remote decisions.

The goal of reaching autonomous production is not a one-stop activity. It has to be planned and executed in a phased manner. The manufacturing industries are watching the steps and stages that have been adopted by the self-driving car industry. Perhaps it will be prudent to first create the basic infrastructure including reskilling and retraining the operators and engineers of the plants. Along with training of manpower, a major cultural shift will be needed in the workforce from the present pattern of silo working to widely collaborative environment. Thereafter, the transformation of the process from the phase of 'autonomous subsystems' to the phase of 'autonomous systems in full control' may be planned and executed. In these transformational phases, the human intervention will be reduced but not dispensed with. In all probability, the total autonomy in cement manufacturing without humans will remain aspirational. But what is essential to achieve is sustainability with lower material footprint and significantly reduced carbon footprint.

Even with embracing AI and autonomous manufacturing, the stiff target of CO_2 emissions by 2050 is unlikely to be achieved, unless the industry advances into carbon capture and sequestration including recycling of CO_2 in the form of novel carbonated non-Portland binders and concrete. These technologies are well researched now but quite a distance away from being viable and scalable for industrial applications. But the developments deserve careful tracking by the cement industry with SWOT analysis. Perhaps, an extension of AI tools and techniques may help in scaling up and integrating the new CCSU technologies faster than expected.

All in all, the intelligent and sustainable production of cement is not a distant goal. The cement sector must achieve this objective sooner than later. The milestones for every five- to ten-year spans for the next three decades need to be set and monitored. Business, management, and production must progress hand-in-hand and there has to be substantial investment in technology development along the pathway.

Anjan Kumar Chatterjee
Editor

Index

Page numbers in **bold** indicate tables, page numbers in *italics* indicate figures.

A

AAS, *see* atomic absorption spectroscopy
ACID (atomicity, consistency, isolation, durability), 85
adaptive weighting, *29*
advanced process control (APC) systems, 94–95, **94**, 138, 143, 161, 358, 388
  and AI, 132–135, 172–176
  implementation steps, *179*
  for thermal process, 176–192
  approach to optimization, *184*
  evaluation and implementation, 192–194
  via digital twins of cement plant, *193*
  future in cement manufacturing, 95–97
AF, *see* alternative fuels
AFE, *see* augmented field engineer
AFR, *see* alternative fuels and raw materials
AI, *see* artificial intelligence
aixPert Optimizer, 279–280, *280*, 291
Akter, S., 328–329
Alborg Portland Cement, 412
algorithms
  categories, *88*
  learning, 60
  overview, 45–46, **87**
  selection, 46, 48–50
  types, 75; *see also* convolutional neural networks (CNN); decision trees; K-means; linear regression; Markov random field (MRF); Monte-Carlo; support vector machine (SVM)
alkali-activated binders, 418–419
alkali-silica reaction (ASR), 416
alternative fuels (AF), 142, 145
  adapting plant and equipment, 154–162
  characteristics, **153**
  chemical analysis, 47
  combustion management, and digital twins, 282–283
  control module, 182–191
    structure, *183*
  firing locations, criteria for selecting, 155–156
  LHV and moisture, **151**
  operational considerations, 153–154
  pyroprocessing, 150–152
    design features of equipment, 156–158
    preheater cyclones, 156
  quality deviations, 212
  rotary drum reactor, 158, *158*
  storing, dosing, and conveying, 152
  types, 152
alternative fuels and raw materials (AFR), 142–143, 145
  carbonate sludge, 147
  and clinker quality, 187
  fly ash, 147–148, 411–412
  industrial waste, 147–150
  and mixing control, 187
  naturally occurring, 146–147
  pyroprocessing, 146–150
    calciner design, 156, *157*
      functional details, *157*
  and quality, 170
  red mud, 148
  slag, 148
    aluminum, 150
    blast furnace, 411
    copper, 149–150
    ferro-alloy, 149
    GGBS, 148, 410, 447
    LD, 149
    non-ferrous metallurgical, 149–150
alternative supplementary cementitious materials (ASCM), 413–418
alumina modulus (AM), 9, 212
aluminum silicate, 54
American National Standards Institution (ANSI), 102
American Society for Testing and Materials (ASTM), 102, 413, 448
ANN, *see* artificial neural networks
ANSI, *see* American National Standards Institution
APC systems, *see* advanced process control systems
APM, *see* asset performance, management
Apollo Guidance Computer, 40–41
artificial intelligence (AI), vii, 81, 83–84, 138, 143, 385–386, 459
  and APC, 132–135, 172–176
  applications, in cement production process, 52–74

basic elements of platform, 96
building block, *174*
components, *42*
definitions, 40–41, 74, 172
and electrical systems, 134–135
essence of, 208–209
major goals, *84*
optimization, **95**
potential for cement manufacturing
process, 50
in process optimization, 386
relevance and limitations, 209–210
strong, 172
subdomains, 174–175
techniques for QC, 199–220
techniques to transform cement
manufacturing, 39–76
technologies, 359
weak, 172
artificial neural networks (ANN), 26, 43, 48,
134–135, 138, 176
application to final product quality,
216–219
principles, *210*
artificial neuron, 44
AR/VR, *see* augmented reality/virtual reality
ASCM, *see* alternative supplementary
cementitious materials
ASR, *see* alkali-silica reaction
asset induction and records, 370
asset performance
approach, in industrial environment,
227–230
insights, 375
management (APM), 227, 399
and EAM, *229*
and ERP, *229*
functions, *228*
with PdM, 254
roles and responsibilities, *229*
strategies and methodologies, 230–232
monitoring
and AI, 244–248
in cement manufacturing, 225–262
near-term opportunities, **252**
structure, *228*
assets, health, connected, 376–380
ASTM, *see* American Society for Testing and
Materials
atomic absorption spectroscopy (AAS), 204
augmented field engineer (AFE), 393, *393*
augmented reality/virtual reality (AR/VR),
401–402
automatic reasoning, 174
automation, 2, 25, 79–80, 230, 458–459

of cement laboratories, 368–369, *368*
limitations of conventional systems,
165–167
Industry 3.0, *166*
autonomy, 80–81
goal of reaching autonomous production,
459–460
steps towards, 81–83, *82*
auto-regression moving average (ARMA), 92
auto sampling, 171

B

bag filters, 235
parameters, **235**
ball mills, 8, 13, 33, 123
gearless drives, 123
process optimization, 360, *360*
basic-oxygen-furnace (BOF), 414
BCT, *see* belite-calciumsulfoaluminate-
ternesite
BDPA, *see* big data and predictive analytics
belite-calciumsulfoaluminate-ternesite (BCT),
429
belite-rich cement (BRC), 425–427, 445
comparative SWOT analysis, **446**
belite, ternary plot, *429*
benchmarking, 375–376, *377*
big data, 327–328, 330
analytics viii, 83, 86, 334
definitions, 83–84
and predictive analytics (BDPA), 329
resources for building capability, 329–335
binders
clinker-based, alternative, 425–430
C-S-H- based, 422–424
Blaine surface area measurement method,
204–205, 208
blast furnace slag, 411
BlendExpert (BLX), 365–366, *366*
mill control, 367–368, *367*
pile control, 366–367, *366*
Blue Planet Ltd, 443, 445
BOF, *see* basic-oxygen-furnace
Bogue's equations, 54
BRC, *see* belite-rich cement
Brundtland report, vii
bucket elevators, 242–243
Bureau of Indian Standards (BIS), 102
burning zone, 301
combustion controller, 181–182
controller, 181
business development
sustainable, 325–326
based on TBL, *326*

business intelligence (BI), 327
business process, connected, 397–399

C

C&D, *see* wastes, construction and
    demolition
cables, intelligent partial discharge
    monitoring, 245–246
calcined clays, 412
calciner
    combustion controller, 183–184
    refractory conditions, inspecting, 395
calcium carbonate, *see* limestone
calcium silicate cement (CSC), 430–439
    average composition, **433**
    CO_2 emissions, reductions, 431–432
    energy savings, 430–431
calcium sulfoaluminate (CSA), 428–430
Calera/Blue Planet, comparative SWOT
    analysis, **446**
Calera cement, 442–443, 447
Calera Corporation, 442, 445, 448
Calera process, 442, *442*
Ca-looping, 444
Caradonna, J. L., 325
    model of sustainability, *326*
Carbicrete, 414
carbon capture and storage (CCS), 22,
    443–444
carbon capture and use (CCU), 22
carbon capture, utilization and storage
    (CCUS), 439, 444, 456
carbon dioxide emissions, 22, *23*, 34–35, 76,
    406, 447
    and AF, 154
    in cement industry, 455
    in global cement sector, *408*
    and Portland cement, 408–409
    reductions within cement industry,
        448, 456
    technology roadmap, *196*
Carbon8 Systems, 445
carbonated steel slag, 414
carbonate sludge, 148
CarbonCure Technologies (CCT), 439–441,
    445, 447
    comparative SWOT analysis, **446**
CCS, *see* carbon capture and storage
CCT, *see* CarbonCure Technologies
CCU, *see* carbon capture and use
CCUS, *see* carbon capture, utilization and
    storage
Celitement GmbH, 424
Celitement®, 422–424, *423*, 445

comparative SWOT analysis, **446**
    performance and durability, 424
Cembureau (European Cement Association),
    324
cement, consumption, per capita, 455
cement fineness, closed loop monitoring and
    control, *311*
cement industry
    technological history, 2, 79–80
    technology roadmap for, 35
cement laboratories, automated, 368–369, *368*
cement manufacture, viii, 1–35
    AI techniques to transform, 39–76
    asset performance monitoring, 225–262
    digital solutions, implementation, 347–403
    and digital twins, 263–291
        options, 280–285
    electrification, 444
    global, 52, 100
    key assets, 232–243, **233**
    maintenance management, 225–262
    and ML, 50–52
    process automation to autonomous
        process, 79–98
    quality control (QC), 200–202
    and soft sensors, 310–319
    sustainable, 99–138
        and IERP, 323–342
        and sensors, 293–320
        thermal energy management, 141–197
        through use of AFR, 145
    thermal process, 143–145, *144*
cement manufacturing groups
    large, 230–232
    small-and medium-sized, 232
cement packing and shipping, 7
cement plants, *52*
    architecture, 33
    central laboratory, 201
    control system, 350, *355*
        essentials, 354–356
        upgrading, 356–358
    current process control infrastructure,
        89–91
    digital solutions, implementation, 347–403
    and digital technologies, 402–403
    digitalization, 349–353
        key components, *350*
    enterprise management systems, 369–372
    and IIoT, 291
    Industry 4.0 system configuration, *169*
    integrated features, 32–33
    LafargeHolcim (LH), 51–52
    main components, 52–61
    process layout, *8*

relevance of AI applications, 52–61
single unit, information flow, *316*
sustainable production, electrical systems,
  99–138
technical performance monitoring,
  230–232
virtual, *see* digital twins
visual impact, 33
cement-producing countries, **4**
cement supply chain, 324
Cemex, 57
CemZero project, 444
CENELAC, *see* European Committee for
  Electro-Technical Standardization
  (CENELAC)
central control room (CCR), 127, 350
CFD, *see* computational fluid dynamics
chatbots, 398
chemical absorption and separation, 443
chromium slag, 150
CIMS, *see* computer integrated
  manufacturing system
Circle Economy, vii
classification, 47
  of SCMs, and AI, 67–71
CLEANKER project, 444
clinker, 10–13, 56, 207
  clinker-to-cement ratio, 409
  and CO_2 emissions, 408–409
  composition, 54
  cooling, 17, 191–192, 234, 302–303
    process optimization, 361
  grinding, **13**
  material chemistry, 12–13
    and ML, 62–71
  microstructure, 67
  production, 9–12, *11*
    and ML, 56–57
  free-lime prediction tool, 58–*62*
  quality, 211
    and AF, 154
    and AFR, 187
  quality deviations, 215
  sampler, 171–172
  volatile constituents, **27**
cloud computing, x, 85, 136–137, 168–169,
  349, 373–374, 458
  security, 374
  storage, viii
cloud data services, 84–85, 249
clustering, 47, **87**
CMMS, *see* computerized maintenance
  management systems
CNN, *see* convolutional neural networks

CO_2 emissions, *see* carbon dioxide emissions
cognitive computing, 208
collaborative operation, in data-driven
  ecosystem, 194–195, *195*
CO_2 ment project, 444
communication, and sensors, 296
communication protocols, 130–131
compressed air systems, optimization,
  126–127
compressive strength, and ML, 62–67, *65*
compressors, 241
  parameters, **242**
computational fluid dynamics (CFD), 26,
  186
computer integrated manufacturing system
  (CIMS), 2, 30, 79–80
computerized maintenance management
  systems (CMMS), 227, 230,
  248–249, 351, 371
computing devices
  advances in, 40–41, *41*
  "AI inside", 41–42
concentrated solar power (CSP), 111–112,
  138
concrete, direct utilization of CO_2 in, 439–445
condition monitoring, 251
  transformer, 245
conductometry, 202
connected
  asset health, 352
  asset insights, 351, 374–385
  business process, 352
  innovation, 352–353
  operation, 352
construction sector, growth, 34
consumption, of cement, per capita, 34
conveyor belt, 240–241
  failure conditions, causes and remedial
    measures, **241**
convolutional neural networks (CNN), 45
copper slag, chemical composition, **149**
CPS, *see* cyber-physical systems
crude steel, global production, **3**
crushers, 122–123
CSA, *see* calcium sulfoaluminate
CSC, *see* calcium silicate cement
CSP, *see* concentrated solar power
culture, *see* data-driven culture, *see*
  organizational culture
cyber-physical systems (CPS), viii, 83
cyclones, 57, *58*
  blockage detection, 57
  preheater, and AF, 156
  samples, volatile constituents, **27**

D

Dais, Sigfried, 457
data
    analysis, 59–60
    analytics, x, 249
        major goals, *84*
    cleansing, 47–48
    collection, 336
    distribution, 47
    feature extraction, 317
    frequency, 59
    fusion, 287
    integration management system, 143
    management, steps and technologies, *286*
    measures and dimensions for developing
        analytics engine, **318**
    metric computation, 317
    mining, 85–86, 327
    modelling, 317
    normalization, 60, 317
    processing infrastructure, and IIoT,
        136–137
    processing, and sensors, 296, 317–318
    science, 85
    sources, 58
    transformation, 60
    types, 75
    understanding, 47
    validation, 60–61, 317
    visualization, 86, 317; *see also* big data
data-driven culture, 334–335
    decision-making, *335*
dataset, quality, 47–48
DCS, *see* distributed control systems
decision trees, **87**, 135, 287
deep learning (DL), 43–44, 89, 96–97,
        134–135, 138, 174, 208, 359, 459
    definitions, 74
demand trends, **3**, 34
depletion of natural resources, 20
detecting events, 59
diagnostic solutions, automated, 398–399, *398*
digital ghost technology, 266
digitalization, x, 459
digital twin aggregate (DTA), 265–266
digital twin environment (DTE), 266
digital twin instance (DTI), 265–266, 269
digital twin prototype (DTP), 265, 268
digital twins, 400–401, *401*, 459
    and AF combustion management,
        282–283
    and cement manufacture, options,
        280–285

in cement manufacturing, 263–291
    comparison of selected tools, **289**
concept
    history, 264–267
    manifestations, 265–266, *267*
    forms, *265*
and grinding of multi-component blended
        cements, 283
and IIoT platforms, 288–289
and Industry 4.0, 263–264
for learning and training, 289–290
and limestone calcination, 282
and limestone mining, 281–282
model-driven, 193–194
and pilot plants, 275–278
and product lifecycle, 268–269, *268*
recent publication classification, **268**
relevance for cement manufacturing,
        278–285
technology in manufacturing
    adoption of, 269–271, *270*
    functionality and structural
        configuration, 271–275
direct separation, 444
dispatched cement, quality deviations,
        215–216
distributed control systems (DCS), ix, 26,
        83, 89–91, 95, 234, 297–298,
        314, 458
    architecture, *314*
DL, *see* deep learning
drone technology, *54*, 394–397
    inspecting refractories, *396*
    photogrammetry, 395
    potential applications, in cement plants,
        397
DTA, *see* digital twin aggregate
DTE, *see* digital twin environment
DTI, *see* digital twin instance
DTP, *see* digital twin prototype
Dubey, R., 331–332, 334
Dunning, Ted, 49–50
dust emission, 21, 23

E

EAF, *see* electric-arc-furnace
EAM, *see* enterprise asset management
edge computing, 84–85, 137, 373
electrical distribution system, automation,
        129–130, *131*
electrical energy performance, 19–20, **20**
    conservation, 124–127
        air conditioner loads, 127

compressed air systems, optimization,
126–127
lighting loads, 127
metering and load profiling, 125
power factor correction, 127
power quality monitoring, 126
voltage imbalance correction, 127
consumption pattern, **125**
electrical systems
AI applications, 134–135
standards and regulations, 101–102
for sustainable cement production, 99–138
electric-arc-furnace (EAF), 414
Electronic Numerical Integrator and
Computer (ENIAC), 40, *41*
electro-thermal phosphorous slag, 150
Elkington, J., 325
emissions, gas analysis, 303–304; *see also*
carbon dioxide emissions; nitrogen
oxides emissions; sulfur dioxide
emissions
EMS, *see* environmental management system
energy balance, and AF, 153–154
energy efficiency, 2, 80
assessment monitoring, 243–244
motors and drives, 121–124, *122*
ENIAC, *see* Electronic Numerical Integrator
and Computer
enterprise asset management (EAM), *229*,
351, 369–371
enterprise resource planning (ERP), ix, 2, 83,
229, 230–231, 327, 351, 371;
see also integrated enterprise
resource planning (IERP)
integration with MES, 30
environmental management system (EMS), 31
environmental sustainability, vii
ERP, *see* enterprise resource planning
European Committee for Electro-Technical
Standardization (CENELAC), 102
expert systems, 138, 278, 358–359
fault diagnosis, *135*
external waste management, 18–19

F

fan power, control, *126*
FastCarb, 414
fault diagnosis, expert system, *135*
feature engineering, 59
feature selection, 60
feeder control, 129–130
ferro-titanium slag, 150
filter bag, online condition monitoring,
384–385, *385*

FLSmidth
connected asset insights, *375*
SiteConnect Mobile app, 374
UptimeGO, *378*
fly ash, 147–148, 411–412
fog computing, 137
Fraunhofer theory, 204
free lime analyzer, 171–172
fuzzification, *163*
fuzzy logic control, 138, 162–163, 278,
358–359
controllers, 91–92, *91*, *133*, *164*

G

gas analysis, **162**
gases, measurement techniques, **303**
gearbox, 239–240
early detection of bearing defect, *384*
failure conditions, causes and remedial
measures, **240**
inline helical, inspection and repair
frequency, **240**
parameters, **239**
Gelernter, David, *Mirror World*, 264
geopolymers, 418–419
GGBS, *see* ground granulated blast furnace
slag
Gibbs Energy Minimization Software
(GEMS), 284, *284*
goal deviation, *29*
greenhouse gas emissions, vii, 100, 253, 406
of global cement industry, 22
grinding, of multi-component blended
cements, and digital twins, 283
ground cement, deviations, influence of SO_3
level, 218
ground granulated blast furnace slag (GGBS),
148, 410, 447
chemical composition, **148**
Gunasekaran, A., 331–332

H

Hadoop framework, 328
hardware and connectivity, 85–86
hCHS, *see* hydraulic calcium hydrosilicates
HCl emissions, *see* hydrogen chloride emissions
HCM system, *see* human capital management
system
health and safety, 50
heavy metals, 23
high-level control (HLC) systems, 51–52
high-pressure grinding rolls, *see* hydraulic roll
presses (HRP)

HLC systems, *see* high-level control systems
horizontal roller mills (HRM), 8, 13, 19, 33
hot meal, 207
   quality deviations, 214
   sampler, 171
HRM, *see* horizontal roller mills
HRP, *see* hydraulic roll presses
human capital management (HCM) system,
   371; *see also* people
human resources, 331–332; *see also* people
hydrate phase assemblage, 283
hydraulic calcium hydrosilicates (hCHS),
   422–423
hydraulic roll presses (HRP), 8, 13, 19, 33,
   123
hydrogen chloride emissions, 23
hyperparameters, 48

I

IaaS, *see* Infrastructure as a Service
IEC, *see* International Electro-Technical
   Commission
IEEE, *see* Institution of Electrical and
   Electronics Engineers
IERP, *see* integrated enterprise resource
   planning
IIoT, *see* industrial internet of things
image analysis/recognition, 41, 50
IMS, *see* information management system
industrial internet of things (IIoT), 83, 97,
   138, 249
   applications, 353
   and cement plants, 291
   and data processing infrastructure,
      136–137
   and digital twins, 290
   platforms for digital twins, 288–289
Industry 4.0, viii, x, 35, 97, 349, 402,
   457–458
   AI tools and techniques, 133–134
   in cement plants
      implementation plan, 167–170
      system configuration, *169*
   and digital twins, 263–264
   process optimization concept, *168*;
      see also advanced process control
      (APC) systems
information density, 279–280
information management system (IMS), 35
information mirroring model, *see* digital
   twins
Infrastructure as a Service (IaaS), 84
innovation, connected, 399–402
in-process solids, analysis, 304–306

Institution of Electrical and Electronics
   Engineers (IEEE), 102
integrated enterprise resource planning
   (IERP), 328–329
   in cement industry
      drivers, **335**
      theoretical model, 340, *341–342*
   developing in cement industry, 335–341
   in sustainable cement production, 323–342
integration, of business, management, and
   production, 30–32
intelligent manufacturing systems, vii
intelligent production management system,
   AI-enabled, 388–390
International Electro-Technical Commission
   (IEC), 101–102
International Energy Agency, viii, 406, 456
   Energy Technology Perspectives 2020, 407,
      447
International Standards Organization (ISO),
   102
internet of things (IoT), viii, 349
   philosophy of management, 252
   platforms, 351
      components, *373*
      foundation, 372–374
IoT, *see* internet of things
ISO, *see* International Standards
   Organization
ISO 9000, 200
   adoption of, 31
ISO 14000, environmental management, 31

K

Kalina, Alexander, 110
Kalina Cycle (KC), 17, 108, 110–111
Kalman filter, 359
Karlsruhe Institute of Technology (KIT),
   422–424
KC, *see* Kalina Cycle
kiln control, 181–182
   smart kilns, online condition monitoring,
      380–381
kiln feed, 9–10, 12, 56; *see also* raw meal
   preparation, 6
   quality deviations, 214
   volatile constituents, **27**
kiln optimization, 26, 361, *362*, 380–381
kilns
   coating, stability index, monitoring, 388
   red spot, *387*
      predicting, 386–387
   shell, temperature, *387–388*
   types, 159

kiln shell scanner, 160–161, *161, 302*
kiln systems, 33
  heat and mass balance, *17*
  plant analytics, *316*
  thermal energy performance, 16–18, **18**
Klien, Alexander, 428
K-means, 47

L

ladle-furnace-refining (LFR), 414
LafargeHolcim Innovation Center (LHIC),
  58, 62
LafargeHolcim (LH), 432
  plants, 51–52
laser granulometry, 204, 208
LD slag, *see* Linz-Donawitz slag
lead-zinc slag, 150
Le Chatelier method, 205
LEILAC project, 444
LFR, *see* ladle-furnace-refining
LH, *see* LafargeHolcim
LHIC, *see* LafargeHolcim Innovation Center
LHV, *see* low heat value
lime saturation factor (LSF), 9, 59, 212, 214,
  366, 425
  variation of raw mix, *366*
limestone, 8, 53–54, 146, 455
  calcination, and digital twins, 282
  and calcined clays cement (LC3), 412–413
  calcining or decarbonation, 10
  compositional grading, **146**
  mining operation
    and digital twins, 281–282
    and ML, 53–54
linear regression, 45–46
Linz-Donawitz slag, 149
  chemical composition, **149**
load control, 130
logistic regression, 47
loss-on-ignition analyzer, 171
low heat value (LHV), of AFs, **151**
LSF, *see* lime saturation factor

M

machine learning (ML), viii, x–xi, 43–45,
  134–135, 138, 173–175, 220, 249,
  287, 359, 385–386, 459; *see also*
  reinforcement learning; supervised
  learning; unsupervised learning
  application development, *44, 88*
  application to lime kiln, 214
  to build free-lime-in-clinker prediction
    tool, 58–61, *62*

  in clinker production, 56–57
  definitions, 74
  essence of, 208–209
  expanding concepts, 87–89
  future for cement manufacturing, 50–52
  implementation process
    at developmental level, 46–48
    at production level, 48–50
  and limestone mining operation, 53–54
  model building, 175
  potential for cement manufacturing
    process, 50–52
  and raw mix design, 54–56
  sound analysis, 71–74
  tool for cement compressive strength,
    62–67
  tooling, 96; algorithms
MAE, *see* mean absolute error
magnesium carbonate-based cements,
  419–422
magnesium silicates, 421–422
magnesium slag, 150
maintenance
  breakdown, 377
  corrective, 248
  planned, 248
  predictive (PdM), 47, 248–249, 251, 370,
    377
    with APM, 254
    with classifiers, configuration, *251*
    ML architecture, 250
    scheme of activities, *250*
  preventive, 248, 370, 377
    near-term opportunities, **253**
  reactive, 377
  reliability-centered (RCM), 377–378
  strategies and practices, advances in,
    248–251
maintenance management, in cement
    manufacturing, 225–262
management information systems (MIS),
    371–372, 388
managers, 333
manufacturing execution system
    (MES), ix, 83
  integration with ERP, 30
Markov random field (MRF), 68–69
material chemistry, 5–13, *6*
material efficiency, 14–16, **16**, 35
material flows, in production
    process, 13–16, *15*
mean absolute error (MAE), 48
membrane separator, 444
MES, *see* manufacturing execution system
MICMAC analysis, 339

microprocessor unit (MPU), 296–297
microsilica, *see* silica fume
Mie theory, 204
mill separators, 236
    parameters, **236**
mirrored spaces model, *see* digital twins
MIS, *see* management information systems
mixing controller, 186–191
ML, *see* machine learning
Modbus, 128, 130–131
    comparison with Profibus
        protocols, **132**
modelling, 25–30
    predictive, 459
    technological spread, *287*
    thermodynamic, 283–284
model predictive control (MPC), ix, 92–94,
        93, 133, 138, 164–165, 173,
        278–279, 358–359
    basic structure, *134*
    for cement kiln process, *92*
    prediction trend, *166*
    principles, *165*
MODICAN, 130
Monte-Carlo, 47
motors
    and drives, 238
    geared, 124
    maintenance schedule, **238**
    protective enclosure, 124
    types, 122–124
MPC, *see* model predictive control
MPU, *see* microprocessor unit
MRF, *see* Markov random field
multi-fuel, process optimization, 360–361
multinational cement enterprises, production
        capacities, **5**

N

NARMAX, *see* non-linear auto-regression
        moving-average model with
        exogenous inputs
National Concrete Masonry Association
        (NCMA), 438
National Electric Manufacturers Association
        (NEMA), 102
National Energy Technology Laboratory
        (NETL), 421
natural language processing (NLP), 87
Near Infrared Borescope (NIR-B), 301
    location, *301*
neural networks (NN), 43, 89, 173, 210,
        287, 386
    in cement manufacturing, *386*

New York State Department of Transportation
        (NYSDOT), 437–438
nickel slag, 150
nitrogen oxides emissions, 17, 23
    and AF, 154, 157–158
    control technologies, 158–160
NLP, *see* natural language processing
NMPC, *see* non-linear model predictive
        control
NN, *see* neural networks
noise pollution, 21–23, 33
non-linear auto-regression moving-average
        model with exogenous inputs
        (NARMAX), 29
non-linear model predictive control (NMPC),
        29–30
Not Only SQL (NoSQL), 85
Novacem™, 420, 422, 447–448
    comparative SWOT analysis, **446**
    production process flow, *421*
NOx emissions, *see* nitrogen oxides emissions

O

O.C.O. Technologies, 445
online analytical processing (OLAP), 85
OPC, *see* Ordinary Portland Cement
Open Platform Communications, 296
optimization; *see also* process optimization
    calciner control module, 182–191
    cooler control module, 191–192
    equations of model, **190**
    kiln control module, 181–182
    mill optimization system, 247–248
    model results, **191**
    of raw mix and fuel mix, *188*
    structured and methodical approach, *178*
ORC, *see* Organic Rankine Cycle
Ordinary Portland Cement (OPC), 410,
        425–426, 430–431
    clinker, CO_2 emissions, during production,
        431
    enthalpy of formation of phases, **426**
    production, 406
        power requirement, **105**, 106
    and Solidia Cement, *435*
Organic Rankine Cycle (ORC), 17, 108
organizational culture, 333–334
output, types, 75
oxyfuel combustion, 444

P

PaaS, *see* Platform as a Service
Paris Agreement, COP-21, 406

partial discharge (PD), 245–246
parts management, 370–371
PCS, *see* plant control system
PD, *see* partial discharge
PDCA cycle, *see* plan-do-check-act cycle
PdM, *see* maintenance, predictive
people
    connected, 352, 391–394, *391*
    health and well-being, 391
    and Industry 4.0, 391
Petuum, 57
PGNA, *see* prompt gamma neutron
    activation
photogrammetry, 395
PID control, *see* proportional-integration-
    derivative control
pilot plants
    building blocks, *277*
    and digital twins, 275–278
    modelling approach, 277–278, *277*
plan-do-check-act (PDCA) cycle, 32
    expanded, *32*
plant control system (PCS), ix
plant operation stability, and AF, 154
plastic, global production, *3*
Platform as a Service (PaaS), 84
PLC, *see* programmable logic controllers
pollutions and emissions, 20–24
polymorphs C_2S, **427**
Portland cements, 207–208
    applications, 33–34
    characteristic features, 24–25
    clinker, 409
    and CO_2 emissions, 408–409
    global production, *3*
    pozzolana, 149
    reactive belite-rich (RBPC), 425–427
    clinker, CO_2 emissions, *427*
    varieties
        processing of clinker into, 7; *see also*
        cement manufacture
    process steps for producing, *24*
power consumption, estimates for cement
    plant, **105**
power distribution system, 113–121, *113*, *114*,
    238
    battery with charger, 120
    cables, 121
    control, automation and information
        system, 127–130
    control system, *128*
    design considerations, 114
    earthing/grounding, 120
    emergency generator, 121

intelligent motor control centres (iMCCs),
    117
load centre substations, 114–115, *115*, **116**
LV distribution
    boards, 117
    transformers, 117
MV switchboards, 116
plant lighting, 120–121
power factor improvement, 120
VFD, 118–119
    energy losses due to harmonics, 119
    LV drives, 118–119
    MV drives, 119
voltage selection, 115, **116**
power supply, 103–107, *103*
    determining requirements, 104–106
    emergency generator, 121
    incoming voltage, 103–104
    sources, 107–113
    renewables, 111–113
    WHR systems, 108–111
    transformers, selection, 106–107
power transformers, 106–107, 237
    dry type, maintenance, **237**
    oil-filled, maintenance, **237**
pozzolans, 412
    natural, 413
prediction controller, AI-based, 185–186
predictive analytics, 327–328
    definition, 327
preheater, with precalciner, 233
primary aluminium, global production, *3*
primary and secondary fuels, composition,
    189
process control, 142–143
    advances in strategy for cement
        manufacturing, 91–95
    infrastructure in cement plants, 89–91, *90*;
        see also advanced process control
        (APC) systems
process engineering, 5–13, *6*
    from raw mix to clinker, *7*
process fans, 124, 236
    parameters, **236**
process instruments, 160–162
    cameras, 160
    gas analyzers, 161–162
    kiln shell scanner, 160–161, *161*
process mining, 328
process optimization, 351
    and AI, 386
    conventional approaches, 162–167
    fuzzy logic control philosophy, 162–163
    Industry 4.0 concept, *168*

solutions, 358–362
process pumps, 241
  parameters, **242**
process signals, 130–132
production forecasting and planning
  challenges, *389*
  operation and maintenance scenarios, *390*
production optimizers, 278
product life cycle, and digital twins, 268–269,
  268
Profibus (Process Fieldbus), 128, 130–131,
  131
  comparison with Modbus protocols, **132**
programmable logic controllers (PLC), viii,
  ix, 28, 83, 89–90, 95, 130–131, 278,
  296–298
prompt gamma neutron activation (PGNA),
  28, 204, 304
proportional-integration-derivative (PID)
  control, 90, *90*, 360
purchased materials and combustibles,
  quality deviations, 212
Purdue University, 437
pyroprocessing, 56–57
  and AF, 150–152, 156–158
  and AFR, 146–150
  connected operation, 385–388
  control modules, *179*
    NOx formation, 159–160, *159*
    structure, *180*; *see also* clinker,
      production

Q

QC, *see* quality control
QMS, *see* quality management system
quality control (QC), 28–29, 33, 200, 363
  and AI techniques, 199–220
  alongside process, 205–208
  in cement manufacturing, 200–202
  data collection methods, 202–205
  decentralized, 201–202
  integrated Robo-Lab, 170–172
  software, 365
quality deviations
  in dispatched cement, 215–216
  and ML, 210–216
quality, of final product, and
  ANN, 216–219
quality management system (QMS), 351,
  363–368
  ISO 9001 principles, *31*
  standard, 31
quality optimization software, 365–368

quarries, 206
  materials from, quality deviations, 211
quarry reclamation, 23

R

radiometric thermal imaging, 246–247, *247*
random forests, 287
raw materials, 53
  chemical composition, **189**
  exploration, *6*
  key features, 8–10
  quality control loop, *171*
raw meal, 206; *see also* kiln feed
  carbonate minerals, 10
  decarbonation or calcination, 12
  fineness, 9
  homogenization, 9
  quality deviations, 213–214
  shale particles, 10
  silicate minerals, 10
raw mix
  design, and ML, 54–56
  optimization, 213
  quality deviations, 212–213
RBPC, *see* Portland cements, reactive
  belite-rich
RCM, *see* maintenance, reliability-centered
RDBMS, *see* relational database management
  system
RDF, *see* refuse-derived fuels
reasoning, 87
red mud, 148
Reference Technology Scenario (RTS), 406,
  455
refractories
  kiln, parameters, **235**
  lining, 234–235
refuse-derived fuels (RDF), 19, 152
regression, 47, **87**, 287; *see also* linear
  regression; logistic regression
reinforcement learning, 47, 88–89, *88*, 96–97,
  174–175, 459
relational database management system
  (RDBMS), 330
relays and protective circuit, maintenance
  schedule, **238**
remote operations-control room (RO-CR),
  391–392, *391*
remote plant operations, 399
remote support, 392–394, *393*
  remote assistance tools, 393–394
remote terminal units (RTU), 128–129, 132
resource circularity, vii

resource conservation, 2, 80
resource efficiency, 35
  in production process, 13–16, *14*
resource sustainability, 80
Rietveld refinement, 28, 203, 306, 433
Rio Tinto, 53, *55*
RMSE, *see* root mean squared error
Robert Bosch GmbH, 457
Robo-Lab, *see* robotic laboratory
robotic laboratory (Robo-Lab), 368–369
  for QC, 170–172
  system configuration, *170*
robots, 28
RO-CR, *see* remote operations-control room
rolling average, 59
root mean squared error (RMSE), 48
rotary kilns, 123–124, 207, 233–234, 301
  variables, fuzzy inference structure, *311*
RTS, *see* Reference Technology Scenario
RTU, *see* remote terminal units
rules-based fuzzy logic, 26

S

SaaS, *see* Software as a Service
sampling, 363–364
  analysis, 365
  design, 336
  importance of, 205–206
  locations, *364*
  preparation, 364–365
SCADA, *see* supervisory control and data acquisition
scanning electron microscopes (SEMs), 67
  multispectral dataset, *68*
Schwenk Zement KG, 424
SCMs, *see* supplementary cementitious materials
Seattle Report on database research, 84
self-diagnostics, 28
SEMs, *see* scanning electron microscopes
sensors, 51
  application to industry, 294
  applications, **295**
    in cement manufacturing, 297–308
    overview, 294–297
  and communication, 296
  and data processing, 296
  definitions, 294
  and IIoT, 294
  in mining, crushing and pre-blending, 307–308
  smart, 296–297, 319
    building blocks, *297*
    motor, 244–245

system architecture, *244*
  soft
    for cement manufacture, 310–319
    design, 308–310, *309*
    development in AI environment, 314–318
    future role in intelligent cement production, 319
    Kalman filter, 359
    in process industry, 308–310
    and sustainable cement manufacturing, 293–320
    temperature, location, 299–302, *300*
    types, **295**
    used in cement production lines, **298**
shutdown planning, 370
silica fume (microsilica), 413
silica modulus (SM), 9, 212
simulation, 25–30
SIRI, 41
Skyonic Corporation, 445
slag, 148–150
  synthetic, 414–415
SM, *see* silica modulus
smart machines, 350, 353–354
  characteristics, *354*
SO₂ emissions, *see* sulfur dioxide emissions
Software as a Service (SaaS), 84
solar photovoltaic (PV) system, 111
solar PV system, *see* solar photovoltaic system
Solidia Cement, 430–431, 445
  application and performance in pavers, 438–439
  average phase composition, **433**
  clinker
    CO₂ emissions, during production, **431**
    nodules, *433*
    production measurements, **432**
  comparative SWOT analysis, **446**
  energy savings, 430–431
  industrial production, 432–434
  and OPC, *435*
Solidia Concrete
  CO₂-cured, microstructure, *437*
  CO₂ reduction potential, compared to OPC, *436*
  curing process, 435–437
  and AI, 437
  performance and durability, 437–438
Solidia Supplementary Cemetititious Material (Solidia SCM™), 415–416
  compressive strength, *417*
Solidia Technologies®, 415, 418, 430–439, 444

SOLPART project, 112
Sorel cements, 419
spectroscopic techniques, 204
SQL, *see* Structured Query Language
SRC, *see* Steam Rankine Cycle
Standardization of systems, 31
    critical function areas, *30*
statistical software, 85
Steam Rankine Cycle (SRC), 108
strength prediction, traditional statistics vs.
        predictive ANN, 216–218
Structured Query Language (SQL), 85
substation control, 129
    radiometric thermal imaging, 246–247
sulfate content, impact on 1-day strength, *219*
sulfoaluminate belite cement, 428–430, 445
    comparative SWOT analysis, **446**
sulfur dioxide emissions, 23
Sumitomo Metal Industries, Kashima Steel
        Works, 110
supercomputers, 40–41, *41*
supervised learning, 46–47, 87–89, 174
supervisory control and data acquisition
        (SCADA), ix, 89, 128–130, 132, 285
supplementary cementitious materials
        (SCMs), 207, 409–418
    classification, and AI, 67–71
    composition, *410*
    traditional, 410–413, 447; *see also*
        alternative supplementary
        cementitious materials (ASCM)
support vector machine (SVM), 46, 68–69
sustainability, vii, 2, 80
    Caradonna model, *326*
    definitions, 325
    environmental, vii
    through use of AFR, 145; *see also* cement
        manufacture, sustainable; resource
        sustainability
sustainable business development, 325–326
    based on TBL, *326*
sustainable firms, *325*
SVM, *see* support vector machine

T

TAD, *see* tertiary air duct
TBL, *see* triple bottom line
technical performance monitoring, in cement
        plants, 230–232
technology roadmaps
    for cement industry, 35, 456
    for low CO₂ emission, *196*
telemetry, 318
tertiary air duct (TAD), 395

text and language processing, 50
TGA, *see* thermogravimetric analysis
thermal energy
    management, data-driven, 141–197
    performance, 16–18
thermography, 160
    infrared, 301
thermogravimetric analysis (TGA), 203
Thomas, C., 324
time plots, 59
time shifting, 59
titration, 202
TLS, *see* Transport Layer Security
Total Organic Carbons (TOC) emission, 23
total quality management (TQM), 324
transport controller, 184–185
    control features, *185*
Transport Layer Security (TLS), 318
triple bottom line (TBL), 325–326, *326*
troubleshooting, remote support, 392–394
Turner-Fairbank Highway Research Center,
        437

U

United Nations Framework Convention on
        Climate Change (UNFCC), 406
United Nations (UN), 325
University of South Florida, 438
unsupervised learning, 47, 87, 174–175
variable frequency drives (VFD), 126–127,
        238–239
    maintenance schedule, **239**

V

vertical roller mills (VRM), 8–9, 13, 33, 123
    case studies, 382–384
    condition monitoring, 381–382, *383*
    process analytics for system, *315*
    process optimization, 361–362, *362*
VFD, *see* variable frequency drives
Vicat apparatus, 205
video monitoring, 160
virtual cement plants, *see* digital twins
VRM, *see* vertical roller mills

W

Wamba, S. F., 328–329
waste burning, 33
waste heat recovery (WHR) systems, 108–111,
        138
    installations, *109*
    ORC based, 109, *110*

wastes
  construction and demolition (C&D),
    carbonated, 414
  industrial, as AFR, 147–150
  pre-combustion treatment, *19*
  recycling, 23
WBCSD, *see* World Business Council for
    Sustainable Development
WHR systems, *see* waste heat recovery
    systems
wireless communication, industrial, 132
World Business Council for Sustainable
    Development (WBCSD)
  Cement Sustainability Initiative (CSI),
    443, 456

"Low-Carbon Transition in Cement
    Industry", 406, *407*, 411, 447
World Commission on Environment and
    Development, Brundtland report, vii

X

X-ray diffraction (XRD), 28, 170, 202–204,
    207–208, 305, 365
  and AF and AFR, 172
X-ray fluorescence (XRF), 170, 202–204,
    207–208, 305, 364–365

Z

Zeobond Pty Ltd, 419